LIBRARY, RO WESTCOTT
REGULATIONS FOR BORROWERS

I. Books are issued on loan for a period of I month
and must be returned to the library promptly.

2. Before books are taken from the Library receipts
for them must be filled in, signed, and handed to a
member of the Library Staff. Receipts for books
received through the internal post must be signed
and returned to the Library immediately.

3. Readers are responsible for books which they
have borrowed, and are required to replace any such
books which they lose or damage. In their own
interest they are advised not to pass on to other
readers books they have borrowed.

4. To enable the Library Staff to deal with urgent
requests for books, borrowers who expect to be
absent for more than a week are requested either to
arrange for borrowed books to be made available to
the PA or Clerk to the Section, or to return them to
the Library for safekeeping during the period of
absence.

SYNTHETIC RUBBERS:
Their Chemistry and Technology

SYNTHETIC RUBBERS:
Their Chemistry and Technology

D. C. BLACKLEY
B.Sc., Ph.D., F.P.R.I.

Reader in Polymer Science and Technology,
London School of Polymer Technology,
The Polytechnic of North London, London, UK

APPLIED SCIENCE PUBLISHERS
LONDON and NEW YORK

APPLIED SCIENCE PUBLISHERS LTD
Ripple Road, Barking, Essex, England
Sole Distributor in the USA and Canada
ELSEVIER SCIENCE PUBLISHING CO., INC.,
52 Vanderbilt Avenue, New York, NY 10017, USA

British Library Cataloguing in Publication Data

Blackley, D. C.
 Synthetic rubbers
 1. Rubber, Artificial
 I. Title
 547.8′47272   TS1895

 ISBN 0-85334-152-4

WITH 26 TABLES AND 81 ILLUSTRATIONS

© APPLIED SCIENCE PUBLISHERS LTD 1983

Printed in Great Britain by Galliard (Printers) Ltd, Great Yarmouth

Preface

This book has its origin in a proposal made a few years ago that I should collaborate with Dr H. J. Stern in the production of a third edition of his well-known text-book entitled *Rubber: Natural and Synthetic*. The suggestion was that I should contribute a series of chapters on synthetic rubbers. Although, in the event, it has not proved possible to publish the full book in the form originally planned, it was apparent that, with some restructuring, the material which I had collected would be valuable as an independent summary of the chemistry and technology of synthetic rubbers. It is in this form that the material is now offered.

The primary purpose of this book is to provide a brief up-to-date survey of the principal types of synthetic rubber which have been and are currently available. Two classes of material are included which are regarded by some as being thermoplastics rather than rubbers, namely, plasticised polyvinyl chloride and the thermoplastic synthetic rubbers. The topics which are covered for each main family of synthetic rubbers are (i) the sources of the monomers, (ii) polymerisation procedures and the effects of important polymerisation variables upon the rubber produced, (iii) the types of rubber currently available commercially, (iv) interesting aspects of the compounding of the rubbers, with special reference to such matters as vulcanisation, reinforcement, protection against degradation, and (where appropriate) plasticisation, and (v) an indication of applications. Where appropriate, an attempt has been made to set the development of the various synthetic rubbers into historical context, and to relate the development to economic, political and commercial factors. In dealing with these diverse aspects of the various synthetic rubbers, an attempt has been made to give broad generalisations of widespread validity, rather than a wealth of detailed information pertaining to particular matters. The author is well aware that such broad generalisations can be misleading in particular instances, but he also knows that it is all too easy to lose the reader in a mass of detail.

The subject of synthetic rubber latices is specifically excluded from this book. It is far too large a subject for inclusion in a book of this type, especially in view of the many additional concepts which would have to be introduced over and above those which are necessary for the treatment of solid synthetic rubbers. In any case, synthetic latices are dealt with in some detail in my book *High Polymer Latices*, thorough revision of which is currently being undertaken.

A difficult decision which faces the author of a broad survey such as this book purports to be concerns the placing of matters of general interest for the subject as a whole. Shall he gather them together in one or more introductory chapters, which will probably be tedious to read because individual materials and processes which exemplify these matters have not yet been encountered? Or shall he insert them as they arise in connection with individual materials and processes? If the latter course were adopted, then the book could have a very fragmented appearance. In fact, a compromise approach has been adopted here. Some general matters, such as the molecular prerequisites for a substance to be rubbery, the basic facts of polymerisation chemistry, and the industrial production of widely used monomers, have been dealt with in the chapters which are essentially introductory to the book as a whole. Other general topics have been dealt with in connection with individual rubbers which exemplify their importance. Examples include the effect of polymerisation temperature upon the structure of rubbers produced by emulsion polymerisation, which is dealt with in connection with styrene–butadiene rubbers; the swelling of elastomer vulcanisates, which is dealt with in connection with acrylonitrile–butadiene rubbers; and the vulcanisation of rubbers by heating with organic peroxides, which is dealt with in connection with ethylene–propylene rubbers. It is hoped that this approach will make for easier reading, at least on the part of the newcomer to the subject.

The audience to whom this book is primarily directed is graduate chemists, physicists and technologists who are employed in the polymer-producing and polymer-using industries and in allied and ancillary industries. Whilst the main intention of this book is to provide a 'free-standing' survey of synthetic rubbers, an important secondary intention is to introduce the reader to the specialised review literature which is available for the various types of synthetic rubber. The volume of primary research literature which now exists for many of the synthetic rubbers is enormous, especially for those rubbers which are used in large tonnage. It would have been quite impracticable to have provided comprehensive lists of references for each type of synthetic rubber. The policy which has been adopted is to provide for each type of synthetic rubber a general bibliography which gives some of the more important review articles and monographs, as well as literature references for some of the specific matters which are discussed in the text. It will become evident to the reader that considerable use has been made in the course of the preparation of this book of several of the articles which appear in the *Encyclopedia of Polymer Science and Technology*, published by John Wiley and Sons. It is appropriate to acknowledge here the usefulness of this encyclopedia, containing as it does authoritative and readable reviews of many of the subjects which will be of interest to readers of this book. The assistance which has been gained from the series of authoritative reviews of individual synthetic rubbers which has appeared over the years in the journal *Rubber Chemistry and Technology*, published by the American Chemical Society, is also acknowledged. Apart from such intrinsic merits as this present book may possess, it will have performed a useful function if it directs its readers to these authoritative articles and reviews, and to the comprehensive bibliographies covering the respective parts of the subject which they give.

An important matter which calls for some comment is the relationship between this book and the book entitled *Rubber Chemistry* which has recently been written by my colleague and friend John Brydson and published by Applied Science Publishers Ltd. There must inevitably be some overlap of subject matter between this book and that by Brydson, but the overlap is not great. Brydson's book is concerned primarily with the chemical aspects of the behaviour of rubbery materials, including natural rubber, whereas this book is concerned with synthetic rubbers as an interesting and useful family of materials. There is, for instance, much greater emphasis here on the variants of a given type of synthetic rubber which are available, and of the consequences of these variants for applications. It is to be hoped that this book will come to be regarded as essentially complementary to that by Brydson, rather than as one which covers the same ground.

There is one further important matter to which any author writing on a technological subject in these days must draw attention. This concerns the health hazards which may accompany the use of some of the materials which he mentions. We are becoming increasingly aware of these hazards. Seldom, if ever, does the reputation of a substance in this respect improve as a consequence of increased knowledge and wider experience. Usually greater knowledge informs us of the existence of a hazard where none was thought to exist, or of the existence of a greater hazard than was believed formerly. For various reasons, not the least of which is the rate at which knowledge and opinion are changing in this field, it has been considered inappropriate to deal with this aspect of the subject in this book. Rather it is considered wiser to give a general warning that some of the materials mentioned here are already believed to be deleterious to health and that the remainder are all potentially liable to become suspect in the light of increasing knowledge. Those who propose themselves to make use of the materials mentioned in this book are therefore strongly advised first to seek up-to-date authoritative opinion on likely health hazards. For those who propose to direct others to use these materials, this precaution is obligatory.

D. C. BLACKLEY

Contents

Chapter 1

Introduction

1.1 RUBBERS NATURAL AND SYNTHETIC

Whatever connotations may have attached to the word 'rubber' in the past, this word has for many years now been used to denote the class of materials which display the property of long-range reversible elasticity. Such materials are also now commonly referred to as 'elastomers'; in fact, the words 'rubber' and 'elastomer' are used almost synonomously.

The property of long-range reversible elasticity can be conveniently defined as 'the ability to be able to sustain large reversible extensions without rupture'. By 'large' in this context is meant 'of the same order of magnitude as the linear dimensions of the material sample being deformed'. The lower limit might reasonably be set at 50 % linear extension, although in practice the maximum linear extensions which rubbery materials can undergo without rupture are commonly in the range 100–500 %. There is no upper limit to the extensibility for the purpose of this definition of a rubber, but in practice the upper limit of reversible elasticity is approximately 1000 % linear extension.

The property of large-scale extension without rupture is not itself sufficient to define a rubbery material. Liquids are capable of enduring almost unlimited extension without rupture; gases certainly are. To qualify as a rubber, the extension must be essentially reversible in the sense that, when the stress which is responsible for the deformation is removed, the material rapidly recovers to approximately its original dimensions. There is inevitably some vagueness in this description because we are seeking to encompass the behaviour of a family of real materials, rather than of ideal materials whose behaviour can be prescribed by definition. By 'rapid' in this context is meant that the recovery takes place over a period of a few seconds at the most. The qualification that the material returns to 'approximately'

1

its original dimensions is necessary because seldom is the return ever to exactly the original dimensions, any more than the recovery ever takes place instantaneously; in fact, the exact extent of the recovery depends upon the time which is allowed for the recovery. By the word 'approximately' in this context is meant that the recovery is such that the linear dimensions of the recovered material do not differ from the corresponding initial dimensions by more than a few per cent.

It is difficult to fit rubbery materials into the normal categories of solid, liquid and gaseous; they are unusual in having affinities with all three of the common states of matter. They are clearly solid in the sense that they feel more-or-less solid and neither flow nor expand spontaneously when left to themselves. They resemble liquids in that they are able to endure large extensions without rupture, although, as has been noted above, unlike liquids the extension is essentially recoverable. The affinity with gases is more subtle, since clearly rubbers do not resemble gases in respect of either very low density, compressibility or tendency to expand spontaneously. The affinity to the gaseous state is revealed by experiments in which the stress required to maintain a strip of rubber at constant extension is measured as a function of temperature, or, alternatively, by experiments in which the extension at constant stress is measured as a function of temperature. Unlike most solid materials, the stress required to maintain a constant extension of a rubber *increases* with increasing temperature if the constant extension is more than a few per cent. Equally unexpectedly—and it is a manifestation of the same phenomenon—the extension at constant stress *decreases* as the temperature increases, provided that the constant stress is such that the extension exceeds a few per cent. If the first of these observations is compared with the fact that the pressure of a gas held at constant volume increases with increasing temperature, then it becomes clear that there is some affinity between rubbers and gases in respect of their thermo-mechanical properties. This affinity is further emphasised by the fact that the elastic properties of rubbers can be explained in general terms by a molecular theory which is in some respects analogous to the kinetic theory of gases.

It will become clear from the final section of this chapter that rubbery materials as normally encountered seldom consist of a single component. Usually they are rather complex mixtures of vulcanising ingredients, fillers, plasticisers, antioxidants, pigments, etc. However, their rubbery character always depends upon the presence in the mixture of one or more base materials which are themselves inherently rubbery in nature, or, in a few cases, become rubbery in the presence of a suitable plasticiser. These base materials are usually, but not invariably, organic in nature. It is with an important sub-group of these base materials, and with the ways in which these base materials are compounded and processed to form useful materials, that this book is concerned.

The base rubbers which almost all rubbery materials contain to a lesser or greater extent are classified into two broad groups, *natural* and *synthetic*. As the title implies, the concern of this book is exclusively with the latter. By the term *natural rubber* is meant a rubber which occurs naturally as a product of the metabolism of certain trees and plants, notably trees of the *Hevea brasiliensis* species. Although not a pure substance, natural rubber is principally a linear *cis*-1,4-polyisoprene of high molecular weight. By the term *synthetic rubber* is understood a rubber which is produced by man from chemical compounds of low molecular weight (say, less than 500). The meaning of the term is not usually extended to cover rubbery materials which are made by the chemical modification of substances which are initially rubbery, such as natural rubber. The term is, however, commonly regarded as including rubbers which have not been produced *directly* from low-molecular-weight compounds. Thus both chlorosulphonated polyethylene and plasticised polyvinyl chloride are conventionally regarded as being synthetic rubbers, even although the high-molecular-weight intermediates from which they are produced (polyethylene and polyvinyl chloride, respectively) are not usually thought of as rubbers.

What were the motivating forces which led to endeavours to manufacture synthetic rubbers? They were probably at least four in number:

1. a desire, if not necessity, to achieve independence in respect of a product which hitherto had been obtainable from natural sources only;
2. a desire to achieve greater ability to meet increased demands;
3. a desire to obtain rubbers which possess properties which are not possessed by the natural product, such as resistance to swelling in oils, resistance to extremes of temperature and resistance to certain deteriorative influences, notably ozone; and
4. curiosity.

Concerning the first of these motivations, it is self-evident that the more routes one has to a desirable product, the better is one able to meet the changes of circumstance which may arise as a consequence of political or economic causes. Concerning the second, it has to be noted that the development of the rubber industry has paralleled very closely the development of transportation powered by the internal-combustion engine. It is extremely doubtful whether the demand for rubber for the manufacture of rubber tyres consequent upon the growth of the auto-motive industry could have been met solely by the producers of natural rubber. Concerning the third of these motivations, the shortcomings of natural rubber in certain applications has led to a requirement for rubbery materials which possess rather different ancillary properties to those possessed by natural rubber. Although the demand for rubbers possessing these properties has never been large compared to the demand for rubber for tyres, the role played by these rubbers has become increasingly important over the years as man has sought to develop his technology in the context of, for example, the aerospace industry. Thus, whereas the early synthetic rubbers were regarded as being substitutes—and 'ersatz' sub-stitutes at that—for natural rubber, several of the newer types permit the application of elastomeric materials in environments where application would scarcely be feasible at all were natural rubber the only rubber available. Concerning the fourth of these motivations, it is the case that many scientists (including prominent scientists such as Michael Faraday) have been very interested in natural rubber and the strange mechanical properties which it possesses. The relationship between rubbery substances on the one hand and low-molecular-weight organic compounds of known structure on the other has often elicited curiosity. The ability to produce rubbery substances from organic compounds of low molecular weight has clearly constituted an important step in the understanding of that relationship.

Consideration of these motivating forces leads immediately to two important observations of a very general nature. The first of these is that synthetic rubbers can be broadly classified into two categories, namely, the *general-purpose types* and the *special-purpose types*. As the names imply, the first category comprises rubbers which are intended for applications, such as the manufacture of tyres and 'general mechanical' products, which call for no properties other than satisfactory mechanical properties at normal temperatures and the ability to retain those properties for satisfactory periods of time under normal service conditions. In the second category are those rubbers which have special properties such as resistance to swelling in oils, and are in consequence intended for specialised applications such as the manufacture of oil seals. Without exception, all the general-purpose synthetic rubbers which have been developed so far have contained only carbon and hydrogen. Likewise, all the special-purpose synthetic rubbers thus far developed have contained other elements besides carbon and hydrogen; indeed carbon and hydrogen are absent from some of the more exotic types.

The second important observation concerns the relationship between the synthetic rubber molecule and the low-molecular-weight compound from which the rubber was manufactured. The molecules of all rubbery materials are in the general class of those known as *polymers*, that is, they are of high molecular weight and comprise one or more small chemical units repeated many times over. The average molecular weights of raw rubbers are often very high, being typically in the range 10^5–10^6. Assuming, as is usually the case, that the molecular weight of the repeat unit is of the order of 50–100, this will mean that each rubber molecule can contain on average between 10^3 and 2×10^4 repeat units per molecule. In the case of a synthetic rubber, the repeat units in the rubber molecule are derived from the low-molecular-weight compounds—called *monomers*—from which the rubber was manufactured. The chemical reaction by which monomer molecules are converted into polymer molecules is known as *polymerisation*. More will be said of these reactions in Chapter 4. It is sufficient to note at this stage that the majority of synthetic rubbers are manufactured by the process of *addition polymerisation*, in which the monomer units become added to one another without any small molecule being eliminated. A few synthetic rubbers are manufactured by *condensation polymerisation*, in which a small molecule is eliminated during each step of the polymer-building reaction. It may also be noted that (a) the polymerisation reaction may involve a mixture of monomers, in which case the product is known as a *copolymer* because it contains more than one type of repeat unit; and (b) the sequences of repeat units in the molecules of raw synthetic rubbers are essentially linear, or, at least, if branching and crosslinking between the essentially linear sequences do occur, then these are regarded as generally undesirable features.

1.2 MOLECULAR REQUIREMENTS FOR A RUBBERY SUBSTANCE

It is appropriate at this stage to summarise the molecular requirements which must be fulfilled if a substance is to have the mechanical properties which we associate with a rubber. Not only should such a summary be helpful in understanding certain of the matters which will be discussed in subsequent chapters; it should also be helpful in giving some insight into the molecular reasons why certain monomers have achieved great importance in connection with the manufacture of synthetic rubbers.

In addition to the obvious requirement of molecular stability, the molecular requirements which have to be fulfilled if a polymer is to be rubbery in nature and to be utilisable as a rubber in practice are five in number:

1.   The segments of the polymer sequence must be capable of moving essentially independently of segments of the same sequence which are some distance away, i.e., the polymer sequence must be flexible.

2.   Strong interaction between neighbouring segments must be absent, and so also must high concentrations of bulky substituents which inhibit segmental motion.

3.   The polymer must not crystallise to any appreciable extent in the unstrained state nor at low extensions.

4.   The molecular weight must be high.

5.   The facility must exist for the introduction of a low concentration of crosslinks between neighbouring polymer molecules.

Considering these points in order, the requirement of independent segmental mobility implies (a) that a sufficient proportion of the main-chain bonds of the polymer sequence have the type of flexibility which is depicted in Fig. 1.1, (b) that the polymer sequence is essentially linear, and (c) if branching does occur, it must not be so extensive as seriously to inhibit segmental mobility. The type of flexibility depicted in Fig. 1.1 is rotational flexibility of one bond about its neighbour. The requirement is that rotation should be possible without undue hindrance, notwithstanding that the angle between the two bonds is fixed within narrow limits. It is by means of essentially independent segmental motions of this type that the rubber molecule as a whole can readily move from one conformation to another. An important consequence of rotational segmental flexibility is that, because of purely statistical considerations, the polymer molecule is more likely to adopt a coiled-up conformation than one in which the molecule is extended. It is this feature, together with the ready inter-conversion of molecular conformations, which endows the bulk material with the property of long-range reversible elasticity.

Molecular features which tend to restrict or inhibit independent segmental mobility must either be absent altogether or else be present in insufficient concentration to destroy the ability of the polymer sequence as a whole to pass readily from one conformation to another. Two important molecular features which have a tendency to restrict independent segmental mobility are (a) attractive forces between segments arising from the presence of chemical groupings which have permanent dipoles associated

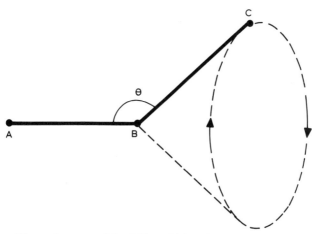

FIG. 1.1. Illustrating type of flexibility which polymer sequence must possess if material is to exhibit rubbery properties in the bulk state. AB and BC are neighbouring segments of the polymer sequence. BC must be able to rotate about AB as shown, without serious hindrance, although the bond angle θ remains essentially unchanged.

with them, and (b) the presence of bulky substituents in the main chain which give rise to steric hindrance.

Another factor which will clearly inhibit independent segmental motions is crystallisation of the polymer segments into a regular lattice, thereby effectively immobilising the segments. Even if the polymer sequence itself is inherently flexible, crystallisation of sections of the sequence can change an otherwise rubbery polymer into a material which is rather more rigid than would be expected. The actual rigidity depends upon the proportion of the polymer which is present in crystalline form. An outstanding example of the effect of segmental crystallisation is to be found in polyethylene. On the basis of inherent segmental mobility, this polymer would be expected to be rubbery at room temperature; in fact, it is a rather rigid plastic because of segmental crystallisation.

The requirement for high molecular weight arises because, if the molecular weight is not sufficiently high, then the polymer molecules are able to move relative to one another in a concerted fashion. The material then behaves as a viscous liquid rather than as a rubber.

However, even if the molecular weight is very high, an uncrosslinked rubber will flow as a very viscous liquid if subjected to a steady stress over a long period of time. It is for this reason that most rubbers have to be lightly crosslinked in order that they may perform satisfactorily in service. The object of introducing the crosslinks is to prevent the occurrence of irreversible viscous flow under sustained stress. But the concentration of crosslinks must not be too high, otherwise the material becomes rigid because independent segmental motions have become seriously restricted. The concentration of crosslinks in most lightly-vulcanised rubbers is of the order of 10^{-5}–10^{-4} mol cm^{-3} (10–100 mol m^{-3}). The actual number of crosslinks in each cubic centimétre of a crosslinked rubber is therefore of the order of 10^{18}–10^{19}.

One further matter has to be considered in connection with the first of the five molecular requirements for rubber elasticity which have been listed above. This concerns the effect of temperature upon the type of bond flexibility and segmental mobility which is necessary if a polymer is to be rubbery in nature. That rotations of the type illustrated in Fig. 1.1 do actually occur is a consequence of the possession of kinetic energy by the polymer molecules and their constituent parts. In fact, the extent of the independent segmental motions required for rubbery character in the bulk material, and the ease with which transitions between molecular conformations connected by such motions can occur, is determined by the balance between the kinetic energy of the segments on the one hand and the resistance which exists to the rotation of one interatomic bond about its neighbour on the other. As the temperature of a rubber is lowered, the kinetic energy of the polymer molecules and their constituent parts is reduced. For this reason, if for no other, the ease of rotation of one interatomic bond about its neighbour is reduced. So too, in consequence, is the ease with which independent segmental motions can occur. One practical consequence is that a rubber becomes less rubbery as the temperature is lowered. (This effect, which is really in the nature of a gradual change of physical state, as will become evident below, must not be confused with the fact, to which reference has been made in Section 1.1, that rubbers in the rubbery state become 'stiffer' as the temperature is *increased*.) As the temperature is progressively lowered, there is eventually reached a small range of temperature over which the material changes from what is evidently essentially a rubber to what is essentially a rigid material. The ability of one interatomic bond to rotate about its neighbour is almost completely lost. So too are the independent segmental motions of the several parts of the polymer chain. The polymer molecules are now virtually frozen in position, and transitions between possible different molecular conformations cannot occur with any ease. Further lowering of the temperature merely leads to further embrittlement of the material. These changes can be reversed by raising the temperature. The transition between behaviour as a rubbery material and behaviour as a rigid material, brought about by change of temperature, is known as the *glass–rubber transition*. The term *vitrification* is also used to denote the change from rubbery behaviour to glassy behaviour brought about by reduction in temperature.

The tendency of a rubber to stiffen when the temperature is lowered can be characterised by its so-called *glass-transition temperature*. In the older literature this was also referred to as the *second-order transition temperature*, but this term is for the present obsolete. This temperature is formally defined as the temperature at which the transition from the glassy state to the rubbery state occurs when the temperature of a glassy polymer is raised, and, conversely, the temperature at which the material changes from a rubber to a glass when the temperature is lowered. However, as has been implied above, the glass–rubber transition does not occur at a single temperature, but over a small range of temperature. It is therefore usual either to quote the middle temperature of this range as the glass-transition temperature, or else the range itself. For technological purposes, the low-temperature stiffening behaviour of a rubber is sometimes characterised by an arbitrary *low-temperature stiffening temperature*. This is defined as the temperature at which the elastic modulus of the rubber, as determined by a standard procedure, is a specified multiple of the corresponding modulus at a specified temperature which is near room temperature.

It must also be noted that some rubbers tend to *crystallise* partially when the temperature is lowered. This also causes stiffening of the material. It is a quite distinct molecular phenomenon from vitrification. It is caused by parts of neighbouring sequences packing together to form structures which have the kind of regularity associated with crystal lattices. Segmental motions within the crystalline regions are almost completely suppressed because the intermolecular attractions constrain the various polymer segments to occupy almost fixed positions in the crystal lattice. If a rubbery polymer crystallises on cooling, it does so over a range of temperature and at a rate which depends upon the temperature. Furthermore, the temperature range over which crystallisation occurs (if it occurs at all) is considerably above the temperature range over which the glass–rubber transition occurs; typically, the difference between the two temperature ranges is approximately 80 °C.

1.3 OUTLINE OF RUBBER PROCESSING TECHNOLOGY

In writing this book, it has been necessary to assume that the reader is familiar in a general way either with the practice of rubber technology, or at least with many of the terms and concepts which are used by the rubber technologist. In particular, it has been necessary to assume that the reader has some knowledge of the technology of natural rubber, so that comparisons between synthetic and natural rubbers in respect of matters such as vulcanisation behaviour can be made without having to discuss the relevant aspects of the technology of natural rubber. However, for the benefit of any readers who are not familiar with the general principles of rubber processing, the following very brief summary is offered in the hope that it will assist in providing a background to the main theme of the book.

A simplified flow diagram which illustrates the basic principles of rubber processing technology is given in Fig. 1.2. The essential molecular problem with which conventional rubber processing technology is concerned can be stated very simply. It is that in order to shape a rubbery material into a useful product, it is necessary that the polymer molecules which comprise the base rubber should be essentially uncrosslinked. Only if this is so is it in general possible to cause the rubbery material to undergo the permanent, irreversible deformation which is an essential feature of the shaping process. But once shaping has been achieved, it is then necessary to insert crosslinks between the polymer molecules in order that no further irreversible deformation can occur during use. The process whereby these crosslinks are inserted is traditionally known as *vulcanisation*, because it is usually necessary to heat the compounded rubber in order to obtain a useful rate of crosslink insertion. However, the crosslinking of rubbers is also referred to as *curing*, because it is a process whereby a *green* material (a term which is sometimes applied to compounded but unvulcanised rubber) is converted into a useful product.

The terms *curing* and *vulcanisation* are often used synonomously, and there is little objection to this. However, they are also sometimes used synonomously with 'crosslinking', and this is open to question because,

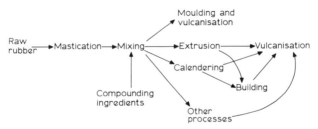

Fig. 1.2.   Flow diagram illustrating basic principles of rubber processing technology.

whilst it is true that in the majority of cases the vulcanisation and curing of rubbers does proceed by way of the formation of crosslinks between hitherto essentially linear polymer molecules, there are cases where the word 'curing' is applied to processes which may well not entail crosslinking. The precise meaning which attaches to these terms is sometimes rather obscure, and caution should be exercised in interpreting them in some cases.

Starting with the raw essentially uncrosslinked rubber, the first step is frequently to soften the rubber by mechanical working and by the application of heat. The softening process is known as *mastication*. It is usually carried out on two-roll mills or in internal mixers. In the case of natural rubber, it is accompanied by a considerable reduction in the molecular weight and gel content of the rubber. (The *gel content* is that fraction of the raw rubber which is lightly crosslinked and therefore insoluble in, though swollen by, normal rubber solvents.) Although some reduction of molecular weight occurs in the case of some synthetic rubbers, this is not generally an important feature of the softening process for synthetic rubbers; indeed the softening is frequently a consequence mainly of such temperature rise as may occur. It is important to have a means of assessing quantitatively the 'plasticity' of a raw rubber and the extent to which it has been increased by mastication. There are various instruments available for making such assessments. One instrument which has been widely used is known as the *Mooney viscometer*. In this instrument, a small cavity is completely filled with the rubber to be tested. The rubber is heated to a controlled temperature (e.g. 100 °C). The cavity also contains a small rotor, and the instrument measures the torque which has to be exerted in order to cause the rotor to rotate at a constant speed. The surfaces of the rotor and cavity are machined in such a way as to minimise slippage at the interface between the rubber and the rotor or cavity. The result of the measurement is quoted as an arbitrary 'Mooney viscosity' reading. In reporting results, it is necessary to state both the temperature to which the rubber was heated and the time which elapsed before the reading was taken.

The next step is the *mixing* or *compounding* of the softened raw rubber. If not done in the same piece of equipment as the softening, it is done in similar equipment, that is, open two-roll mills or internal mixers. A wide variety of substances are frequently compounded with rubbers. They are usually classified as follows:

1. vulcanising ingredients, or *curatives*;
2. fillers;
3. softeners, extenders, plasticisers and processing aids;

4.   antioxidants and other anti-degradants;
5.   pigments and colours;
6.   others, such as fungicides, re-odourants, etc.

Before considering in outline the function of each of these groups of ingredients, it should be mentioned that some confusion may arise because terms such as 'rubber compound' and 'compounding' are used where strictly the terms 'rubber mix' and 'mixing' respectively should be used. Indeed, there is (rightly) a tendency to prefer the latter terms to the former. But sanctification by common usage will probably ensure the continuing use of the former terms, to the confusion of newcomers to rubber processing technology.

As the name implies, vulcanising ingredients comprise those substances which are added in order that crosslinking may occur subsequently when the compounded rubber is heated to an appropriate temperature. Sulphur is the principal vulcanising agent which is used for most rubbers which contain copolymerised dienes, such as butadiene and isoprene. The ability of sulphur to act as a vulcanising agent for natural rubber was discovered more than a century ago. It is still the primary vulcanising agent for natural rubber and for most other hydrocarbon diene rubbers and related rubbers. In order that economic rates of vulcanisation can be achieved, it is necessary that the rubber compound should contain vulcanisation accelerators and activators. Not only does the presence of these substances bring about an enhanced rate of vulcanisation, but also the mechanical properties of the vulcanisate are generally improved, as is the resistance to deterioration during ageing. A wide variety of vulcanisation accelerators is available today. Almost without exception they are organic compounds. The most widely-used vulcanisation activators are either combinations of a metal oxide (e.g., zinc oxide) and a fatty acid (e.g., stearic acid), or a heavy-metal soap (such as zinc laurate). Whilst sulphur and appropriate accelerators and activators are commonly used for the vulcanisation of diene rubbers, other vulcanising agents are also available for these rubbers. Furthermore, some synthetic rubbers require the use of non-sulphur vulcanising systems anyway. Examples include the use of metal oxides for vulcanising certain halogen-containing rubbers, and the use of organic peroxides for vulcanising certain olefin and silicone rubbers. These matters will be discussed in greater detail when the various classes of synthetic rubber are described.

Fillers are usually inorganic powders of small particle size. They are incorporated for various purposes. Some, typified by whiting and ground chalk, are incorporated primarily to cheapen and stiffen the final product. Others, typified by the carbon blacks, china clays, silicas and silicates, are incorporated primarily to stiffen and *reinforce* the product. By 'reinforcement' in this context is meant enhancement of properties such as tear strength, abrasion resistance and tensile strength. Thus fillers tend to be classified into the two broad groups of *reinforcing* and *non-reinforcing*. But the distinction between the two groups is not clear-cut, and there are many fillers available which are intermediate in character. It should also be borne in mind that the carbon blacks embrace a large range of types, some of which show little tendency to reinforce rubber whilst others are very highly reinforcing indeed. A further complication is that, in the amounts in which they are usually added (say, between 30 and 300 parts by weight per 100 parts by weight of rubber†), particulate fillers are able profoundly to alter the flow behaviour of unvulcanised rubbers. In particular, they can reduce the elasticity of the unvulcanised rubber. But whatever may be the nature of the alteration, it can have important implications for the subsequent processing of the unvulcanised rubber.

The third group of compounding ingredients listed above comprises substances whose functions are not always precisely defined. The substances in this group are generally liquids or resinous materials which mix intimately with the rubber thereby modifying its flow behaviour. Some of these materials are added primarily as extenders and cheapeners for the rubber. Others are added primarily as softeners in order to enable large quantities of fillers to be incorporated. Others are added primarily to reduce the temperature at which the rubber becomes brittle, thereby extending the lower end of the temperature range over which the material is usable as a rubber. Such additives in effect reduce the glass-transition temperature of the rubber by increasing the separation between its molecules, thereby reducing the attractive interactions between the various parts of its molecules. Still other materials in this group are added primarily because of their effects upon the flow behaviour of the compounded unvulcanised rubber. Because of the complexity of this subject, and because the effect of any given substance in this class depends not only upon the substance itself but also upon the nature of the rubber to which it is added, it would be misleading to give specific examples of these substances here.

Substances in the fourth of the above groups are added in order to

---

† In subsequent pages the abbreviation 'pphr' will be used consistently to denote the cumbersome expression 'parts by weight per 100 parts by weight of rubber'.

increase the ability of the vulcanised product to resist the various deteriorative influences to which it may be subject during use. These influences almost always include oxygen, but other common influences are heat, sunlight and ozone. Thus almost all rubber compounds include an antioxidant, but some may also include substances, such as antiozonants, which are intended to give protection against specific deteriorative influences.

In the fifth group are substances which are added in order to impart a suitable colour to the vulcanised product. The sixth group contains a wide diversity of substances which are added in order to achieve special effects.

The plasticity of compounded rubbers can be assessed quantitatively by the same procedures which are used to assess the plasticity of raw and masticated rubbers. In particular, the Mooney viscosity has been widely used for assessing the plasticity of compounded rubbers. The *vulcanisation behaviour* of compounded rubbers can be assessed by determining the decrease in plasticity with time of heating at a fixed temperature. In the past, the Mooney viscometer has been widely used for this purpose. However, for some years now it has become customary to use instruments known as *oscillating disc rheometers* for this purpose. In principle, these instruments are similar to the Mooney viscometer, except that the rotor is caused to oscillate slowly at a constant low frequency and at a small but definite angular amplitude, rather than to rotate at a constant speed. Again, what is measured is the torque which has to be exerted in order to sustain the motion of the rotor.

Having accomplished the mixing process, it is now necessary to shape and vulcanise the rubber article. These processes are achieved in virtually one step if shaping is carried out by *moulding*. Moulding is a process in which the unvulcanised rubber is forced into a mould cavity of the same shape as that of the desired product, and is then vulcanised *in situ* in the mould cavity. The vulcanisation may not necessarily be complete; moulding is in some cases followed by a *post-curing* step in which the moulding is further heated—usually in air—in order to improve the properties of the rubber. An important aspect of the post-curing process is that often further crosslinking occurs. Moulding processes for rubbers can be classified under three broad headings:

1.   compression moulding,
2.   transfer moulding,
3.   injection moulding.

In *compression moulding* processes, a suitable blank is placed in the open heated mould cavity, which is then closed and maintained closed for a time which is sufficient to ensure adequate vulcanisation at the particular moulding temperature. In *transfer moulding* processes, the rubber is pre-heated in a chamber prior to being forced through a channel into the mould cavity. *Injection moulding* is a kind of continuous transfer moulding, in which the unvulcanised rubber compound is injected into the mould cavity according to a predetermined cycle which comprises injection, vulcanisa-tion, mould opening, ejection of moulding and closing of mould.

A second type of shaping process which is used in rubber processing technology is *extrusion*. In this process, the unvulcanised rubber com-pound is forced through a die, the shape of which is the principal determinant of the cross-section of the extrudate. Extrusion is used for the manufacture of products such as rubber tubing, insulated cables and weather sealing strip. It is also used for the production of tubes and other types of extrudate (e.g., tyre treads) which will be used subsequently for the construction of other products. But in all cases the shaping process is essentially distinct from the subsequent vulcanisation.

A third type of shaping process which is used in rubber processing technology is *calendering*. In this process, the unvulcanised rubber is formed into a thin sheet by passing through heated steel rollers. The sheet may either be vulcanised as such, or be used for the building of other articles (e.g., football bladders). But, as in the case of extrusion, the shaping process is essentially distinct from the subsequent vulcanisation.

There are, of course, other processes which are used for shaping unvulcanised rubber besides the three broad types described above. Two processes which should be mentioned are the *frictioning* and *skim-coating* of textiles. These are really variants of calendering, but are sufficiently distinctive to merit separate mention. The term 'skim-coating' refers to a process whereby a thin layer of rubber is calendered on to the surface of a textile fabric. In order to ensure satisfactory adhesion between rubber and textile, it may be necessary first to treat the textile with a very thin layer of rubber which is of such a consistency that, under the conditions of treatment, the rubber partially penetrates the interstices of the fabric. This process is known as 'frictioning' because the friction generated by calendering rolls rotating at different speeds is used to drive the rubber compound into the fabric. A further process is *spreading*. This again is a process for applying a thin layer of rubber to a textile fabric. In this case, the rubber is applied in admixture with a rubber solvent, the mixture being commonly known as a *dough*. Spreading is achieved by passing the fabric between a roller and a *doctor blade*. The dough is maintained as a 'bank' behind the doctor blade, and is partially driven into the interstices of the fabric by the action of the doctor blade.

One final comment is necessary in order to complete this brief outline of rubber processing technology. It is that whereas some rubber products are entirely or mainly rubber, others are composite materials. Thus rubber tubing, weather sealing strip, O-ring seals and shoe soles are examples of the former. The outstanding example of the latter is the rubber tyre, which is a composite of rubber, steel and textile fabric. Other examples of rubber composites include hoses and power-transmission belts. Under the broad heading of the *building processes*, to which passing reference has been made above, and which are noted in Fig. 1.2, is included operations in which the various components of the composite are brought together either to form essentially the finished product (apart from vulcanisation) or something which can be converted into the final product by moulding.

Chapter 2

Historical Development of Synthetic Rubbers

2.1 EARLY ENDEAVOURS IN THE MANUFACTURE OF SYNTHETIC RUBBERS

The earliest recorded accounts of the production of synthetic rubbers by polymerisation refer to isoprene (2-methyl-1,3-butadiene) (I), a substance which is closely related to the repeat unit in the molecule of natural rubber.

$$CH_2\!\!=\!\!C\!\!-\!\!CH\!\!=\!\!CH_2$$
$$|$$
$$CH_3$$

(I)

Williams[1] appears to have been the first to isolate this compound from the products obtained by the destructive distillation of natural rubber in the absence of air, and to show that it becomes viscous in the presence of air. Furthermore, when this viscous liquid was heated, it was found to change into a spongy, rubbery substance. These latter observations would today be interpreted in terms of peroxidation of the isoprene followed by polymerisation of the isoprene under the influence of heat, the peroxidised isoprene acting as a catalyst. That isoprene could be converted into a rubbery solid by treatment with hydrochloric acid was demonstrated by Bouchardat[2] and by Tilden.[3] Wallach[4] showed that a similar change occurred spontaneously when isoprene was exposed to light in sealed tubes.

The first patent to be granted for the manufacture of synthetic rubber was to Matthews and Strange in 1910 for the production of a rubbery substance from isoprene by treating the latter with sodium metal. This important discovery was made accidentally in the course of endeavouring to dry isoprene by means of sodium metal. A similar discovery was made almost simultaneously in Germany by Harries and the Bayer Company.

17

At about the same time, interest was being shown by Earle and Kyriakides[5] and by Spence and associates,[6] both groups working in the U.S.A., in the possibility of manufacturing synthetic rubber from 2,3-dimethyl-1,3-butadiene (II). This monomer was principally of interest because it could be prepared fairly readily from acetone by reduction of the latter to pinacol followed by dehydration:

$$CH_3-\underset{\underset{O}{\|}}{\overset{\overset{CH_3}{|}}{C}} \; + \; \underset{\underset{O}{\|}}{\overset{\overset{CH_3}{|}}{C}}-CH_3 \xrightarrow[Mg/Hg]{[H]} CH_3-\underset{\underset{OH}{|}}{\overset{\overset{CH_3}{|}}{C}}----\underset{\underset{OH}{|}}{\overset{\overset{CH_3}{|}}{C}}-CH_3$$

pinacol

$$CH_2=\underset{}{\overset{\overset{CH_3}{|}}{C}}----\underset{}{\overset{\overset{CH_3}{|}}{C}}=CH_2$$

$-2H_2O$ acid catalyst

(II)

Considerable interest was shown in the production of synthetic rubber from dimethylbutadiene in Germany during World War I. Cut off from supplies of natural rubber by the British blockade, it was necessary to seek alternative materials. The product obtained by polymerisation of dimethyl-butadiene was known as 'methyl rubber'. Two grades principally were produced: methyl rubber H intended for hard-rubber goods, and methyl rubber W intended for soft-rubber goods. The H grade was produced by allowing the monomer to stand in partially-filled vessels containing air for some 6–10 weeks at approximately 30 °C. A small amount of polymer was added to accelerate the polymerisation. The product of the polymerisation was hard and was converted to a more elastic material by prolonged milling. The W grade was produced by heating the monomer without a catalyst for 3–6 months at about 70 °C in double-walled pressure vessels. The product was again rather tough and needed to be softened by milling before it could be used as a rubber. A certain amount of a third type (methyl rubber B) was also produced using sodium metal as a polymerisation catalyst. The methyl rubbers were very poor materials compared with natural rubber when used as bases for the rubber compounds in use at the time. However, in passing judgement upon these early attempts to produce synthetic rubber, it must be noted that no reinforcing carbon blacks were available at the time. Indeed, it has been remarked[7] that 'had carbon black attained at the time of the war, as it has today, the position of a recognised rubber compounding ingredient, and had it been available in Germany, the story of synthetic rubber might have been very different from what it was'. Certainly it is the case that several of the general- and special-purpose synthetic rubbers developed subsequently owe their commercial exploitation to the possibility of having otherwise very unsatisfactory mechanical properties enhanced by compounding with carbon black.

At the end of World War I, development work on synthetic rubber ceased in Germany until about 1926. During the late 1920s, interest shifted from dimethylbutadiene to 1,3-butadiene (III) as a monomer because a

$$CH_2{=}CH{-}CH{=}CH_2$$

(III)

more satisfactory rubber was obtained from the latter and because it was more readily avilable than isoprene (which might otherwise have been used as a monomer). Butadiene was produced from acetylene, which was in turn available from calcium carbide or from methane by the electric-furnace process. Polymerisation was effected using sodium metal as catalyst. From this fact originated the word 'Buna' as a generic name for synthetic rubbers produced in Germany at that time and subsequently. The word was obtained by combining the first two letters of 'butadiene' with the first two letters of the German word for sodium, 'Natrium'. The early Buna rubbers were never very satisfactory as general-purpose synthetic rubbers.

Two important developments occurred in the U.S.A. in the years immediately following World War I. These developments are associated with the names of Ostromislensky and Maximoff.[8] The first was the use of butadiene as a monomer, because of its ready availability from ethyl alcohol (by a type of process which will be described in Chapter 3). The second was the introduction of the technique known as *emulsion polymerisation* for converting the monomer to synthetic rubber. Hitherto, most endeavours at producing synthetic rubbers had used the process known as *bulk polymerisation*, that is, the monomer was polymerised in bulk with the addition of at most a catalyst. In the emulsion polymerisation process, the monomer is polymerised in the presence of water, a surface-active substance such as soap, and a polymerisation initiator. The polymer is obtained in the form of a *latex*, that is, a stable dispersion of minute particles of polymer in an essentially aqueous medium. The process will be described in greater detail in Chapter 4. It is sufficient to note here that, compared to bulk polymerisation, the emulsion process offered the advantages of greater rate of polymerisation, improved control of the reaction, greater versatility and a superior product. Until comparatively recently, emulsion polymerisation has been the principal process for producing synthetic rubbers from their monomers. In recent years, however, increasing quantities of synthetic rubber have been made by a process known as *solution polymerisation*, in which, as the name implies, the monomer is polymerised in solution in a suitable solvent.

So far, attention has been confined to general-purpose synthetic rubbers. The most significant early event in the development of special-purpose synthetic rubbers may well have been the announcement in 1931 by the Du Pont Company that they had succeeded in developing a new synthetic rubber which they first called 'Duprene', but whose name was later changed to 'Neoprene'.[9] This synthetic rubber is produced by the polymerisation of 2-chloro-1,3-butadiene (IV), commonly known as 'chloroprene'. This

$$CH_2=C-CH=CH_2$$
$$|$$
$$Cl$$

(IV)

monomer is a kind of analogue of isoprene, in that the methyl group of the latter is replaced by chlorine. It is the presence of this chlorine atom in each repeat unit of the rubber molecule which gives the product its unusual properties, such as moderate resistance to swelling in hydrocarbon oils and resistance to deterioration by heat and ozone.

At about the same time as polychloroprene was being developed, an American chemist by the name of Patrick was attempting to find ways of utilising the large amounts of olefin gases, such as ethylene and propylene, which were being disposed of by combustion at oil refineries. He recovered the gases by chlorination to dichlorides. In the course of his investigations of the reactions of these dichlorides, he discovered that they interact with sodium di- and polysulphides to give rubbery materials.[10] It appears that Patrick was not initially investigating the possible manufacture of synthetic rubber, but rather that of anti-freeze mixtures for use in the radiators of internal-combustion engines. In the course of his experiments, he tried a mixture of ethylene dichloride, sodium polysulphide and a trace of acid. He found that a rubbery material was formed in the radiator, thereby choking it. On trying to remove the rubbery material with petrol, he found the material to be almost completely unaffected. In this way, not only were the polysulphide rubbers themselves discovered, but also their most interesting property, namely, that of extreme resistance to swelling in hydrocarbon fuels. These rubbers were given the generic name 'Thiokol'. They are in effect addition polymers of alkylene polysulphides (V), although the reaction by which they are prepared is a condensation polymerisation, not an addition polymerisation, e.g.:

$$n\text{ClCH}_2\text{CH}_2\text{Cl} + n\text{Na}_2\text{S}_x \longrightarrow -(\text{CH}_2\text{CH}_2\text{S}_x)_n^- + 2n\text{NaCl}$$

(V)

Although polysulphide rubbers have rather poor mechanical properties as well as a bad odour, their interesting and important property of very high resistance to swelling in hydrocarbon oils has ensured continuing commercial interest in them. The formation of rubbery products by reaction between ethylene dichloride and potassium sulphide had also apparently been observed previously during the latter part of World War I by chemists looking for better ways to make mustard gas (2,2'-dichlorodiethyl sulphide), but the reaction product was never successfully developed at that stage.[11]

A few years after the development of polychloroprene and the polysulphides came the development of an important family of synthetic rubbers derived by copolymerisation of butadiene and acrylonitrile (VI).

$$\text{CH}_2=\text{CH}$$
$$|$$
$$\text{CN}$$

(VI)

The weight percentage of acrylonitrile in the copolymers varied over the range 25–40 %. Marketed under a variety of trade names such as 'Buna N', 'Perbunan', 'Ameripol', 'Butaprene', 'Chemigum', 'Hycar' and 'Paracril', these synthetic rubbers offered a most useful combination of good mechanical properties and excellent resistance to swelling in hydrocarbon oils. Furthermore, properties such as oil resistance and tendency to stiffen at low temperatures could be usefully varied by varying the ratio of acrylonitrile to butadiene in the rubber molecule.

A very important class of general-purpose synthetic rubbers comprises those which are produced by copolymerising styrene (VII) and butadiene. The initial development work on styrene–butadiene copolymers as general-purpose synthetic rubbers was carried out in Germany in the early 1930s.

$$\text{CH}_2=\text{CH}$$

(VII)

Having become convinced that the most promising butadiene-based synthetic rubbers were those produced by emulsion polymerisation rather than those produced by bulk polymerisation using sodium metal as catalyst, German workers then endeavoured to improve the mechanical properties of the rubber by copolymerising the butadiene with a second monomer. Copolymers of butadiene with styrene and with acrylonitrile were investigated. Emulsion-polymerised copolymers of styrene and butadiene were found to have better mechanical properties than butadiene homopolymers. Copolymers of acrylonitrile and butadiene were found to have interesting oil-resistance properties, as has already been noted above. The rubbers obtained from butadiene and styrene were designated as 'Buna S' and those from butadiene and acrylonitrile as 'Buna N'. The Buna S rubbers were tough materials which had to be softened by heating in air, as well as by the addition of plasticising oils, before they could be processed on conventional rubber-manufacturing equipment. Interest was shown in emulsion-polymerised styrene–butadiene rubber in the U.S.A. during the 1930s, and commercial arrangements were made whereby American rubber and oil companies could have access to German materials and technology. However, up until the late 1930s the principal American interest was in Buna N as a special-purpose oil-resistant rubber, rather than in Buna S as a general-purpose replacement for natural rubber. (Buna N was introduced into the U.S.A. in 1937 following disruption of the supply of chloroprene from the Du Pont Company owing to an explosion at their plant.)

This brings the story of the development of synthetic rubbers up to the outbreak of World War II. Because of the consequences of the momentous events of that conflict for the development of synthetic rubbers, it is appropriate to regard the period of World War II as constituting the second major phase of the historical development of synthetic rubbers. It is to this phase that the following section of this chapter will be devoted. The final section of this chapter will be concerned with developments in the manufacture of synthetic rubbers since World War II; these advances are conveniently regarded as comprising a third phase of the development of synthetic rubbers.

2.2   DEVELOPMENTS IN SYNTHETIC RUBBER MANUFACTURE DURING WORLD WAR II

The outstanding development in synthetic rubber manufacture during World War II was undoubtedly that of emulsion-polymerised styrene–

butadiene rubber. This development took place in both Germany and the U.S.A. In both countries it was prompted by a shortage of natural rubber. The shortage was of critical significance for the Allies, because during the early part of the war the Japanese were able to overrun many of the natural-rubber-producing areas in the Far East.

We have already noted that the initial development work on styrene–butadiene synthetic rubbers was carried out in Germany in the early 1930s. Furthermore, although these rubbers were known in the U.S.A., the Americans were until the late 1930s more interested in acrylonitrile–butadiene rubbers as special-purpose oil-resistant rubbers than in styrene–butadiene rubbers as general-purpose replacements for natural rubbers. Indeed, such styrene–butadiene rubber as was required in the U.S.A. tended to be imported from Germany.

The situation in the U.S.A. changed dramatically with the outbreak of World War II in September 1939. Imports of synthetic rubber from Germany ceased. The fall of France in May 1940 made the possibility of a shortage of natural rubber increasingly likely. From experience gained in producing and using the pilot-plant quantities of styrene–butadiene rubber which were being manufactured at the outbreak of World War II, it became clear that this type of rubber was the most promising candidate upon which to concentrate as a replacement for natural rubber in tyres and inner tubes, that is, in those rubber products which were of critical importance for the pursuance of the war. The version of this rubber which was made with a 25/75 styrene/butadiene charge ratio became known as GR–S (Government Rubber–Styrene).† One important respect in which the technology of the GR–S rubbers was an improvement on that of the German Buna S rubbers from which they were developed was that GR–S did not require to be softened by heating in air. Softening by this means was regarded as unsatisfactory because it was thought that it might lead to oxidative crosslinking with consequent deterioration in dynamic mechanical properties. In the case of the GR–S rubbers, control of plasticity was achieved by the inclusion of mercaptan modifiers in the emulsion polymerisation recipe.

Rubber was designated as a strategic and critical material by the U.S. President on 28th June 1940. On the same day, the Rubber Reserve Company was set up by the U.S. Reconstruction Finance Corporation. According to Dunbrook,[12] the Rubber Reserve Company 'took over the

† Acrylonitrile–butadiene rubber was subsequently known as 'GR–A' (Government Rubber–Acrylonitrile) and chloroprene rubber as 'GR–M' (Government Rubber–Monovinylacetylene).

responsibility of laying the foundations of the synthetic-rubber industry and was charged with the responsibility of producing the raw materials necessary for the manufacture of synthetic rubber, the actual manufacture of the synthetic rubber and the sale and use of the synthetic rubber'. The ensuing synthetic rubber programme was a cooperative activity involving several rubber-manufacturing and oil companies. Because of the urgency of the need, arrangements were made for full exchange of technical information between the various companies and other institutions concerned with the synthetic rubber programme, and for the cross-granting of licences in respect of patents for particular processes owned or controlled by the various participating parties. In this way, provision was made for full and effective interchange of technical information amongst the several private organisations concerned in the programme, and also for patent protection. Arrangements for patent protection covered not only patents in existence at the inception of the programme, but also those which arose as a consequence of developments made in the course of the execution of the programme.

Following the conquest by the Japanese of many of the rubber-growing areas of the Far East, the rubber supply situation became very critical in the summer of 1942. It was estimated that unless new supplies of rubber—either natural or synthetic—could be obtained, stocks would be exhausted before the end of the summer of 1943. A 'Rubber Survey Committee' was appointed by the U.S. President on 6th August 1942 to study the situation. This committee recommended reorganisation and consolidation of the governmental agencies then dealing with the supply of rubber. All sensible means for conserving, reclaiming and producing rubber were urged, including a significant increase in the production of styrene–butadiene rubber. A 'Rubber Director' was appointed, his job being to oversee the implementation of these recommendations. Special powers were vested in him in respect of matters relating to rubber. Largely because of the success of the synthetic rubber programme, the office of the Rubber Director was dissolved in 1944, and the special powers terminated.

The success of the U.S. World War II synthetic rubber programme is an outstanding example of what can be achieved by cooperation in a critical situation such as arises from a war-time emergency. An enormous amount of development work was carried out in the laboratories of private companies and of universities. Some of this work was concerned with deepening understanding of the fundamentals of the process of emulsion polymerisation. Some was concerned with the more practical aspects of the emulsion copolymerisation of styrene and butadiene. Some was concerned with the processing and production technology of styrene–butadiene rubbers. Of the information which was acquired in the course of the execution of the American synthetic rubber programme, a certain amount became publicly available shortly after the end of World War II by way of publication in American scientific and technical journals, such as the *Journal of the American Chemical Society*, the *Journal of Polymer Science*, and *Industrial and Engineering Chemistry*. No doubt much of the information still remains buried in the files of the organisations which participated in the programme. It is known that at least some of the unpublished information has been lost for ever as a consequence of the flooding of a building in the basement of which the information was stored.

One other important synthetic rubber development which occurred during World War II must also be mentioned. This concerns the so-called 'butyl' rubber, a copolymer of isobutene(VIII) with a minor amount of

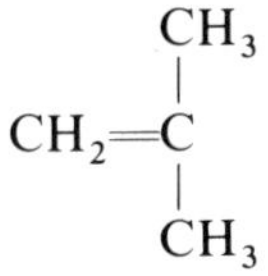

$$CH_2\!\!=\!\!\underset{\displaystyle \underset{CH_3}{|}}{\overset{\displaystyle \overset{CH_3}{|}}{C}}$$

(VIII)

isoprene(I). This rubber was known during the war as 'GR–I' (Government Rubber–Isobutylene). As will be described subsequently (Chapter 7), it is made by a type of polymerisation (cationic polymerisation) which is otherwise little used for the production of synthetic rubbers. Butyl rubber was first announced in 1940. It has two properties which made it of special interest for certain applications:

1.   It has relatively low permeability to gases, and was therefore of immediate interest for the manufacture of inner tubes for use with tyres.
2.   It displays good resistance to certain deteriorative influences, notably to cracking by ozone and to embrittlement on ageing, thereby making it of interest for cable insulation and for articles such as rubber hoses, grommets and weather-seals.

Like the styrene–butadiene rubbers developed during World War II, butyl rubber is still made and sold today, although, of course, there have been significant improvements in the types available.

2.3 DEVELOPMENTS IN SYNTHETIC RUBBER MANUFACTURE SINCE WORLD WAR II

The American cooperative programme which led to the successful development of styrene–butadiene rubber during World War II came to an end shortly after the end of the war. Since that time, the development of synthetic rubbers in non-communist countries has been determined principally by normal commercial factors such as demand, supply, and cost and availability of raw materials. Two additional factors have also had an influence:

1. the tendency of governments to 'stockpile' essential materials such as rubber during times of crisis, as happened in the U.S.A. during the Korean war; and
2. the need for special-purpose synthetic rubbers in small quantities to meet exacting requirements in connection with various high-technology ventures such as aerospace programmes.

With regard to the second of these factors, it should be noted that, because high-technology ventures are frequently financed by governments and governmental agencies, it is the case that many of the more exotic special-purpose synthetic rubbers which have appeared since the end of World War II have been developed either in government laboratories or by way of government contracts placed with other laboratories. It seems likely that few of these newer types of special-purpose synthetic rubber would have been developed in the West had normal commercial considerations been operative.

As may be inferred from what has been said in the preceding paragraph, an extremely diverse range of special-purpose synthetic rubbers has been developed since the end of World War II. This range has extended from inorganic elastomers intended for use at very high temperatures to elastomeric hydrogels intended for bio-medical applications such as soft contact lenses. It would be quite inappropriate to describe these materials here, although reference will be made to some of them in subsequent chapters. What will be attempted here is an outline of some of the more important developments in the area of general-purpose synthetic rubbers.

A series of important developments in the manufacture of emulsion-polymerised styrene–butadiene rubbers has been exploited since the end of World War II. Of these, the first and most important was the discovery that the emulsion polymerisation reaction could with advantage be carried out at much lower temperatures (c. 5 °C) than hitherto (c. 50 °C). Special types of polymerisation initiator are required for this reaction, and these will be described briefly subsequently (Chapter 5). What is important here is to note the advantages which accrue from carrying out polymerisation at low temperature. It is found that the polymerisation temperature affects the microstructure of the polymer which is formed (that is, the way in which the successive monomer units become enchained in the polymer sequence), and also the molecular weight and gel content of the polymer. One practical consequence is that some of the mechanical properties of vulcanisates prepared from the rubber are improved, notably the resistance to abrasion which is so important in tyres.

Styrene–butadiene rubber produced by low-temperature polymerisation has a higher molecular weight than does that produced at higher temperatures, and also lacks the low-molecular-weight 'tail' of the latter. Consequently, it is possible in effect to replace the low-molecular-weight 'tail' by a cheap hydrocarbon oil. This is the principle of the 'oil-extended' synthetic rubbers. The principal motivation in the development of these materials was the desire to produce cheaper rubbers without seriously compromising the mechanical properties of vulcanisates obtained from them. A further development of rather lesser importance has been that of carbon-black masterbatching. This was made possible by the fact that the emulsion polymerisation reaction gives the polymer in the form of a latex, which, as has been noted earlier, is a dispersion of minute particles of the rubber in an aqueous medium. It is therefore possible to blend the latex with an aqueous dispersion of carbon black, and then to co-coagulate the two to give an intimate masterbatch of rubber and carbon black.

Another important development in the technology of styrene–butadiene rubbers also depended upon the fact that the emulsion polymerisation reaction initially gives the polymer in the form of a latex. Hitherto this circumstance had been regarded as something of a disadvantage in that, insofar as the rubber was to be used in the dry form, it has to be separated from the latex by some such process as coagulation. In the years immediately following World War II, rubber scientists and technologists began to explore very seriously the possibility of using the product of emulsion polymerisation as a replacement for natural rubber latex. Ways were found of increasing the solids content of styrene–butadiene rubber latex to a level equal to or greater than that of concentrated natural rubber latex without at the same time causing the viscosity of the latex to rise to an unacceptable level. For many years now, concentrated styrene–butadiene rubber latices have been used for the production of latex foam rubber, either alone or in admixture with natural rubber latex. A further significant development has been the discovery that the properties of synthetic latices (of those containing other types of rubber as well as styrene–butadiene rubber), and of the polymer which they contain, can be usefully modified by incorporating minor amounts (up to 5% by weight) of unsaturated carboxylic acids in the monomer mixture from which the latex was prepared. An example of such a monomer is acrylic acid (IX). The resultant

$$CH_2{=}CH$$
$$|$$
$$CO_2H$$

(IX)

latices and rubbers are described as being 'carboxylated'. The advantages which carboxylation confers upon the latex include (a) enhanced latex colloidal stability, (b) improved tolerance to the addition of inorganic fillers, (c) improved adhesion to paper and textile fibres, and (d) the possibility of the copolymer participating in novel crosslinking reactions by way of the copolymerised carboxylic acid functionality. The subject of synthetic rubber latices is a very extensive and specialised aspect of the general subject of synthetic rubbers. Because of their specialised nature, synthetic rubber latices have been excluded from the scope of this present book. However, in reviewing the development of synthetic rubbers, it is necessary to note that a significant contribution in the post-World-War II period has been the exploitation of synthetic rubber latices as latices.

Perhaps the most exciting development in the area of general-purpose synthetic rubbers since the end of World War II is that of the stereospecific polyisoprenes—the 'synthetic natural' rubbers. (Of almost equal interest is the discovery of methods of manufacturing stereospecific polybutadienes.) In 1956, two methods were disclosed for the polymerisation of isoprene to essentially cis-1,4-polyisoprene. In one of them,[13] polymerisation is effected in solution by means of lithium metal or by alkyllithiums. As will be seen subsequently (Chapter 4), the reaction is an anionic polymerisation. Control of the reaction conditions and of the purity of the reactants is essential if a satisfactory product is to be obtained. The second method[14] uses a catalyst of the so-called 'Ziegler–Natta' type, that is, a complex formed by reacting an organometallic compound with a transition-metal compound. Both these processes were developed in the U.S.A. It has been suggested that the search for processes of this kind was motivated by fears of further world conflict or unrest which would again cut off supplies of natural rubber from the Far East. There was deep concern that the U.S.A.

should never again be in a position of complete dependence upon foreign sources of natural rubber.

In the years immediately following the announcement of practicable routes to 'synthetic natural' rubber, it was widely forecast that these discoveries would herald the demise of the natural-rubber-producing industry; 20 years was typical of the predicted time-scale over which this demise would occur. In fact, no such demise has occurred nor is likely to occur. Several factors have been responsible for the failure of this prophecy. These include (a) the sharp rise in world petroleum prices and in the prices of all products derived from petroleum, (b) the fact that the rubber tree produces polyisoprene from renewable starting materials, and (c) the fact that the rubber tree produces polyisoprenes by a process which does not tend to pollute the environment (although, of course, the subsequent processing of the product obtained from the tree may cause pollution).

All of the synthetic rubbers noted so far, with the exception of butyl rubber and the polysulphide rubbers, are made either wholly or mainly from diene monomers. With the ready availability of the simple olefin monomers ethylene (X) and propylene (XI) and the development of new

$$CH_2{=}CH_2 \qquad\qquad CH_2{=}CH$$
$$|$$
$$CH_3$$

(X)   (XI)

methods of polymerising them, it became inevitable that useful elastomers would eventually be produced from them. The development of ethylene–propylene copolymers which can be used as general-purpose synthetic rubbers is yet another important landmark in the story of the development of synthetic rubbers since World War II. Like butyl rubber, these materials show excellent resistance to heat, light and ozone. In order that they may be vulcanisable by means of conventional sulphur/accelerator systems, it is necessary that small amounts of dienes be included in the monomer mixture from which they are prepared, thus imparting a minor degree of unsaturation to the polymer.

There are three other developments which must be mentioned in this general review. The first concerns that of materials which are often known as 'thermoplastic rubbers'. These are materials which are rubbery at room temperature but which on heating melt to viscous liquids which can be moulded and otherwise processed. The change is quite reversible; on cooling, the melt reverts to a solid material having rubbery properties. The reason why the change is reversible is that the development of rubber properties does not depend upon the occurrence of chemical crosslinking reactions, but depends instead upon the occurrence of certain physical changes. Many types of thermoplastic rubber have been developed. This subject will be discussed in some detail in a later chapter (Chapter 10). It is sufficient to note here that one commercially-important type is a styrene–butadiene–styrene 'block' copolymer (that is, the molecule comprises a 'block' of styrene units followed by a 'block' of butadiene units followed by another 'block' of styrene units), and that these copolymers are manufactured by solution polymerisation.

The second development is that of powdered rubbers. These are rubbers in a finely-powdered form which facilitates mixing of the rubber with other compounding ingredients. Although some interest has been shown subsequently in the production of natural rubber in powder form, this development was initially made by producers of synthetic rubbers.

The third development is that of fluid rubbers. These are high-viscosity liquids which are sufficiently fluid to be capable of being cast into moulds, and which can then be vulcanised to give solid elastomeric materials. At the time of writing, there has been little successful commercial exploitation of this idea, except in the field of cast urethane elastomers. However, the idea is an interesting one, and it may be that the future will see greater use being made of it.

It may be asked what developments remain to be made in connection with general-purpose synthetic rubbers. There is certainly one outstanding development which has yet to be made, and that concerns the manufacture of stereospecific polydiene rubbers by emulsion polymerisation. Although it has been found possible to effect the stereospecific emulsion polymerisation of butadiene using catalysts such as rhodium chloride, it has unfortunately not so far been possible to produce useful rubbery polymers by this means. Thus rhodium chloride catalysis produces a *trans*-1,4-polybutadiene of rather low molecular weight, which possesses no rubbery properties at all. All attempts to synthesise a reasonably linear high-molecular-weight *cis*-1,4-polybutadiene by an emulsion polymerisation reaction seem so far to have met with failure. The successful attainment of this objective would be of considerable industrial interest for at least two reasons: (a) it would provide a means of producing a 'synthetic natural' rubber latex by emulsion polymerisation, and (b) it might provide a more convenient means of manufacturing stereospecific polydienes for use as dry synthetic rubbers.

An outstanding problem in connection with the production of synthetic rubbers of all types which seems certain to arise before many decades have elapsed is that of alternative sources to petroleum for monomers. One suggestion which has been made is that monomers might be obtained from the products of destructive distillation of natural rubber as starting materials. This scarcely seems a feasible possibility. A more likely route would be from ethyl alcohol as starting material, the alcohol being obtained from, for example, cane sugar by fermentation.

GENERAL BIBLIOGRAPHY

Whitby, G. S., Davis, C. C. and Dunbrook, R. F. (Eds.) (1954). *Synthetic Rubber*, John Wiley and Sons, New York. (Especially Chapters 1, 2, 7 and 8.)

Cooper, W. (1966). Elastomers, synthetic. In: *Encyclopedia of Polymer Science and Technology*, Vol. 5, John Wiley and Sons, New York, pp. 406f.

D'Ianni, J. D. (1961). Newer synthetic rubbers derived from olefins and diolefins, *Rubb. Chem. Technol.*, **34**, 361.

Dunbrook, R. F. Series of papers entitled: 'Contributions of Organic Chemistry to the War Effort—Synthetic Rubber', *India Rubber World* (1947), **117**(2), 203; (1947), **117**(3), 355; (1948), **117**(4), 486; (1948), **117**(5), 617; (1948), **117**(6), 745.

Saltman, W. M. (Ed.) (1977). *The Stereo Rubbers*, John Wiley and Sons, New York. (Especially Chapter 1.)

REFERENCES

1. Williams, G. (1860). *Proc. Roy. Soc.*, **10**, 516; *Phil. Mag.*, **80**, 245.
2. Bouchardat, G. (1875). *Compt. Rend.*, **80**, 1446; (1879), **89**, 1117; (1875), *Bull. Soc. Chim.*, **24**, 108.
3. Tilden, W. A. (1884). *J. Chem. Soc.*, **45**, 411.
4. Wallach, O. (1887). *Ann.*, **239**, 48.
5. Kyriakides, L. P. (1945). *Chem. Eng. News*, **23**, 531; (1914), *J. Amer. Chem. Soc.*, **36**, 530, 657, 663, 980.
6. Semon, W. L. (1943). *Chem. Eng. News*, **21**, 1613.
7. Whitby, G. S. and Kata, M. (1933). *Ind. Eng. Chem.*, **25**, 1204.
8. U.S. Rubber Co. Private communication to R. F. Dunbrook.
9. Carothers, W. H., Williams, I., Collins, A. M. and Kirby, J. E. (1931). *J. Amer. Chem. Soc.*, **53**, 4203.
10. Patrick, C. J., U.S. Patent 1 890 191, 1932.
11. Fisher, H. L. and Moskowitz, M. Private communication to R. F. Dunbrook.
12. Dunbrook, R. F. (1954). In: *Synthetic Rubber*, Whitby, G. S., Davis, C. C. and Dunbrook, R. F. (Eds.), John Wiley and Sons, New York, p. 42.
13. Stavely, F. W., *et al.* (1956). *Ind. Eng. Chem.*, **48**, 778.
14. Horne, S. E. *et al.* (1956). *Ind. Eng. Chem.*, **48**, 784.

Chapter 3

Monomers for Synthetic Rubber Production

3.1 INTRODUCTION

The purpose of this chapter is to describe briefly industrial processes for the manufacture of the more common of the olefinic monomers which are used in the production of synthetic rubbers. Some indication of the properties of these monomers will also be given. These processes have been gathered together in one chapter for convenience in order to avoid unnecessary subsequent repetition or cross-referencing, and also to avoid interruption of the subsequent development of the main theme of this book which is the production, properties and technology of the synthetic rubbers themselves.

Some introductory comment is desirable concerning the types of monomer which have found widespread application in the manufacture of synthetic rubbers. Of course, the monomers which are actually used in large quantity probably represent a compromise between desirability and availability. Clearly, however desirable a particular monomer may be, it will not find wide application unless an economic synthetic route is available for its manufacture. But, notwithstanding this, it is possible to make some very broad generalisations concerning the common monomers which are now used for the production of synthetic rubbers. The common monomers which are used can be conveniently classified into two broad groups as follows:

1. those which impart flexibility to the polymer sequence of the rubber molecule; and
2. those which do not impart flexibility but do confer some other desirable property or properties.

In Chapter 1, it was noted that an essential prerequisite for a polymer to behave as a rubber is that it should comprise linear sequences which are flexible in the manner indicated in Fig. 1.1. Bulky substituents which would inhibit free rotation about main-chain bonds should be absent, as also should strongly polar substituents which would give rise to strong attractions between neighbouring segments of the polymer chains. In fact, the main-chain flexibility of most synthetic rubbers derives from the presence in the polymer sequence of units of either the type (XII) or the type (XIII). In both cases, the substituents are either hydrogen, or fairly small non-polar groups, or occasionally a moderately polar group, or occasionally (e.g., in units derived from *n*-butyl acrylate) groups of atoms which although partially polar also contain moieties which are able to keep the polymer segments apart and thus reduce the attractions between the polar parts of the whole substituent group. It may seem odd at first sight that a sequence of units like (XII) which contain carbon–carbon double bonds should be flexible when it is well known that rotation about such double bonds is difficult. Whilst it is true that carbon–carbon double bonds are themselves rotationally rigid at normal temperatures, the presence of the carbon–carbon double bonds increases the rotational flexibility of the adjacent carbon–carbon single bonds, essentially because the number of substituents is reduced. The double bond in the unit of the type (XII) usually has the *cis* configuration with respect to the remainder of the polymer sequence, although this is not an invariable rule; in one important family of synthetic rubbers in which the main-chain flexibility derives from units of type (XII), the double bond has predominantly the *trans* configuration. Units of types (XII) and (XIII) are usually obtained by the addition polymerisation of monomers of types (XIIa) and (XIIIa), respectively. Amongst the other types of repeat unit which confer mainchain flexibility upon a polymer sequence, and which are utilised in the production of synthetic rubbers, are included the types (XIV) and (XV). The substituents are either hydrogen or relatively small non-polar groups. Units of the oxyethylene type (XIV) are obtained by the addition polymerisation of the corresponding ethylene oxide derivative (XIVa). Units of type (XV) are obtained by the condensation polymerisation of the corresponding dihalogen alkane and an alkali-metal polysulphide of appropriate sulphur rank (XVa). Lest it should be thought that all flexible polymer sequences are based upon main chains which are either exclusively carbon or at least contain carbon, mention should be made of an important example of a unit whose main chain does not contain carbon. This is the siloxane unit (XVI) upon which is based the silicone rubbers. The

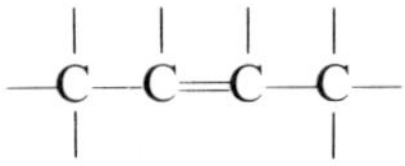

$$-\overset{|}{\underset{|}{C}}-\overset{|}{C}=\overset{|}{C}-\overset{|}{\underset{|}{C}}-$$

(XII)

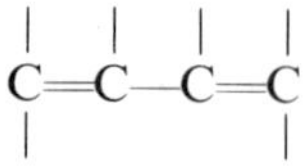

$$\overset{|}{C}=\overset{|}{C}-\overset{|}{C}=\overset{|}{C}$$

(XIIa)

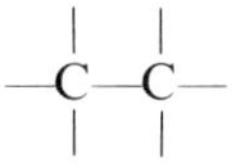

$$-\overset{|}{\underset{|}{C}}-\overset{|}{\underset{|}{C}}-$$

(XIII)

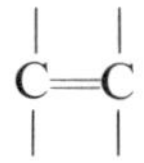

$$\overset{|}{C}=\overset{|}{C}$$

(XIIIa)

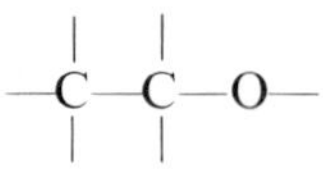

$$-\overset{|}{\underset{|}{C}}-\overset{|}{\underset{|}{C}}-O-$$

(XIV)

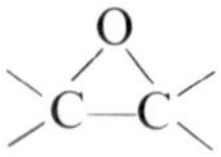

$$\overset{O}{\overset{/\backslash}{>C-C<}}$$

(XIVa)

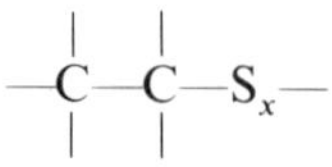

$$-\overset{|}{\underset{|}{C}}-\overset{|}{\underset{|}{C}}-S_x-$$

(XV)

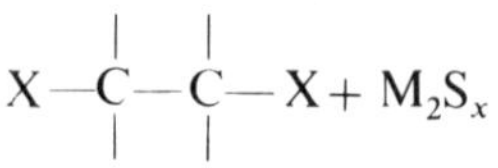

$$X-\overset{|}{\underset{|}{C}}-\overset{|}{\underset{|}{C}}-X + M_2S_x \quad X = \text{halogen}$$

(XVa)

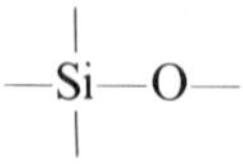

$$-\overset{|}{Si}-O-$$

(XVI)

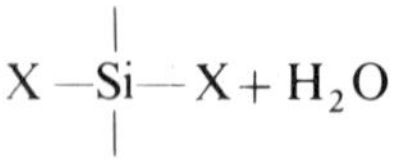

$$X-\overset{|}{\underset{|}{Si}}-X + H_2O \quad\quad X = \text{halogen}$$

(XVIa)

substituents are small alkyl, alkenyl or phenyl groups. Sequences of units of this type are obtained by condensation polymerisation between the corresponding dihalogen silane and water (XVIa).

Amongst the various repeat units which modify the properties of flexible polymer sequences but which do not themselves contribute to the flexibility (at least at normal temperatures), are included the types (XVII) and (XVIII). The aryl-substituted units of type (XVII), derived from monomers of type (XVIIa), inhibit the segmental mobility of polymer sequences and can give rise to the presence of rigid domains within a rubbery matrix. Units of the type (XVIII), derived from monomers of type (XVIIIa), are used to impart polarity to the polymer as a whole, thereby increasing the resistance of the rubber to swelling in hydrocarbon fluids.

It remains to note one further type of unit which is used as the main component of synthetic polymers which are used as the basis for rubbery materials, but which are not themselves rubbery at normal temperatures. This is the unit of type (XIX), obtained by addition polymerisation of the monomer (XIXa), the substituents being hydrogen. Although polymers

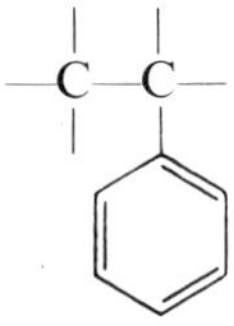

$$-\overset{|}{\underset{|}{C}}-\overset{|}{\underset{}{C}}-$$

(XVII)

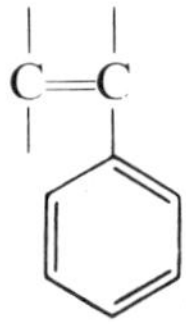

$$\overset{|}{\underset{|}{C}}=\overset{|}{\underset{}{C}}$$

(XVIIa)

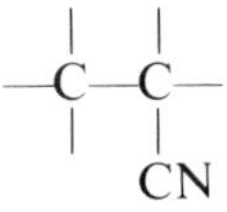

$$-\overset{|}{\underset{|}{C}}-\overset{|}{\underset{|}{C}}- \\ CN$$

(XVIII)

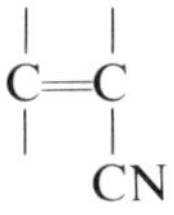

$$\overset{|}{\underset{}{C}}=\overset{|}{\underset{|}{C}} \\ CN$$

(XVIIIa)

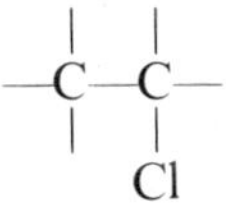

$$-\overset{|}{\underset{|}{C}}-\overset{|}{\underset{|}{C}}- \\ Cl$$

(XIX)

$$\overset{|}{\underset{}{C}}=\overset{|}{\underset{|}{C}} \\ Cl$$

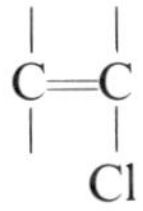

(XIXa)

comprising sequences of the unit (XIX) are rigid and non-rubbery at normal temperatures, they become flexible and more-or-less rubbery if they are compounded with sufficient of a plasticiser of low molecular weight. In this case, the segmental flexibility and mobility which is an essential prerequisite for rubbery properties is not an inherent property of the polymer sequence itself, but is a property which develops when the polymer repeat units are separated from one another laterally by the presence of the plasticiser molecules.

3.2 BUTADIENE (1,3-BUTADIENE)

Two types of processes have been used for the large-scale production of butadiene:

1. those which use petroleum-derived C_4 hydrocarbons as the starting material; and
2. those which use ethyl alcohol as the starting material.

Processes of the first type fall into three broad classes:

(a) those in which butadiene is produced from n-butenes;
(b) those in which butadiene is produced from n-butane, and
(c) those in which butadiene is produced directly by the steam cracking of naphtha and gas–oil petroleum fractions.

Other processes which have been used for, or proposed for, the production of butadiene include:

3.   the Reppe processes, for which the starting materials are acetylene and formaldehyde; and

4.   a process for which acetaldehyde is the starting material.

The earliest large-scale processes for the production of butadiene were those which use ethyl alcohol as the starting material; indeed, butadiene from petroleum sources only became readily available in 1939. Furthermore, because of the great need to make petroleum available for aviation fuel during World War II, over 50% of the butadiene produced in the U.S.A. during the years 1940–45 was made from ethyl alcohol. However, the economics of alcohol processes being very unfavourable compared to petroleum processes, alcohol processes were abandoned after the cessation of World War II, although they have been revived for short times during and after the Korean War. The various synthesis routes which have been commonly used for the production of butadiene have been summarised by Hofmann[1] as shown in Table 3.1. Also given in this table is an indication of the relative masses of starting materials which are required to produce a given mass of butadiene by each route.

3.2.1 Production of Butadiene from n-Butenes

The n-butenes are obtained from the C_3–C_4 streams from the cracking of petroleum crudes. Propane and isobutane are removed by fractionation, and isobutene is removed by absorption in c. 65% sulphuric acid. The stream is then subjected to catalytic dehydrogenation using a catalyst such as mixed calcium–nickel phosphate stabilised with chromium oxide. 1,3-Butadiene is produced by reactions such as:

$$CH_2{=}CH{-}CH_2{-}CH_3 \longrightarrow CH_2{=}CH{-}CH{=}CH_2 + H_2$$
1-butene

$$CH_3{-}CH{=}CH{-}CH_3 \longrightarrow CH_2{=}CH{-}CH{=}CH_2 + H_2$$
2-butene

Typically, a mixed stream containing 1-butene and 2-butene is preheated to c. 650 °C, mixed with superheated steam, and then passed through a fixed-bed catalyst. Contact times are short, being of the order of 0·2 s. The hot exit gases are cooled, compressed, then treated with light naphtha absorber oil in order to remove propanes and pentanes. Further purification of the butadiene is necessary before it is suitable for polymerisation.

TABLE 3.1

SYNTHESIS ROUTES FOR BUTADIENE, AS SUMMARISED BY HOFMANN.[1] THE NUMBERS IN PARENTHESES GIVE THE RELATIVE MASSES OF STARTING MATERIAL NECESSARY TO PRODUCE 100 PARTS BY WEIGHT OF BUTADIENE

Acetylene	Acetylene	Ethyl alcohol	n-Butane	n-Butene	n-Butane	Petroleum distillate
(155)	(150)	(250)	(170–190)	(120–140)	(170–190)	(2 300–3 000)
$\rightarrow$	$\rightarrow$	$\rightarrow$	$\rightarrow$	$\rightarrow$	$\rightarrow$	$\rightarrow$
Acetaldehyde	Acetaldehyde	Butadiene	n-Butene	Butadiene	Butadiene	C_4-fraction
$\rightarrow$	$\rightarrow$		$\rightarrow$			$\rightarrow$
Acetaldol	Ethyl alcohol		Butadiene			Butadiene
$\rightarrow$	$\rightarrow$					
Butylene glycol	Butadiene					
$\rightarrow$						
Butadiene						

The ratio of steam to n-butenes is in the range 15 to 20:1. The purpose of mixing with steam is to reduce the partial pressure of the butenes. Not only does this favour the dehydrogenation reaction (Le Châtelier's principle); it also discourages polymerisation of the butadiene which is formed.

Processes have also been developed for the production of butadiene from n-butenes by oxidative dehydrogenation. Air can be used as the oxidant, the nitrogen functioning as a diluent.

3.2.2 Production of Butadiene from n-Butane

Butadiene can be produced from n-butane by a single-step dehydrogenation reaction using a chromium–aluminium catalyst:

$$CH_3-CH_2-CH_2-CH_3 \longrightarrow CH_2=CH-CH=CH_2 + 2H_2$$
n-butane

The temperature must be in excess of 600 °C, being typically 620 °C. Pressures are low (typically 3 lb in^{-2} absolute pressure) in order to favour the dehydrogenation. This process is known as the Houdry process. There is also a process—known as the Phillips process—for the dehydrogenation of n-butane to butadiene in two steps, the first being dehydrogenation to n-butenes.

3.2.3 Production of Butadiene by Steam Cracking of Naphtha Petroleum Fraction

This is now a process of great importance, especially in Europe. The thermal cracking of hydrocarbons in the naphtha petroleum fraction by heating at 700–900 °C in the presence of steam is now an important source of various olefins and diolefins, such as ethylene and butadiene. No catalyst is used. Residence times are short. Steam–hydrocarbon ratios are generally in the range 0·2–0·8. The product of the cracking process depends upon both the nature of the feedstock and the cracking conditions.

3.2.4 Production of Butadiene from Ethyl Alcohol

The ethyl alcohol for these processes has usually been obtained by fermentation. In a typical process, ethyl alcohol is first dehydrogenated to acetaldehyde by passing at 250 °C over a supported copper catalyst containing a little chromium oxide. A mixture of acetaldehyde and ethyl alcohol is then passed at 350 °C over a tantalum oxide catalyst supported on silica gel. The chemistry of the process is said to be as follows:

$$C_2H_5OH \longrightarrow CH_3.CHO + H_2$$

$$2CH_3.CHO \longrightarrow CH_3.CHOH.CH_2.CHO \xrightarrow{-H_2O} CH_3.CH=CH.CHO$$
$$\text{acetaldol} \qquad\qquad\qquad \text{crotonaldehyde}$$

$$CH_3.CH=CH.CHO + CH_3.CH_2OH \longrightarrow$$
$$CH_3.CH=CH.CH_2OH + CH_3.CHO$$
$$\text{crotyl alcohol}$$

$$CH_3.CH=CH.CH_2OH \xrightarrow{-H_2O} CH_2=CH-CH=CH_2$$

Single-step processes for the production of butadiene from ethyl alcohol have also been developed, the overall reaction being:

$$2C_2H_5OH \longrightarrow CH_2=CH-CH=CH_2 + 2H_2O + H_2$$

3.2.5 The Reppe Process

Acetylene and formaldehyde are passed over a copper(I) acetylide catalyst to form 2-butyne-1,4-diol. Typical reaction conditions are 100 °C and 4·5 atm pressure. The 2-butyne-1,4-diol is then hydrogenated to 1,4-butanediol using a copper/nickel catalyst, the reaction conditions being 120 °C and 300 atm. The final step is removal of water by passing over trisodium orthophosphate at 280 °C and 1 atm pressure. The reactions which occur are as follows:

$$CH{\equiv}CH + 2H.CHO \longrightarrow HOCH_2-C{\equiv}C-CH_2OH$$
$$\text{2-butyne-1,4-diol}$$

$$\Big\downarrow {+2H_2}$$

$$CH_2=CH-CH=CH_2 \xleftarrow{-H_2O} \underset{\substack{CH_2 \quad CH_2 \\ \diagdown O \diagup}}{CH_2-CH_2} \xleftarrow{-H_2O} HOCH_2.CH_2.CH_2.CH_2OH$$
$$\text{tetrahydrofuran} \qquad\qquad\qquad \text{1,4 butanediol}$$

3.2.6 Production of Butadiene from Acetaldehyde

Acetaldehyde is first condensed with itself to form acetaldol. This is then reduced to 1,3-butanediol, which is dehydrated to butadiene:

$$2CH_3.CHO \longrightarrow CH_3.CHOH.CH_2.CHO$$

acetaldol

$$\downarrow [H]$$

$$CH_2{=}CH{-}CH{=}CH_2 \xleftarrow{-2H_2O} CH_3.CHOH.CH_2.CH_2OH$$

1,3-butanediol

The acetaldehyde required for this process can be obtained from acetylene by reaction with water in the presence of mercury salts.

Butadiene is a colourless gas at normal temperatures and pressures. Its boiling point at 760 mm of mercury is $-4\cdot4\,°C$ and its freezing point is $-109\,°C$. In the liquid state, it has a very low density ($0\cdot645\,kg\,dm^{-3}$ at $0\,°C$).

3.3 STYRENE

Styrene is produced from ethylbenzene, which in turn is produced by alkylating benzene with ethylene:

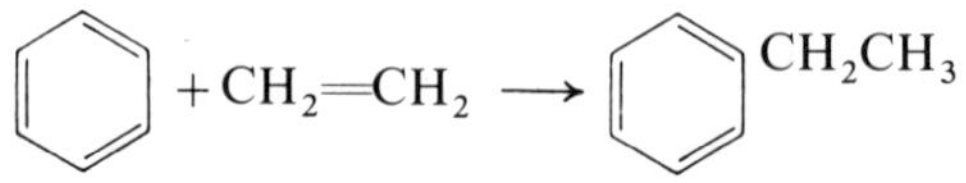

benzene   ethylbenzene

The alkylation can be carried out in the liquid phase as a Friedel–Crafts reaction using aluminium chloride as catalyst. Typical reaction conditions are 85–95 °C and a slightly elevated pressure. A typical mole ratio of ethylene to benzene is 0·6:1; this ratio is chosen in order to maximise the yield of monoethylbenzene and to minimise the formation of higher alkylbenzenes. A small quantity of ethyl chloride may be present in the ethylene feed; the ethyl chloride breaks down into ethylene and hydrogen chloride, and the latter acts as a promoter for the aluminium chloride catalyst.

Two types of reaction are used for obtaining styrene from ethylbenzene:

1.   dehydrogenation;
2.   oxidation.

Of these reactions, dehydrogenation is overwhelmingly the more important; almost all the styrene produced at the present time is obtained by dehydrogenation of ethylbenzene.

3.3.1 Production of Styrene by Dehydrogenation of Ethylbenzene

The reaction is as follows:

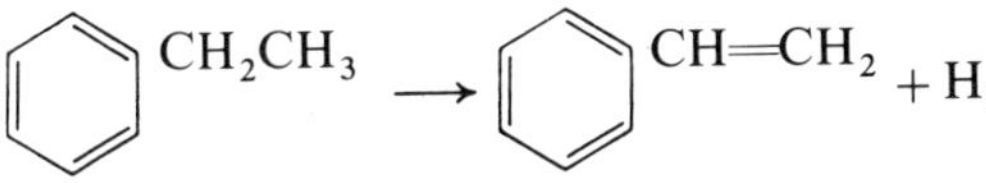

The reaction is strongly endothermic. Processes differ in the way in which the heat necessary to sustain the reaction is supplied. In the Dow 'adiabatic' cracking process, the heat is provided by mixing the ethylbenzene with super-heated steam prior to contacting with a catalyst, of which the main constituent is typically iron(III) oxide with minor amounts of potassium hydroxide and chromium(III) oxide. The feed mixture is introduced at a temperature of $c.$ 650 °C. The pressure is typically 5–15 lb in^{-2} in order to encourage dehydrogenation and discourage polymerisation. Because of the high temperature, some thermal cracking of the hydrocarbons occurs.

A second process for the dehydrogenation of ethylbenzene to styrene is known as the 'isothermal' process. Developed by B.A.S.F., heat is provided indirectly via a tubular reactor. Temperatures are maintained in the range 580–610 °C, and little thermal cracking occurs. The catalysts used are similar to those used for the adiabatic process. The yield of styrene based on ethylbenzene is rather higher than for the adiabatic process.

3.3.2 Production of Styrene by Oxidation of Ethylbenzene

In one such process, ethylbenzene is oxidised to acetophenone with air in the liquid phase. The acetophenone is then hydrogenated to 1-phenylethyl alcohol in the presence of a copper–chromium–iron catalyst. The 1-phenylethyl alcohol is then dehydrated to styrene over a titanium dioxide catalyst:

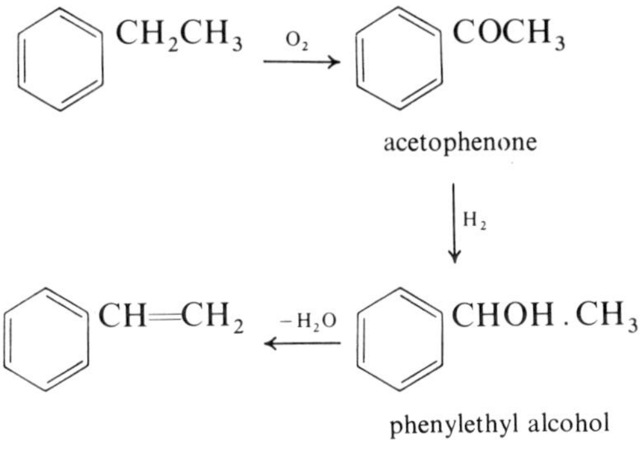

In a second process, ethylbenzene is first hydroperoxidised with air, and the hydroperoxide then reacted with propylene to form 1-phenylethyl alcohol. The latter is then dehydrated to styrene:

$$\text{C}_6\text{H}_5\text{—CH}_2\text{CH}_3 \xrightarrow{\text{O}_2} \text{C}_6\text{H}_5\text{—CH(OOH).CH}_3$$

$$\downarrow \text{CH}_3.\text{CH}\!=\!\text{CH}_2$$

$$\text{C}_6\text{H}_5\text{—CH}\!=\!\text{CH}_2 \xleftarrow{-\text{H}_2\text{O}} \text{C}_6\text{H}_5\text{—CHOH.CH}_3 + \text{CH}_3.\text{CH}\!-\!\text{CH}\diagdown_{\text{O}}\diagup$$

Styrene is a colourless, mobile liquid which boils at 145 °C (760 mm of mercury) and freezes at −31 °C. It has a characteristically 'aromatic' odour. It is very sparingly soluble in water (0·032 % at 25 °C), and this has important implications in respect of its emulsion polymerisation. The solubility of water in styrene is also very low (0·070 % at 25 °C), but this is of little significance in relation to polymerisation behaviour. Styrene is completely miscible with most common organic solvents.

The changes in the price of styrene which have occurred in recent years provide an interesting illustration of the complex interrelationships of supply and demand which often exist in the modern industrial world. Because of the recent introduction of anti-lead legislation in respect of petrol to be used in road vehicles in the U.S.A., the demand for aromatic hydrocarbons for use in petrol has increased. Because of limited production capacity, insufficient aromatic hydrocarbons were for a period available for styrene production. The consequences were acute shortage and a sharp rise in price. Inevitably these consequences have been reflected in the price and availability of products, such as synthetic rubber, which are manufactured from styrene.

3.4   ACRYLONITRILE

The following are amongst the processes which have been developed for the production of acrylonitrile:

1.   reaction between acetylene and hydrogen cyanide;
2.   reaction between ethylene oxide and hydrogen cyanide;

3. reaction between propylene and ammonia in the presence of oxygen;
4. reaction between propylene and nitric oxide; and
5. the lactonitrile process.

Of these, methods 1, 2 and 3 are especially important.

3.4.1 Production of Acrylonitrile by Reaction between Acetylene and Hydrogen Cyanide

The chemistry which underlies this process is straightforward addition of hydrogen cyanide to acetylene:

$$CH{\equiv}CH + HCN \longrightarrow CH_2{=}CH.CN$$

A typical catalyst for the reaction is copper(I) chloride dissolved in aqueous ammonium chloride solution and acidified with hydrochloric acid. Typical reaction conditions are 70–90 °C and 2–5 lb in^{-2} pressure. Yields are 75–80 % on the acetylene and 85–90 % on the hydrogen cyanide. Many by-products are formed. In order to reduce the formation of by-products, anhydrous catalyst systems based upon copper(I) chloride have also been developed. In these systems, copper(I) chloride is dissolved in an aliphatic or aromatic nitrile, such as adiponitrile or benzonitrile. Addition of a small amount of an anhydrous strong acid, such as hydrogen chloride, increases the activity of the catalyst.

3.4.2 Production of Acrylonitrile by Reaction between Ethylene Oxide and Hydrogen Cyanide

Again, the underlying chemistry is straightforward:

$$\underset{\text{ethylene oxide}}{CH_2\underset{\diagdown O \diagup}{—}CH_2} + HCN \longrightarrow \underset{\text{ethylene cyanohydrin}}{\underset{OH \quad CN}{\overset{|\qquad|}{CH_2—CH_2}}}$$

$$\Big\downarrow {\scriptstyle -H_2O}$$

$$CH_2{=}CH.CN$$

The addition of hydrogen cyanide to ethylene oxide can be carried out as a liquid-phase reaction in which an alkaline catalyst (e.g., an alkaline earth hydroxide) is used. The reaction is strongly exothermic, and it is therefore necessary to employ a solvent, such as ethylene cyanohydrin or water, in order to moderate the reaction. The dehydration step can also be carried out in the liquid phase by heating to 200–350 °C in the presence of alkaline catalysts such as sodium formate or calcium oxide. One-step vapour-phase processes have also been developed.

Ethylene oxide is usually produced by the direct oxidation of ethylene using a silver catalyst:

$$CH_2{=}CH_2 + \tfrac{1}{2}O_2 \longrightarrow \underset{\diagdown O \diagup}{CH_2{-}CH_2}$$

It can also be obtained from ethylene by the so-called 'chlorohydrin process':

$$CH_2{=}CH_2 + H_2O + Cl_2 \longrightarrow \underset{\underset{OH\quad Cl}{|\qquad |}}{CH_2{-}CH_2} + HCl$$

ethylene chlorohydrin

$$\downarrow NaOH$$

$$\underset{\diagdown O \diagup}{CH_2{-}CH_2}$$

3.4.3 Production of Acrylonitrile by Reaction Between Propylene and Ammonia in the Presence of Oxygen

The reaction is sometimes known as 'ammoxidation'. Its approximate stoichiometry can be represented as follows:

$$CH_3 . CH{=}CH_2 + NH_3 + \tfrac{3}{2}O_2 \longrightarrow CH_2{=}CH . CN + 3H_2O$$

Typical reaction conditions are a temperature of 450 °C and a pressure in the range 2–3 atm. Suitable catalysts include the bismuth, tin and lead salts of molybdic and phosphomolybdic acids; bismuth phosphomolybdate is particularly suitable. The catalyst can be used alone or supported on an inert carrier such as silica. Contact times are in the range 10–25 s. The presence of water is found to be beneficial for the formation of acrylonitrile. Recovery of the acrylonitrile is a comparatively simple process; the hot effluent gases are passed into a cool column at a temperature below 23 °C, and the resultant aqueous solution is fractionally distilled. Commonly known as the Sohio process, this method for making acrylonitrile is of special interest because of the cheapness of the starting materials.

3.4.4 Production of Acrylonitrile by Reaction Between Propylene and Nitric Oxide

The overall reaction can be represented approximately as:

$$4C_3H_6 + 6NO \longrightarrow 4CH_2{=}CH.CN + N_2 + 6H_2O$$

3.4.5 Production of Acrylonitrile by the Lactonitrile Process

The starting materials are acetaldehyde and hydrogen cyanide. These are first reacted together to form lactonitrile (which is isomeric with ethylene cyanohydrin). This is then dehydrated to acrylonitrile by heating to 600–620 °C in the presence of orthophosphoric acid and steam. The reactions which occur are as follows:

$$CH_3.CHO + HCN \longrightarrow CH_3.\underset{\overset{|}{OH}}{CH}.CN \xrightarrow{-H_2O} CH_2{=}CH.CN$$

lactonitrile

This process is sometimes known as the Knapsack process.

Acrylonitrile is a mobile, colourless liquid which boils at 77·3 °C (760 mm of mercury) and freezes at c. -84 °C. Its solubility in water is relatively high, being 7·5 % at 20 °C; this has implications for its behaviour in emulsion polymerisation reactions. The solubility of water in acrylonitrile at 20 °C is 3·1 %. Acrylonitrile is completely miscible with most polar and non-polar solvents.

3.5   ISOPRENE (2-METHYL-1,3-BUTADIENE)

The small quantities of isoprene required for the manufacture of butyl rubber were obtained by the steam cracking of naphtha. However, isoprene produced in this way was not available in sufficient quantity to support the large-scale manufacture of synthetic polyisoprene rubber. It was necessary to develop other methods of production in order to obtain isoprene in sufficient quantity and of sufficient quality for this purpose. Amongst the processes which are in current use, or have been developed to the pilot stage, are the following:

1.   the propylene dimer process;
2.   dehydrogenation of isopentane and/or 2-methyl butenes;
3.   the isobutene–formaldehyde process; and
4.   the acetone–acetylene process.

3.5.1 Production of Isoprene by the Propylene Dimer Process

Propylene is dimerised to 2-methyl-1-pentene, which is then isomerised to 2-methyl-2-pentene. This is then pyrolysed in a cracking furnace to give isoprene:

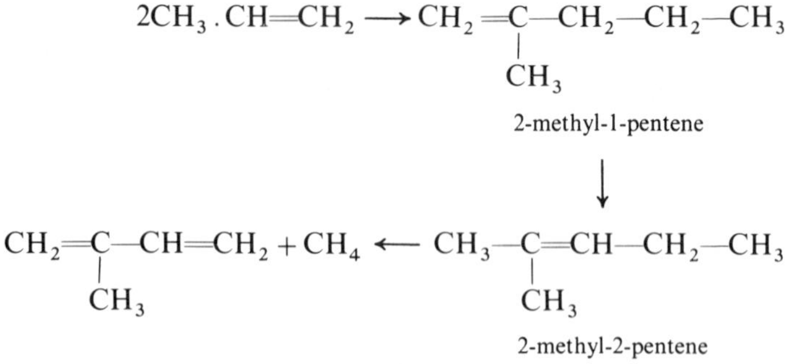

$$2CH_3 . CH{=}CH_2 \longrightarrow CH_2{=}\underset{\underset{CH_3}{|}}{C}{-}CH_2{-}CH_2{-}CH_3$$

2-methyl-1-pentene

$$\downarrow$$

$$\underset{\underset{CH_3}{|}}{CH_2{=}C}{-}CH{=}CH_2 + CH_4 \longleftarrow CH_3{-}\underset{\underset{CH_3}{|}}{C}{=}CH{-}CH_2{-}CH_3$$

2-methyl-2-pentene

Catalysts for the dimerisation step include aluminium tri-*n*-propyl, dilute orthophosphoric acid, silica–alumina, silica–alumina plus nickel, and a combination of alumina and molybdenum trioxide. The preferred catalyst seems to be aluminium tri-*n*-propyl. The following reaction mechanism has been proposed for propylene dimerisation using this catalyst:

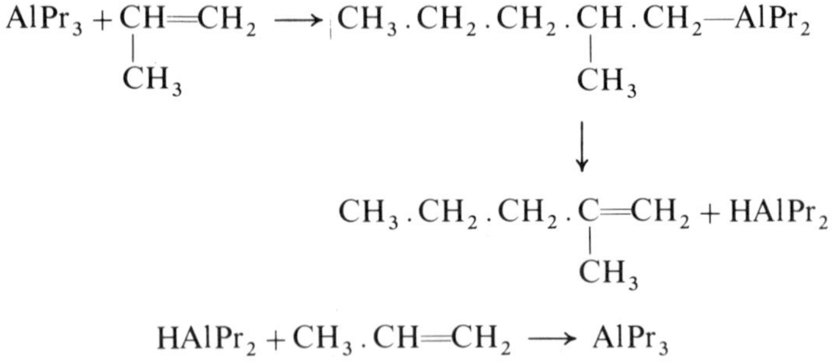

$$AlPr_3 + \underset{\underset{CH_3}{|}}{CH{=}CH_2} \longrightarrow {}_|CH_3 . CH_2 . CH_2 . \underset{\underset{CH_3}{|}}{CH} . CH_2{-}AlPr_2$$

$$\downarrow$$

$$CH_3 . CH_2 . CH_2 . \underset{\underset{CH_3}{|}}{C}{=}CH_2 + HAlPr_2$$

$$HAlPr_2 + CH_3 . CH{=}CH_2 \longrightarrow AlPr_3$$

In this reaction scheme, Pr denotes the radical $-CH_2CH_2CH_3$. This reaction mechanism shows clearly that the aluminium tri-*n*-propyl catalyst is regenerated. Typical reaction conditions for the dimerisation are 150–250 °C and 3000 lb in^{-2} pressure. The propylene must be free from polar compounds and oxygen because these react with aluminium tri-*n*-propyl.

Isomerisation of 2-methyl-1-pentene to 2-methyl-2-pentene is achieved by passing over an acidic fixed-bed catalyst at a temperature in the range 150–300 °C. The final pyrolysis step is carried out at a temperature in the range 650–800 °C using, for example, hydrogen bromide as catalyst and a very short pyrolysis time (in the range 0·05–0·3 s).

3.5.2 Production of Isoprene by Dehydrogenation of Isopentane and/or 2-Methyl Butenes

This process is analogous to the Houdry process for the production of butadiene from n-butane. The reactions which occur are:

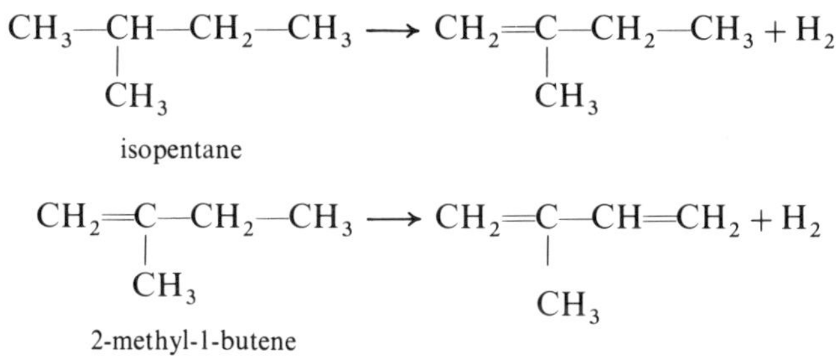

$$CH_3\!-\!\underset{\underset{CH_3}{|}}{CH}\!-\!CH_2\!-\!CH_3 \longrightarrow CH_2\!=\!\underset{\underset{CH_3}{|}}{C}\!-\!CH_2\!-\!CH_3 + H_2$$

isopentane

$$CH_2\!=\!\underset{\underset{CH_3}{|}}{C}\!-\!CH_2\!-\!CH_3 \longrightarrow CH_2\!=\!\underset{\underset{CH_3}{|}}{C}\!-\!CH\!=\!CH_2 + H_2$$

2-methyl-1-butene

Typically, the reaction is carried out at a temperature in the range 540–620 °C and under a partial vacuum of 22–24 in of mercury. A metal-oxide catalyst, e.g., chromia–alumina, is used. The process yields isoprene of high purity (99 %).

3.5.3 Production of Isoprene by the Isobutene–Formaldehyde Process

The underlying chemistry of this process is as follows:

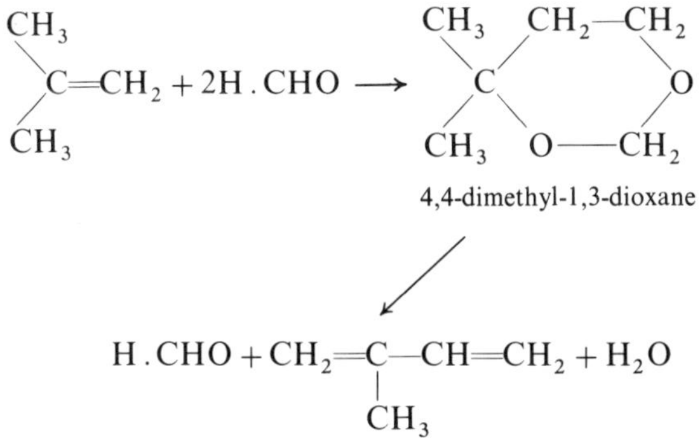

$$\underset{\underset{CH_3}{}}{\overset{CH_3}{}}\!\!C\!=\!CH_2 + 2H.CHO \longrightarrow$$

4,4-dimethyl-1,3-dioxane

$$H.CHO + CH_2\!=\!\underset{\underset{CH_3}{|}}{C}\!-\!CH\!=\!CH_2 + H_2O$$

The reaction between isobutene and formaldehyde is effected at 95 °C under a few atmospheres pressure and using an acidic catalyst. The 4,4-dimethyl-1,3-dioxane is separated from the reaction products by distillation *in vacuo*. The decomposition of the latter to isoprene and formaldehyde is effected by heating to a temperature below 400 °C at atmospheric pressure in contact with a moving-bed catalyst.

3.5.4 Production of Isoprene by the Acetone–Acetylene Process

In this process, isoprene is produced by the following series of reactions:

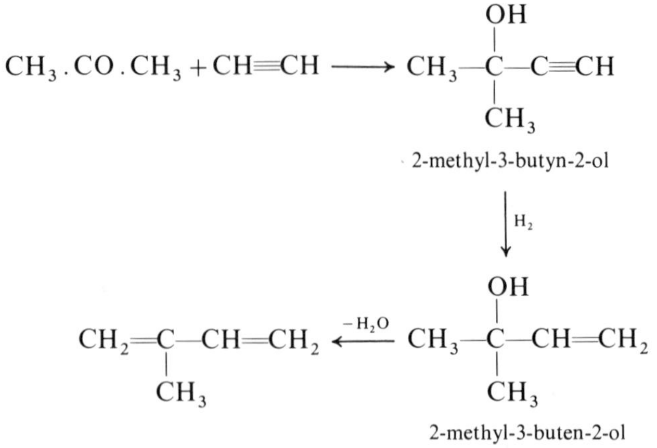

$$CH_3.CO.CH_3 + CH\equiv CH \longrightarrow CH_3-\overset{\overset{\displaystyle OH}{|}}{\underset{\underset{\displaystyle CH_3}{|}}{C}}-C\equiv CH$$

2-methyl-3-butyn-2-ol

$$\big\downarrow H_2$$

$$CH_2=\overset{}{\underset{\underset{\displaystyle CH_3}{|}}{C}}-CH=CH_2 \xleftarrow{-H_2O} CH_3-\overset{\overset{\displaystyle OH}{|}}{\underset{\underset{\displaystyle CH_3}{|}}{C}}-CH=CH_2$$

2-methyl-3-buten-2-ol

The initial reaction between acetone and acetylene is effected at 10–40 °C and 285 lb in^{-2} pressure in liquid ammonia using a catalyst. The hydrogenation is effected at 30–80 °C and 70–140 lb in^{-2} hydrogen pressure using a supported palladium catalyst. The final dehydration is achieved by passing the 2-methyl-3-buten-2-ol over high-purity alumina at 260–300 °C under atmospheric pressure. It has been claimed that isoprene produced by this process is cost-competitive with butadiene.

Isoprene is a very volatile, colourless, mobile liquid which boils at 34·1 °C (760 mm of mercury) and freezes at −146 °C. Liquid isoprene has a low density (0·681 kg dm^{-3} at 20 °C). Its solubility in water is low.

3.6   CHLOROPRENE (2-CHLORO-1,3-BUTADIENE)

Processes which have been developed for the production of chloroprene include:

1. production from acetylene; and
2. production from butadiene.

Of these processes, the first was the more important in earlier years, but in more recent years the latter has become the dominant process.

3.6.1 Production of Chloroprene from Acetylene

The chemistry which underlies this process is as follows:

$$2CH\!\equiv\!CH \longrightarrow CH_2\!=\!CH\!-\!C\!\equiv\!CH$$

monovinylacetylene

$$\downarrow HCl$$

$$CH_2\!=\!\underset{\underset{Cl}{|}}{C}\!-\!CH\!=\!CH_2 \longleftarrow [CH_2Cl\!-\!CH\!=\!C\!=\!CH_2]$$

4-chloro-1,2-butadiene
('chlorallene', 'isochloroprene')

Monovinylacetylene is produced by passing acetylene through an aqueous catalyst solution which contains copper(I) chloride, hydrochloric acid, and a 'solubilising' chloride of another monovalent cation such as potassium. A catalyst composition corresponding approximately to $M^+[CuCl_2]^-$ seems to be required for optimum activity. The temperature of the catalyst solution is in the range 55–65 °C. The monovinylacetylene, which is a colourless liquid with boiling point 3·5 °C (760 mm of mercury), is then dried and purified by fractionation. A major impurity is divinylacetylene, which is formed by reaction between monovinylacetylene and further acetylene under the influence of the copper(I) chloride catalyst:

$$CH_2\!=\!CH\!-\!C\!\equiv\!CH + CH\!\equiv\!CH \longrightarrow CH_2\!=\!CH\!-\!C\!\equiv\!C\!-\!CH\!=\!CH_2$$

divinylacetylene

The next step is to react monovinylacetylene with an equivalent weight of hydrogen chloride in the presence of an aqueous solution of copper(I) chloride at 35–45 °C. The initial product of the reaction is believed to be 4-chloro-1,2-butadiene, but this spontaneously isomerises to chloroprene. The exit gases are condensed, the water separated and returned to the reactor, and the organic layer dried. Unreacted monovinylacetylene is recovered by stripping under nitrogen; the chloroprene is separated by vacuum distillation.

3.6.2 Production of Chloroprene from Butadiene

This process comprises three essential steps, namely (a) chlorination of butadiene, (b) isomerisation of the product, and (c) dehydrochlorination.

The underlying chemistry of the process can be summarised as follows:

$$CH_2{=}CH{-}CH{=}CH_2 + Cl_2 \longrightarrow CH_2Cl{-}CH{=}CH{-}CH_2Cl$$

1,4-dichloro-2-butene

$$\downarrow$$

$$CH_2{=}\underset{\underset{Cl}{|}}{C}{-}CH{=}CH_2 \xleftarrow{\;-HCl\;} CH_2Cl{-}CHCl{-}CH{=}CH_2$$

3,4-dichloro-1-butene

The initial chlorination is carried out in the vapour phase, with reaction conditions adjusted to prevent over-chlorination. The preferred temperature range is 290–330 °C. A mixture of products is obtained. Those of interest for the subsequent production of chloroprene are the *cis*- and *trans*-isomers of 1,4-dichloro-2-butene and 3,4-dichloro-1-butene. Isomerisation of the 1,4-dichloro-2-butenes to the desired isomer is achieved by heating to *c*. 100 °C with a catalyst such as copper metal or copper(I) chloride, and arranging the experimental conditions such that the desired isomer is continuously separated. The dehydrochlorination step is achieved by heating the 3,4-dichloro-1-butene with 5–15% sodium hydroxide solution at 80–110 °C.

Chloroprene is a colourless, mobile, volatile liquid which boils at 59·4 °C (760 mm of mercury) and freezes at *c*. −131 °C. Its solubility in water is low, but it is miscible with most organic solvents.

3.7   ETHYLENE

Major sources of ethylene are the catalytic cracking of crude petroleum oil and the pyrolysis of saturated hydrocarbons such as ethane and propane present in natural gas. A minor amount of ethylene has been produced by the dehydration of ethyl alcohol; however, it is now the case that ethylene is an important source of industrial ethyl alcohol.

The separation of ethylene from the mixed gases produced by the cracking of hydrocarbons is effected by fractionation, adsorption, or solvent-extraction procedures.

Ethylene is a colourless gas which can be condensed to a liquid of boiling point −103·7 °C (760 mm of mercury).

3.8   PROPYLENE

Propylene is also produced by the cracking of other hydrocarbons from petroleum sources. In a typical process, a petroleum distillate fraction boiling in the range 200–430 °C is preheated to about 400 °C and then passed over a silica–alumina catalyst at a temperature of about 480 °C. A small proportion of the feed becomes converted to propylene. This is concentrated into a mixture which comprises mainly propylene and propane. The final separation is by fractional distillation. Acidic components which boil in the same range as propylene are removed by absorption in an alkali, such as soda-lime, prior to the final distillation. Propylene is also produced by the thermal cracking of light naphtha at high temperatures (700–760 °C), moderate pressures, and short contact times (c. 1 min).

Propylene is a colourless gas which can be condensed to a liquid of boiling point −47 °C (760 mm of mercury).

3.9   ISOBUTENE (ISOBUTYLENE)

Isobutene occurs in the C_4 gas fraction (the so-called 'C_4 cut') which is obtained by various petroleum refinery cracking processes such as the catalytic or thermal cracking of gas oil for the production of high-octane petrol, the thermal reforming of petrol to increase its octane number, and the low-pressure cracking of various hydrocarbon mixtures for the purpose of olefin production. The C_4 fractions obtained by these methods contain 10–35 % of isobutene. The isobutene is separated by absorption in c. 65 % aqueous sulphuric solution. The absorbed isobutene is subsequently liberated from the acid extract by treatment with steam. The product is then washed with alkali and then water, after which it is compressed and condensed.

Like butadiene, isobutene is a gas at normal temperatures and pressures. It condenses to a liquid of very low density ($0.600 \, kg \, dm^{-3}$ at −15 °C) which boils at −6.9 °C (760 mm of mercury) and freezes at −105 °C.

3.10   ACRYLIC MONOMERS

Under this general heading are included the alkyl esters of acrylic acid (IX) and methacrylic acid (XX). These esters are sometimes produced by

esterification of the acids by the appropriate alcohol, and sometimes
directly. They can also be produced by the transesterification of other

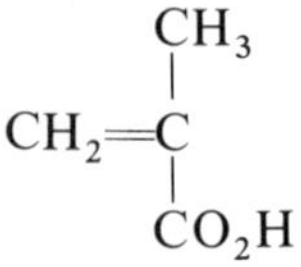

$$CH_2\!=\!\underset{\underset{CO_2H}{|}}{\overset{\overset{CH_3}{|}}{C}}$$

(XX)

acrylic esters. Processes for the production of the acids or their esters
include:

1.   the ethylene cyanohydrin process for acrylates;
2.   production of acrylates from acetylene;
3.   the β-propiolactone process for acrylates; and
4.   the acetone cyanohydrin process for methacrylates.

3.10.1 The Ethylene Cyanohydrin Process for Acrylates
The underlying chemistry of this process is as follows:

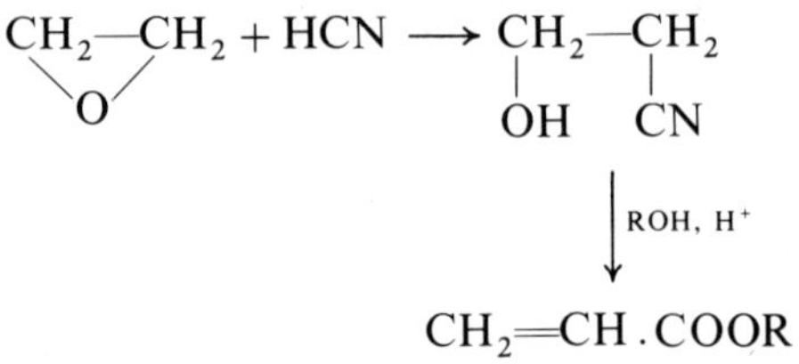

$$\underset{O}{CH_2\!-\!CH_2} + HCN \longrightarrow \underset{OH \quad CN}{CH_2\!-\!CH_2}$$

$$\downarrow \text{ROH, H}^+$$

$$CH_2\!=\!CH.COOR$$

The conversion of ethylene cyanohydrin to the alkyl acrylate is achieved by
reaction with the appropriate alcohol under strongly acid conditions.

3.10.2 Production of Acrylates from Acetylene
In the Reppe process, acetylene is reacted with carbon monoxide, water and
an alcohol in the presence of nickel carbonyl under mild conditions. The
reaction is in effect a carbonylation of acetylene by means of nickel
carbonyl. The reaction which occurs can be represented as follows:

$$4CH\!\equiv\!CH + 4ROH + 2H^+ + Ni(CO)_4 \longrightarrow$$

$$4CH_2\!=\!CH.COOR + Ni^{2+} + 2H$$

In a second process, acrylic acid is produced by reacting acetylene,
carbon monoxide and water at supra-atmospheric pressure and elevated
temperature in the presence of tetrahydrofuran and a nickel(II) catalyst:

$$CH\!\equiv\!CH + CO + H_2O \longrightarrow CH_2\!=\!CH.COOH$$

3.10.3 The β-Propiolactone Process for Acrylates

The starting materials for this process are ketene and formaldehyde. These are first reacted to give β-propiolactone, which is then polymerised to poly-β-propiolactone. This can then be depolymerised to acrylic acid by heating to 100–150 °C with water in the presence of a little sodium carbonate. Alternatively, it can be converted directly to acrylic acid esters by passing with the appropriate alcohol through activated charcoal at c. 250 °C. The reactions which underly this process are as follows:

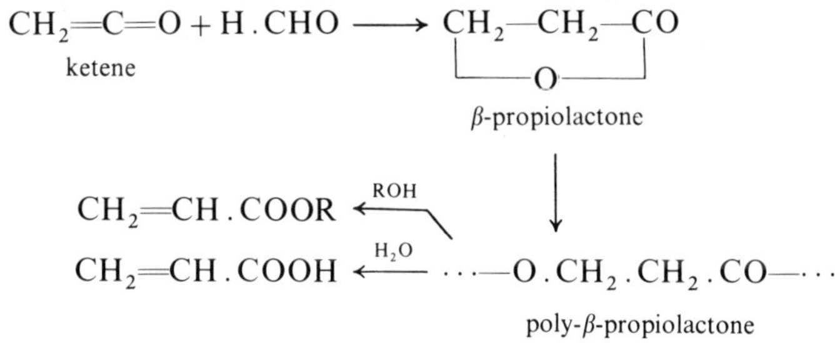

$$CH_2{=}C{=}O + H.CHO \longrightarrow \underset{\underset{\text{β-propiolactone}}{\displaystyle \lfloor\!-\!O\!-\!\rfloor}}{CH_2{-}CH_2{-}CO}$$

$$CH_2{=}CH.COOR \xleftarrow{\text{ROH}}$$

$$CH_2{=}CH.COOH \xleftarrow{\text{H}_2\text{O}} \cdots{-}O.CH_2.CH_2.CO{-}\cdots$$

poly-β-propiolactone

3.10.4 The Acetone Cyanohydrin Process for Methacrylates

The starting materials for this process are acetone and hydrogen cyanide. These are reacted together to give acetone cyanohydrin, which is then treated with 98 % sulphuric acid to give methacrylamide bisulphate. The latter is not usually isolated as such, but is treated with an appropriate alcohol under acid conditions to give the methacrylic acid ester. The reactions are as follows:

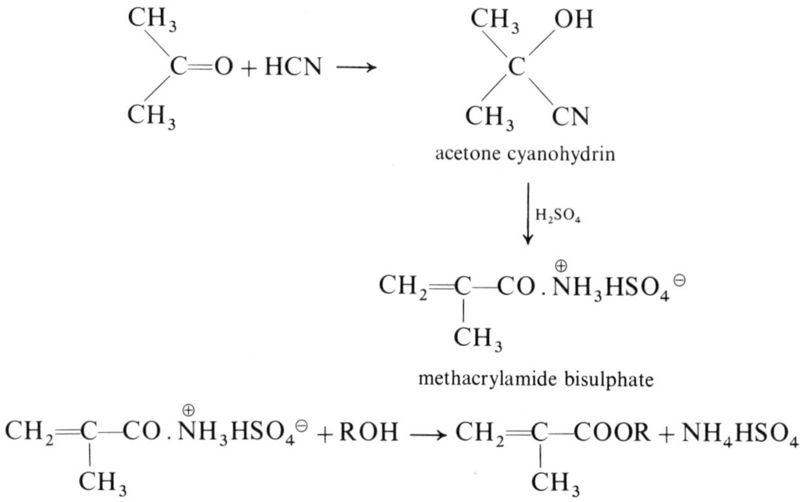

$$\underset{CH_3}{\overset{CH_3}{>}}C{=}O + HCN \longrightarrow \underset{CH_3}{\overset{CH_3}{>}}\underset{CN}{\overset{OH}{C}}$$

acetone cyanohydrin

$$\Big\downarrow \text{H}_2\text{SO}_4$$

$$CH_2{=}\underset{\underset{CH_3}{|}}{C}{-}CO.\overset{\oplus}{N}H_3HSO_4{}^{\ominus}$$

methacrylamide bisulphate

$$CH_2{=}\underset{\underset{CH_3}{|}}{C}{-}CO.\overset{\oplus}{N}H_3HSO_4{}^{\ominus} + ROH \longrightarrow CH_2{=}\underset{\underset{CH_3}{|}}{C}{-}COOR + NH_4HSO_4$$

The lower alkyl esters of acrylic and methacrylic acid are sweet-smelling, colourless, mobile liquids of moderate volatility. In accordance with expectation, the boiling point increases with increasing size of the alkyl group, and the solubility in water decreases.

3.11 VINYL CHLORIDE

There are two other important vinyl monomers which are of relevance for the manufacture of synthetic rubbers. The first of these, vinyl chloride (XXI), is used to produce resinous polymers and copolymers which, when suitably plasticised, can be used as synthetic rubbers. The second, vinyl acetate (XXII), is copolymerised with ethylene to give synthetic rubbers.

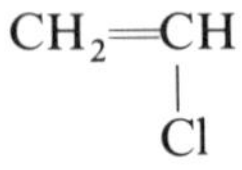

$$CH_2=CH \qquad\qquad CH_2=CH$$
$$| \qquad\qquad\qquad |$$
$$Cl \qquad\qquad\qquad OCOCH_3$$

(XXI) 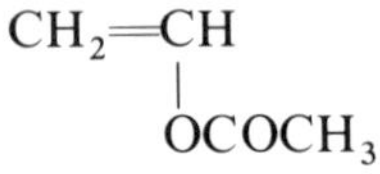   (XXII)

Processes for the production of vinyl chloride include:

1. reaction between acetylene and hydrogen chloride;
2. dehydrochlorination of ethylene dichloride; and
3. oxychlorination of ethylene.

3.11.1 Production of Vinyl Chloride by Reaction Between Acetylene and Hydrogen Chloride

The reaction is straightforward addition of hydrogen chloride to acetylene:

$$CH{\equiv}CH + HCl \longrightarrow CH_2=CHCl$$

Almost equimolar proportions of the two reactants are passed over a mercury(II) chloride catalyst supported on active carbon and contained in a series of reactors which are designed to ensure effective removal of the heat of reaction. The reaction temperature is in the range 80–150 °C. The reaction goes almost to completion. The unreacted hydrogen chloride, together with other volatile impurities, is subsequently removed by distillation. The monomer is then dried and distilled either under pressure or at atmospheric pressure using a refrigerated system.

3.11.2 Production of Vinyl Chloride by Dehydrochlorination of Ethylene Dichloride

The ethylene dichloride is obtained by reaction between ethylene and chlorine. Dehydrochlorination is effected by pyrolysis at 300–600 °C using a contact catalyst and elevated pressure. The underlying chemistry of the process is as follows:

$$CH_2\!\!=\!\!CH_2 + Cl_2 \longrightarrow CH_2Cl\!\!-\!\!CHCl \longrightarrow CH_2\!\!=\!\!CHCl + HCl$$
$$\text{ethylene dichloride}$$

The hydrogen chloride generated in the course of the dehydrochlorination reaction can be used to produce further vinyl chloride by reaction with acetylene.

3.11.3 Production of Vinyl Chloride by the Oxychlorination of Ethylene

This is in effect an alternative process for the production of ethylene dichloride from ethylene. Instead of reacting the ethylene with chlorine, it is reacted with a mixture of hydrogen chloride and oxygen. The reaction which takes place can be represented as follows:

$$CH_2\!\!=\!\!CH_2 + 2HCl + \tfrac{1}{2}O_2 \longrightarrow CH_2Cl\!\!-\!\!CH_2Cl + H_2O$$

A catalyst such as copper chloride, potassium chloride, or lanthanum chloride is required. Vinyl chloride is then obtained by pyrolysing the ethylene dichloride as described in Section 3.11.2 above. The hydrogen chloride produced in the dehydrochlorination step can then be used for further oxychlorination.

Vinyl chloride is a gas at normal temperatures and pressures. It condenses to a liquid which boils at $-13\cdot4$ °C (760 mm of mercury) and freezes at -154 °C. The solubility of vinyl chloride in water depends upon the temperature and pressure; at 50 °C and 600 cm of mercury pressure, the solubility is approximately 1 %.

3.12 VINYL ACETATE

There are two industrial processes for the production of vinyl acetate:

1.   by reaction between acetylene and acetic acid; and
2.   from ethylene.

3.12.1 Production of Vinyl Acetate by Reaction Between Acetylene and Acetic Acid

The reaction is straightforward addition of acetic acid to acetylene:

$$CH\equiv CH + CH_3.COOH \longrightarrow CH_2\!\!=\!\!CH.OCOCH_3$$

Ethylidene diacetate is formed as a by-product by the further addition of acetic acid to vinyl acetate:

$$CH_2\!\!=\!\!CH.OCOCH_3 + CH_3.COOH \longrightarrow CH_3CH(OCOCH_3)_2$$
ethylidene diacetate

The addition of acetic acid to acetylene can be carried out either by a liquid-phase reaction or by a vapour-phase reaction. In liquid-phase processes, acetylene is passed into liquid acetic acid under normal pressure. The temperature is kept low (less than 70 °C) because high temperatures favour the formation of ethylidene diacetate. The preferred temperature is 40–50 °C. Mercury(II) compounds are used as catalysts, e.g., a mixture of mercury(II) oxide, iron(II) oxide and sulphuric acid. The mechanism of the reaction using mercury(II) catalysts is thought to be as follows:

$$CH\equiv CH + HgSO_4 \longrightarrow$$

$$CH\!\!=\!\!CH \quad \xrightarrow{+CH_3COOH} \quad CH_2\!\!=\!\!CH.OCOCH_3 + HgSO_4$$
$$\diagdown\diagup$$
$$HgSO_4$$

mercury
sulphate–acetylene
complex

In the vapour-phase process, a mixture of acetylene and acetic acid is passed through a catalyst bed comprising zinc acetate and activated carbon at a temperature in the range 180–210 °C.

3.12.2 Production of Vinyl Acetate from Ethylene

In this process, ethylene is reacted with sodium acetate in the presence of palladium chloride and acetic acid. The reactions which underly this process are complex, and have been represented as follows:

$$CH_2\!\!=\!\!CH_2 + 2CH_3CO_2Na + PdCl_2 \xrightarrow{CH_3COOH}$$
$$CH_2\!\!=\!\!CH.OCOCH_3 + Pd + 2NaCl + CH_3COOH$$

As with the acetylene–acetic acid process, the reaction can be carried out either in the liquid phase or in the vapour phase. In general, vinyl acetate obtained by the ethylene process contains fewer impurities that affect polymerisation activity than does vinyl acetate from acetylene.

Vinyl acetate is a clear, mobile liquid which boils at 72·7 °C (760 mm of mercury) and freezes at − 100 °C. It is moderately soluble in water (2·3 % at 20 °C), and completely miscible with common organic solvents. It hydrolyses fairly rapidly in contact with water and acids or alkalis. The products of the hydrolysis are acetaldehyde and acetic acid or alkali acetate (according to whether the hydrolysis was catalysed by an acid or an alkali). The rate of hydrolysis depends upon the pH of the aqueous phase, and is a minimum at pH 4·44. The acetaldehyde which is produced forms by tautomeric rearrangement of the vinyl alcohol which is the initial product of the hydrolysis:

$$CH_2\!\!=\!\!CH.OH \rightleftharpoons CH_3.CHO$$

Vinyl alcohol is so unstable relative to acetaldehyde that the equilibrium lies almost entirely on the side of acetaldehyde.

GENERAL BIBLIOGRAPHY

Waddams, A. L. (1962). *Chemicals from Petroleum*, Shell Chemical Co. Ltd, London.

Whitby, G. S., Davis, C. C. and Dunbrook, R. F. (Eds.) (1954). *Synthetic Rubber*, John Wiley and Sons, New York, Chapters 3, 4, 5, 6, 23 and 24.

Saltman, W. M. (1965). Butadiene polymers. In: *Encyclopedia of Polymer Science and Technology*, Vol. 2, John Wiley and Sons, New York, pp. 678 f.

Kirschenbaum, I. (1978). Butadiene. In: *Kirk–Othmer Encyclopedia of Chemical Technology*, 3rd edn., Vol. 4, John Wiley and Sons, New York, pp. 313 f.

Hofmann, W. (1964). Nitrile rubber, *Rubb. Chem. Technol.*, **37** (2[2]), 1.

Coulter, K. E. and Kehde, H. (1970). Styrene polymers (monomers). In: *Encyclopedia of Polymer Science and Technology*, Vol. 13, John Wiley and Sons, New York, pp. 135 f.

Bamford, C. H. and Eastmond, G. C. (1964). Acrylonitrile polymers. In: *Encyclopedia of Polymer Science and Technology*, Vol. 1, John Wiley and Sons, New York, pp. 374 f.

Bean, A. R., Himes, G. R., Holden, G., Houston, R. R., Langton, J. A. and Mann, R. H. (1967). Isoprene polymers. In: *Encyclopedia of Polymer Science and Technology*, Vol. 7, John Wiley and Sons, New York, pp. 782 f.

Hargreaves, C. A. and Thompson, D. C. (1965). 2-Chlorobutadiene polymers. In: *Encyclopedia of Polymer Science and Technology*, Vol. 3, John Wiley and Sons, New York, pp. 705 f.

Johnson, P. R. (1976). Polychloroprene rubber, *Rubb. Chem. Technol.*, **49**, 650.

Raff, R. A. V. (1967). Ethylene polymers. In: *Encyclopedia of Polymer Science and Technology*, Vol. 6, John Wiley and Sons, New York, pp. 275 f.

Jezl, J. L. and Honeycutt, E. M. (1969). Propylene polymers. In: *Encyclopedia of Polymer Science and Technology*, Vol. 11, John Wiley and Sons, New York, pp. 597 f.

Buckley, D. J. (1965). Butylene polymers. In: *Encyclopedia of Polymer Science and Technology*, Vol. 2, John Wiley and Sons, New York, pp. 754 f.

Coover, H. W. and Wicker, T. H. (1964). Acrylic ester polymers. In: *Encyclopedia of Polymer Science and Technology*, Vol. 1, John Wiley and Sons, New York, pp. 246 f.

Brighton, C. A. (1971). Vinyl chloride polymers (introduction). In: *Encyclopedia of Polymer Science and Technology*, Vol. 14, John Wiley and Sons, New York, pp. 305 f.

Lindemann, M. K. (1971). Vinyl ester polymers. In: *Encyclopedia of Polymer Science and Technology*, Vol. 15, John Wiley and Sons, New York, pp. 531 f.

REFERENCE

1. Hofmann, W. (1964). *Rubb. Chem. Technol.*, **37** (2[2]), 1.

Chapter 4

Outline of Polymerisation Methods

4.1 INTRODUCTION

It has already been noted in Chapter 1 that polymers are substances of high molecular weight whose molecules comprise one or more small chemical units repeated many times over, and that the chemical reaction by which these molecules are produced from the corresponding monomers is known as polymerisation. Polymerisation reactions are classified into two broad groups:

1. *addition polymerisations*, in which the monomer units become attached to one another without any molecule being eliminated; and
2. *condensation polymerisations*, in which a small molecule is eliminated during each step of the polymer-building reaction.

Although synthetic rubbers are produced by both addition and condensation polymerisation reactions, very much more synthetic rubber is produced by the latter type of reaction than by the former. Furthermore, addition polymerisation is a very general type of reaction in which, in principle, a wide variety of unsaturated and cyclic compounds can participate. However, in fact it is the case that most of the synthetic rubber which has been manufactured and is being manufactured today is produced by way of addition polymerisation reactions which involve the opening of carbon–carbon double bonds. Commonly referred to as 'addition polymerisation' without qualification, this reaction can be represented very generally as follows:

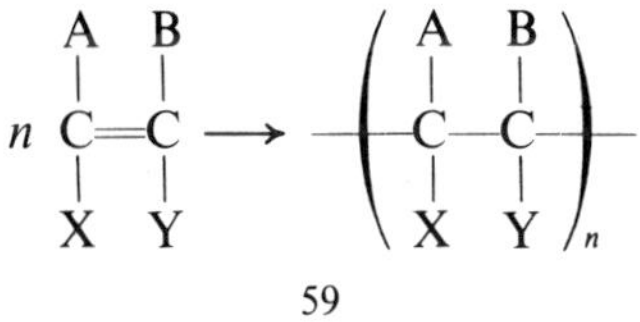

$$n \underset{\substack{|\\X}}{\overset{\substack{A\\|}}{C}} = \underset{\substack{|\\Y}}{\overset{\substack{B\\|}}{C}} \longrightarrow \left(\underset{\substack{|\\X}}{\overset{\substack{A\\|}}{C}} - \underset{\substack{|\\Y}}{\overset{\substack{B\\|}}{C}} \right)_n$$

59

where the letters A, B, X and Y denote the many and varied substituents which can surround the double bond. Thus the reaction can be thought of as one in which one of the bonds in each carbon–carbon double bond breaks, and the broken bonds on adjacent monomer units then link together to form bonds which in consequence link together the monomer units themselves.

Before proceeding to consider polymerisation reactions in more detail it is necessary to introduce the idea of the *functionality* of a monomer. This is defined as the maximum number of polymer-building bonds which a single monomer molecule can form if it reacts as fully as possible in a particular type of polymerisation reaction. It is thus a property not only of the monomer itself but also of the polymerisation reaction in which it is participating. The functionality of a monomer of the type $CH_2{=}CHX$ with respect to addition polymerisation by opening of the carbon–carbon double bond is two (assuming the substituent X to be inert with respect to addition polymerisation). Such monomers are commonly said to be *difunctional*. The maximum number of polymer-building bonds which such a monomer can form is clearly two. Likewise the functionality of a monomer of the type $CH_2{=}CH.R.CH{=}CH_2$ with respect to addition polymerisation is said to be four, because this is the maximum number of polymer-building bonds which it can form in an addition polymerisation. Monomers of this type are commonly said to be *tetrafunctional*.

4.2   ADDITION POLYMERISATION BY OPENING OF CARBON–CARBON DOUBLE BONDS

4.2.1   Outline of Mechanism of Addition Polymerisations by Opening of Carbon–Carbon Double Bonds

An important feature which many addition polymerisations involving the opening of carbon–carbon double bonds have in common is that high-molecular-weight polymer is present from the very earliest stage of the reaction; in fact, at all stages in the reaction, the reaction mixture comprises monomer and polymer of high molecular weight. This contrasts with condensation polymerisations, which proceed by a 'step growth' mechanism in which the polymer molecules grow slowly throughout the whole course of the reaction. Thus in many addition polymerisations the final polymer molecule forms very quickly—in a matter of seconds, in fact. This observation has led to the formulation of a very general reaction mechanism in which (i) centres of abnormally high reactivity are formed within the monomer, (ii) these active centres propagate themselves from one monomer molecule to another, thereby causing the monomer molecules to become chemically united into a polymer chain, and (iii) the active centres become destroyed. If an active centre is denoted generally by R*, then the first stage of the mechanism is the formation of R*. The second stage can then be represented quite generally as a succession of reactions of the type:

$$R^* + \overset{|}{\underset{|}{C}} = \overset{|}{\underset{|}{C}} \longrightarrow R - \overset{|}{\underset{|}{C}} - \overset{|}{\underset{|}{C}}{}^*$$

$$R - \overset{|}{\underset{|}{C}} - \overset{|}{\underset{|}{C}}{}^* + \overset{|}{\underset{|}{C}} = \overset{|}{\underset{|}{C}} \longrightarrow R - \overset{|}{\underset{|}{C}} - \overset{|}{\underset{|}{C}} - \overset{|}{\underset{|}{C}} - \overset{|}{\underset{|}{C}}{}^*$$

$$R - \left(\overset{|}{\underset{|}{C}} - \overset{|}{\underset{|}{C}} \right)_i \overset{|}{\underset{|}{C}} - \overset{|}{\underset{|}{C}}{}^* + \overset{|}{\underset{|}{C}} = \overset{|}{\underset{|}{C}} \longrightarrow R - \left(\overset{|}{\underset{|}{C}} - \overset{|}{\underset{|}{C}} \right)_{i+1} \overset{|}{\underset{|}{C}} - \overset{|}{\underset{|}{C}}{}^*$$

In the third stage, the high chemical reactivity associated with the active centre is lost, and so the active centre ceases to propagate and the polymer molecule which remains is of normal reactivity.

The description of the reaction mechanism given in the previous paragraph is very general. It raises the immediate question as to the nature of the abnormal reactivity of the propagating centres. Experience has shown that the active centres can be of three broad types:

1. carbon free radicals, in which the propagating centre is a carbon atom which has only seven outer electrons, one of which (the one responsible for the high reactivity) is unpaired;

2. carbonium cations, in which the propagating centre is a carbon atom which has only six outer electrons, and in consequence has a net positive charge; and

3. carbon anions (carbanions), in which the propagating centre is a carbon atom which has a full complement of eight outer electrons but only three of the four valencies are satisfied, and thus the carbon atom has a net negative charge.

The conventional representations of these three types of active centre are shown below, together with their respective electronic structures:

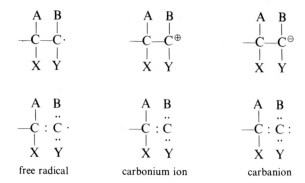

free radical   carbonium ion   carbanion

Addition polymerisations which occur by way of these three types of propagating centre are known respectively as *free-radical* addition polymerisations, *cationic* addition polymerisations, and *anionic* polymerisations. It is also necessary to point out that there are very important addition polymerisation reactions in which the propagating centre is not a free, independent entity as implied by the above structures, but is part of a coordination complex. This can have profound implications for the structure of the polymer which is produced. An important example is addition polymerisation by way of coordinated anionic centres; as will be seen in due course, this is a reaction by means of which some of the newer important types of synthetic rubber are produced.

4.2.2 Mechanism of Free-radical Addition Polymerisation
Until comparatively recently, most synthetic rubber was produced by free-radical addition polymerisation. It is therefore necessary to consider the mechanism of this reaction in some detail, and then to proceed to consider the various ways in which the reaction can be carried out.

Initiation
The free radicals from which the propagating centres are derived can be generated by a wide variety of reactions, of which the thermal decomposition of dibenzoyl peroxide is typical. The details of this reaction are rather complicated, but the initial stages of the reaction can be represented as follows:

$$\langle \bigcirc \rangle CO.OO.OC \langle \bigcirc \rangle \longrightarrow 2 \langle \bigcirc \rangle CO.O\cdot$$

$$\langle \bigcirc \rangle CO.O\cdot \longrightarrow \langle \bigcirc \rangle \cdot + CO_2$$

Denoting by $R\cdot$ the radicals which are formed, and the monomer by M, the next stage of the reaction is interaction between $R\cdot$ and M to form a species, denoted by $M_1\cdot$, which contains one monomer unit and an active free radical at one of the carbon atoms of the carbon–carbon double bond of the monomer:

$$R\cdot + M \longrightarrow M_1\cdot$$

where $M_1\cdot$ denotes the species $RM\cdot$. The totality of the reactions which lead to the formation of the $M_1\cdot$ species is known collectively as *initiation*.

Propagation
The polymer sequence is formed by a succession of radical addition reactions involving monomer molecules:

$$M_1\cdot + M \longrightarrow M_2\cdot$$
$$M_2\cdot + M \longrightarrow M_3\cdot$$
$$M_i\cdot + M \longrightarrow M_{i+1}\cdot$$

These reactions are collectively known as *propagation*.

If we examine the propagation reaction in a little more detail, we see that each step presents several possibilities. Consider first the polymerisation of a *vinyl* monomer, $CH_2 : CHX$, and the case in which each propagation step is such as to lead to the formation of the polymer sequence:

$$\cdots -\overset{\overset{\displaystyle H}{|}}{\underset{\underset{\displaystyle H}{|}}{C}}-\overset{\overset{\displaystyle H}{|}}{\underset{\underset{\displaystyle X}{|}}{C\dagger}}-\overset{\overset{\displaystyle H}{|}}{\underset{\underset{\displaystyle H}{|}}{C}}-\overset{\overset{\displaystyle H}{|}}{\underset{\underset{\displaystyle X}{|}}{C\dagger}}-\cdots$$

(XXIII)

Taking into account the fact that polymer chains of different lengths, if not of detailed structures, will be attached to either side of the unit —CH$_2$— CHX—, it follows that the carbon atoms marked † are asymmetric. Any one of them can therefore have one of two spacial configurations as regards the groups which are attached to them, namely, D and L. The polymer sequence which is formed by a succession of propagation steps may therefore comprise (i) a succession of units all of which have the same configuration (all D or all L), (ii) a succession of units each of which has the mirror-image configuration to its predecessor and successor, or (iii) a succession of units whose configurations are randomly D and L. Vinyl polymers of these three types are known respectively as (i) *isotactic*, (ii) *syndiotactic*, and (iii) *atactic*. The properties of a vinyl polymer depend, amongst other things, upon its structure in respect of the sequential arrangement of D and L units in the polymer chain.

However, the polymer sequence (XXIII) shown above is not the only possibility. Successive monomer units may also add to the polymer sequence to give the pairs of units shown below:

$$
\begin{array}{ccccc}
 & H & H & H & H \\
 & | & | & | & | \\
\cdots -C & -C & -C & -C & - \cdots \\
 & | & | & | & | \\
 & H & X & X & H
\end{array}
\qquad\qquad
\begin{array}{ccccc}
 & H & H & H & H \\
 & | & | & | & | \\
\cdots -C & -C & -C & -C & - \cdots \\
 & | & | & | & | \\
 & X & H & H & X
\end{array}
$$

(XXIV)   (XXV)

Because of the relative dispositions of the X groups in the three types of sequence (XXIII), (XXIV) and (XXV), the three pairs of units shown are said to be joined together in *head-to-tail*, *head-to-head*, and *tail-to-tail* modes respectively.

Turning now to consider the polymerisation of a 1,3-*diene* monomer, that is, a monomer based upon the structure C=C—C=C, we find the number of possibilities for the outcome of each propagation step to be greatly increased. In the following diagram, an attempt has been made to enumerate these possibilities for the case of the polymerisation of 1,3-butadiene, and to show the relationships between them. In the first place, each butadiene unit can become enchained either as a 1,4 unit or as a 1,2 unit. If the former, then each double bond in the enchained unit can have either the *cis* configuration or the *trans* configuration as regards the disposition of the four substituents attached to the double bond. If the mode of enchainment is 1,2, then the monomer has in effect reacted as the

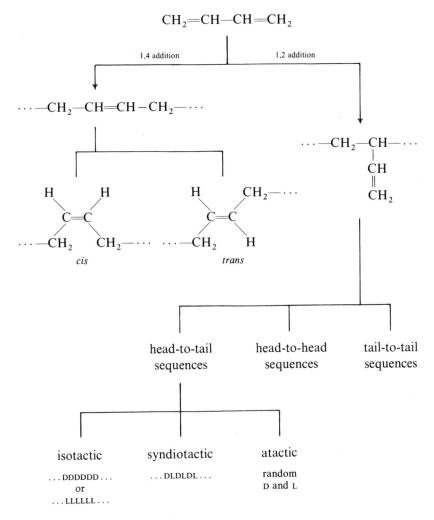

vinyl monomer CH_2: $CH(CH:CH_2)$. All the various configurations and sequences of configurations discussed above in connection with vinyl monomers then become possibilities.

The possibilities become still more numerous in the case of the polymerisation of substituted 1,3-dienes such as isoprene and chloroprene. There is now a third distinct mode of enchainment of the monomer unit, namely, by 3,4 addition. (In the case of butadiene, of course, the result of 3,4 addition is identical with that of 1,2 addition.)

Termination

We pass now to consider reactions which lead to the destruction of radical activity (although not necessarily to destruction of the radical itself). These reactions are known collectively as *termination*. They include:

1. interaction between pairs of propagating chains to give non-radical polymer species by either combination or disproportionation:

$$M_i \cdot + M_j \cdot \begin{array}{c} \xrightarrow{\text{combination}} M_{i+j} \\ \xrightarrow{\text{disproportionation}} M_i + M_j \end{array}$$

where M_i denotes a 'dead' polymer sequence containing i monomer units;

2. interaction between a propagating chain and some other molecule in the reaction system leading to the formation of a stable (and therefore unreactive) free radical, e.g.:

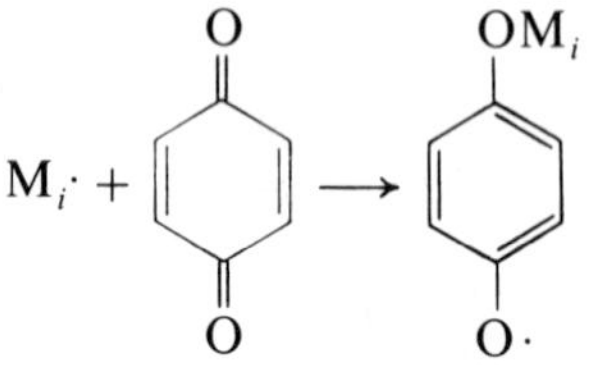

Transfer reactions

Initiation, propagation and termination do not exhaust the possible reactions which may occur in a free-radical addition polymerisation, although, of course, they are the only types of reaction which are essential for the formation of polymers of high molecular weight by this mechanism. There is a fourth group of reactions of considerable importance, known collectively as *transfer* reactions. As the name implies, they are reactions in which the radical activity is not destroyed but is transferred from one propagating chain to another chemical species. The outcome is then a 'dead' polymer chain, together with an active radical centre elsewhere through which propagation can continue. Such reactions have little effect upon the rate of polymerisation, but they can have an important effect upon the molecular weight and structure of the polymer produced.

Transfer reactions can be conveniently classified as:

1. transfer reactions to monomer, in which the reaction

$$M_i\cdot + M \longrightarrow M_i + M_i\cdot$$

occurs as an alternative to propagation;

2. transfer reactions to initiator, I, which can be represented as

$$M_i + I \longrightarrow M_i + I\cdot$$

3. transfer reactions to other molecules which have been deliberately added to, or are adventitiously present in, the polymerisation system; and

4. transfer reactions to 'dead' polymer.

A few comments are necessary concerning transfer reactions of types (3) and (4). The two most common types of 'other molecule' which are deliberately added to free-radical polymerisation reaction systems are (a) monomer diluents and (b) molecular-weight regulators or *modifiers*. If a polymerisation reaction is carried out in, say, ethylbenzene as solvent, then occasional transfer reactions of the type

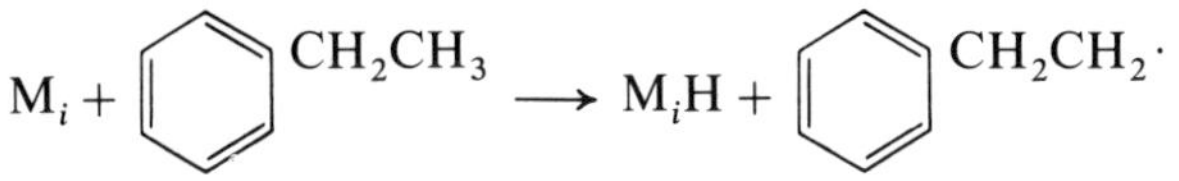

$$M_i + \text{(benzene ring)}\,CH_2CH_3 \longrightarrow M_iH + \text{(benzene ring)}\,CH_2CH_2\cdot$$

will occur. And if substances such as carbon tetrachloride or dodecyl mercaptan (typical molecular-weight regulators) are present in the reaction system, then the following reactions will occur:

$$M_i\cdot + CCl_4 \longrightarrow M_iCl + \cdot CCl_3$$

$$M_i + C_{12}H_{25}SH \longrightarrow M_iH + C_{12}H_{25}S\cdot$$

(In these examples, the 'dead' polymer has been written as M_iH or M_iCl rather than as M_i, in order to show clearly the fate of the various atoms involved in the reaction.)

Turning now to transfer-to-polymer reactions, these are important because they lead to *branching* of the polymer sequence, and ultimately to the *crosslinking* of polymer chains. Taking poly-1,4-butadiene as an example, a typical transfer-to-polymer reaction is one which involves the abstraction of a hydrogen atom from a carbon atom adjacent to a carbon–carbon double bond as follows:

$$\cdots -CH_2-CH=CH-CH_2\cdot + \cdots -CH_2-CH=CH-CH_2-\cdots$$

$$\downarrow$$

$$\cdots -CH_2-CH=CH-CH_3 + \cdots -\dot{C}H-CH=CH-CH_2-\cdots$$

Propagation from the radical centre formed in this reaction will clearly lead to the formation of a branch in the polymer sequence *to* which the transfer reaction occurred. In the case of a diene polymer, branching and crosslinking can also occur by way of propagation through the pendant vinyl group which results from 1,2 addition and also (although it is less likely) through the carbon–carbon double bond which results from 1,4 addition:

$$
\sim\!\!\cdot + \sim\!\!CH_2-CH\!\!\sim \qquad \sim\!\!CH_2-CH\!\!\sim
$$
$$
\hspace{2.5cm} | \hspace{3cm} |
$$
$$
\hspace{2.5cm} CH \quad \longrightarrow \quad \sim\!\!CH
$$
$$
\hspace{2.5cm} \| \hspace{3.2cm} |
$$
$$
\hspace{2.5cm} CH_2 \hspace{3cm} \cdot CH_2
$$

It is important to notice that transfer-to-polymer reactions, and also propagation reactions through enchained double bonds, will be favoured by an increasing concentration of polymer molecules. It is therefore to be expected that these reactions will become increasingly important as the monomer becomes progressively converted to polymer.

One final comment is necessary concerning termination and transfer reactions in free-radical addition polymerisations. These are the reactions by which a propagating polymer chain loses its reactivity and becomes a 'dead' polymer molecule. These are therefore the reactions which control the number of monomer units which are present in the final polymer molecule, and which therefore control the molecular weight of the final polymer molecule. But these reactions are essentially random in nature, in the sense that the probability that a particular propagating polymer chain in a given polymerisation system becomes involved in a termination or transfer reaction is for practical purposes independent of the length of the propagating polymer chain. An important consequence of the random nature of termination and transfer is that a given free-radical polymeris-ation reaction produces a polymer whose molecules are not all of the same size. The distribution of molecular sizes produced by free-radical addition polymerisation is in fact broad, even if transfer reactions are absent. Transfer reactions tend to broaden the distribution still further, especially if transfer-to-polymer reactions occur. Transfer-to-polymer reactions produce a 'dead' polymer molecule whose average molecular size is less than what it would have been in the absence of the transfer reaction. But they also produce a propagating species which is macromolecular from its inception and which will probably grow to a size which is considerably larger than the average size of the polymer molecule which would be formed in the absence of the transfer-to-polymer reaction. If the branching which is the usual concomitant of transfer-to-polymer reactions occurs to such an extent that crosslinked structures of indefinitely large size are produced, then the distribution of molecular sizes produced by the polymerisation reaction can be very broad indeed.

4.2.3 Ways in which Free-radical Addition Polymerisation can be Effected

The most obvious way of carrying out the reaction is to dissolve the initiator (if one is to be used) in the monomer and then to establish whatever conditions (e.g., elevated temperature) are necessary in order to ensure that polymerisation occurs at a convenient rate. Polymerisations effected in this way are known as *bulk polymerisations*. This procedure offers the advantage that the polymer is obtained in a comparatively uncontaminated condition and in such a form that there is little from which to separate it other than residual monomer. The disadvantages of the process include the extremely high viscosity which develops as the reaction proceeds, the difficulty of dissipating the heat of polymerisation with the consequent problem of controlling the polymerisation temperature, and the prevalence of reactions involving dead polymer molecules, especially in the latter stages of the reaction.

The procedure known as *suspension polymerisation* offers a way of overcoming some of the disadvantages of bulk polymerisation. In this procedure, the reaction system is broken up into a large number of small droplets dispersed in an inert medium. In the case of the suspension polymerisation of hydrocarbon monomers, the obvious choice for the dispersion medium is water. The whole system remains fluid throughout the course of the reaction, because the viscosity is determined mainly by that of the dispersion medium. Heat dissipation is no longer a problem, because the dispersion medium acts as an effective heat-transfer medium. The characteristics of the reaction are essentially those of a bulk polymerisation; indeed, suspension polymerisations can be fairly described as 'micro-bulk' polymerisations. The disadvantages of suspension polymerisation include those which accrue from the presence of high concentrations of 'dead' polymer molecules, those which accrue from the presence of the dispersion medium and any dispersion stabiliser necessary to ensure adequate dispersion stability, and those which are inherent in the process itself. Included amongst the latter is a tendency for the monomer/polymer 'beads' to stick together to form a semi-solid mass during the course of the reaction.

An alternative approach to the problem of high viscosity in the later stages of the reaction is that of *solution polymerisation*. In this procedure, the polymerisation is carried out in a solvent for the monomer (which may or may not be a solvent for the polymer also). The reaction system can remain reasonably fluid if conditions are suitably adjusted. Reactions involving 'dead' polymer molecules tend to occur less than in bulk or suspension reactions, because the concentration of polymer molecules is reduced. However, for a given rate of initiation, the rate of polymerisation is reduced because the concentration of monomer is reduced. The other disadvantages of solution polymerisation come under three broad headings: (a) the cost of the solvent, (b) the problem of subsequent solvent recovery, and (c) the effect of transfer-to-solvent reactions in reducing the molecular weight of the polymer produced

The fourth way in which free-radical addition polymerisation can be effected is known as *emulsion polymerisation*. Until comparatively recently, this method for effecting polymerisation has been of overwhelming importance for the production of synthetic rubbers. It is still a very important reaction, and for this reason a separate section is devoted to a description of its mechanism (Section 4.2.4). We merely note here the way in which the reaction is carried out, together with its principal advantages and disadvantages.

As practised industrially, emulsion polymerisation bears a superficial resemblance to suspension polymerisation. The polymerisation is carried out in a reaction system which comprises monomer and a dispersion medium (usually aqueous) in which the monomer is either virtually insoluble, or at most sparingly soluble. Unlike suspension polymerisation, a soap-like substance is present, together with a water-soluble initiator. Other substances may also be present, but these are the essential components. The monomer(s), dispersion medium, soap, initiator, and other ingredients are brought together and heated to an appropriate temperature under agitation. Polymerisation normally occurs quite smoothly, and the reaction lends itself readily to continuous operation. Furthermore, the dispersion medium usually being water, the use of expensive and potentially hazardous solvents is eliminated, as is the problem of solvent recovery. As has been noted in Chapter 2, the end-product of the reaction is a stable fine-particle polymer latex. The reaction system remains perfectly fluid over the entire extent of the reaction. Heat dissipation is no problem. Provided that the reaction recipe has been correctly formulated, there is very little tendency for the particles to coalesce as the reaction proceeds. Apart from the controllability of the reaction and the inherent ease of carrying it out, the dominant advantage of the emulsion polymerisation reaction is that it provides a method whereby polymer of very high molecular weight can be obtained at relatively fast rates of polymerisation. This feature is of particular significance for the production of polymers to be used as synthetic rubbers, because the development of satisfactory elastomer properties requires that the primary polymer molecular weight be high. It is significant that any attempt to increase the rate of polymerisation in bulk, suspension or solution polymerisations is invariably accompanied by a reduction in the molecular weight of the polymer produced. This is an inescapable kinetic consequence of the reaction mechanism. It arises essentially from the fact that the propagation reaction is unimolecular with respect to propagating radicals, and is kinetically of first order with respect to the concentration of these radicals, whereas the termination reaction is bimolecular with respect to propagating radicals, and is kinetically of second order with respect to the concentration of these radicals. Thus any attempt to increase the rate of polymerisation by increasing the concentrations of propagating radicals is inevitably accompanied by a disproportionate increase in the rate of termination. Emulsion polymerisation offers a way out of this impasse.

There are several disadvantages which attend the emulsion polymerisation process. An obvious one is that, if the polymer is required in dry bulk form as opposed to latex form, it is necessary to separate the polymer from the dispersion medium. A second disadvantage is that the polymer, even when separated from the latex, has associated with it relatively large quantities of non-polymer substances such as soap and soap derivatives. Other disadvantages include a tendency for the reaction to be rather variable, especially when carried out as a batch reaction, and, as compared with bulk polymerisation, a large reduction in effective reactor volume because of the presence of the dispersion medium. Regarding this latter point, it may be noted that the monomer phase of a typical industrial reaction may comprise as little as 40 % of the total volume of the reaction system; it is therefore necessary to provide a reactor volume which is more than twice that of the monomer itself.

It is necessary to draw attention to one very important shortcoming of the emulsion polymerisation process which it shares in common with other procedures for effecting free-radical addition polymerisation. This concerns the very limited extent to which it is normally possible to control and vary the exact mode in which the propagation reaction occurs. Thus, for instance, the free-radical polymerisation of butadiene invariably produces a polymer which contains *cis*-1,4 units, *trans*-1,4 units and 1,2 units. Despite the strenuous efforts which have been made, there has been little progress in developing free-radical systems (and, in particular, emulsion systems) which produce so-called *stereoregular* or *stereospecific* polymers, that is, polymers in which a high proportion of the repeat units has a single isomeric configuration. Polymers of this type have been in production for several years now, but, with one important exception (that of chloroprene rubber), they are invariably produced by an ionic polymerisation mechanism.

4.2.4 Outline of Mechanism of Emulsion Polymerisation

At first sight, it is tempting to suppose that the polymerisation in an emulsion polymerisation reaction occurs in the emulsified droplets of monomer. If this were the case, then the reaction would be indistinguishable from a suspension polymerisation. That this is not usually the case is indicated by several pieces of evidence, including:

1. the fact that the latex particles produced are usually much smaller than the droplets of emulsified monomer initially present;
2. the fact that it is not necessary for the monomer to be emulsified in the dispersion medium, nor, indeed, for it to be in direct contact with the dispersion medium at all;
3. evidence from the composition of copolymers produced by emulsion copolymerisation; and
4. evidence from certain other features of the reaction, such as the nature of the dependence of the rate of polymerisation upon the amount of soap present in the reaction system.

If we examine the conversion–time curve for a typical emulsion polymerisation reaction (Fig. 4.1), we see that the reaction divides into three stages or 'intervals', as they are commonly known:

1. Interval I, which is a stage during which the rate of polymerisation increases progressively up to a maximum;
2. Interval II, in which the rate of polymerisation remains constant at the maximum reached at the end of Interval I; and
3. Interval III, in which the rate of polymerisation gradually falls to zero.

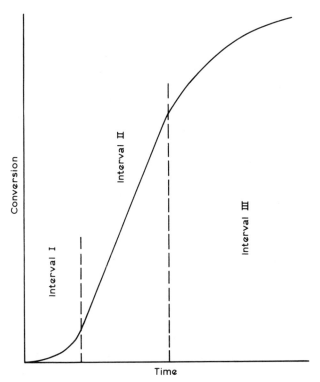

FIG. 4.1. Conversion–time curve for a typical emulsion polymerisation reaction, showing delineation of Intervals I, II and III.

All the available evidence supports the view that the polymerisation during Intervals II and III is taking place within a very large and essentially constant number of individual particles or reaction loci (typically 10^{15} cm^{-3} of reaction system), and that Interval I is the stage of the reaction in which those loci are formed. Interval I is therefore referred to as the stage of *locus nucleation* or *particle nucleation*. In Intervals II and III, the number of reaction loci is essentially constant because no further locus nucleation is taking place, and because, if the system is colloidally stable, little agglomeration of particles occurs. In Interval II, the rate of polymerisation is constant because unreacted monomer is still present as a separate monomer droplet phase. The concentration of monomer in the reaction loci is therefore held constant (at a level which many believe is essentially the level at which it would be maintained if no polymerisation were occurring), and so we have polymerisation taking place in a fixed number of reaction loci under conditions such that the monomer concentration remains constant within each locus. Under these conditions, and given a fixed concentration of propagating radicals in each locus, the rate of polymerisation will not vary with extent of reaction; the rate of polymerisation will therefore appear to be kinetically of zero order with respect to monomer concentration. The transition from Interval II to Interval III marks the disappearance of the separate monomer droplet phase which hitherto has maintained the monomer concentration in the reaction loci at a constant level. The rate of polymerisation now falls as reaction proceeds, because of the depletion of monomer within the reaction loci.

The above gives a brief resumé of the mechanism of emulsion polymerisation which probably attracts general consent today. However, when we pass from the broad generalisations of the preceding paragraph to consider the mechanism in a little more detail, we enter into realms of controversy concerning which the last word is far from yet having been written.

Consider first the mechanism of locus nucleation. Until comparatively recently, the view was held almost universally that this occurred by a process known as *micellar nucleation*, at least in the case of the hydrocarbon monomers of low water-solubility from which many synthetic rubbers are made. According to this view, the soap which is present in the emulsion polymerisation system fulfils the crucial role of providing the sites for locus nucleation. Soap molecules are surface-active in nature, that is, their molecules comprise regions of markedly different solubility behaviour, the one part tending to be water-soluble, and the other part tending to be expelled from an aqueous environment into a contiguous non-polar phase. In aqueous solution, they therefore tend to be adsorbed at air–water and monomer–water interfaces. Adsorption at the air–water interface is manifested by reduction in the surface tension of the aqueous solution. However, once such interfaces are filled up with adsorbed soap molecules (and in a typical emulsion polymerisation system, very little soap will be requird to saturate these interfaces), the molecules tend to aggregate together in such a way as jointly to satisfy the diverse solubility tendencies of their several parts. These aggregates are known as *micelles*. Once the air–water interface of an aqueous solution is filled up with adsorbed soap molecules, the surface tension of the solution becomes insensitive to further additions of soap to the solution. In fact, the point at which the surface tension becomes independent of soap concentration is frequently taken as diagnostic of the commencement of micelle formation. Typically, micelles are formed from 50–100 soap molecules. In the aqueous phase of a typical emulsion polymerisation, there are of the order of 10^{18} micelles cm^{-3}. It is known that, because micelles have non-polar interiors, they are able to imbibe within their interiors small amounts of other non-polar molecules, such as hydrocarbon monomers. This process is known as *solubilisation*. According to the theory of micellar initiation, free radicals generated in the aqueous phase from the water-soluble initiator (or possibly surface-active oligomer radicals formed by aqueous solution polymerisation) diffuse into the interiors of some of the micelles and there initiate polymerisation of the monomer molecules which have been solubilised within the micelle. The concentration of monomer in those micelles (or particles, as we must now begin to call them) then falls, and more monomer is taken up from the monomer droplet phase via the aqueous phase. Further polymerisation occurs, and, by continuation of the processes of polymerisation and monomer imbibition, the nucleated micelle grows into what is best described as a latent latex particle. In so growing, the surface area of the particle increases, and thus residual monomolecular soap from the aqueous phase tends to be adsorbed at the growing particle surface. This soap is then replenished by dissociation of those micelles which have not as yet been nucleated by the absorption of a radical from the aqueous phase. As can be seen by comparing the numbers given above for the concentrations of micelles and ultimate reaction loci in a typical emulsion polymerisation reaction system, very few (approximately one in every 10^3) micelles become reaction loci. This, then, is an outline of the theory of micellar initiation.

In recent years, an alternative mechanism for locus nucleation has been proposed and has received considerable attention. This is the mechanism known as *homogeneous nucleation*. According to this mechanism, nucleation occurs by way of solution polymerisation of the monomer in the aqueous phase, followed by precipitation of the low-molecular-weight polymer molecules (called *oligomers*) when they reach a certain critical size. Precipitation will tend to occur because the larger is an oligomer molecule the less soluble will it be in a liquid, such as water, in which the high-molecular-weight version is completely insoluble. In principle, the solubility limit might be reached either because the molecule has grown to the critical size during the propagation step, in which case the precipitated polymer molecule will already contain an active free radical, or because the size has increased abruptly as the result of mutual termination by combination within the aqueous phase, in which case the precipitated polymer molecule will not contain a free radical. The function of the soap is seen as providing colloidal stability for the reaction loci which form by precipitation, and possibly also some stability for the growing oligomeric radicals, preventing them from precipitating prematurely. According to the theory of homogeneous nucleation, the presence of the soap micelles is irrelevant to the nucleation mechanism; the soap micelles merely provide a reservoir of soap molecules to be drawn upon as needed for the stabilisation of the nucleating and growing reaction loci.

It has long been recognised that homogeneous nucleation must occur in emulsion polymerisation reactions whose initial aqueous phases do not contain soap micelles. Such reactions are known, and they occur particularly readily if the monomer is fairly soluble in water. What is still unresolved is how the loci are nucleated if micelles are present initially, especially if the monomer is very sparingly soluble in water. The prevailing view at the time of writing seems to favour micellar nucleation in such cases, but it is possible, of course, that reaction loci are formed by both mechanisms simultaneously.

Turning now to consider briefly what happens in Intervals II and III of the reaction—that is the stages in which most of the polymerisation occurs—the first point to notice is that each reaction locus contains very few propagation radicals. It is not the same propagating radicals which are present throughout the whole course of the reaction. Loci are continually acquiring radicals from the aqueous phase (in which radicals continue to be generated by decomposition of the initiator) and losing them by mutual termination within loci and by diffusion out into the aqueous phase. According to the 'classical' view of emulsion polymerisation reactions, each locus contains, on average, one half of a propagating radical, that is to say, at any instant one half of the loci contain one propagating radical and the other half contain no radicals at all. The actual average number of radicals per reaction locus is probably rather higher than 0·5, and tends to increase as the reaction proceeds. However, the average number is undoubtedly of the order of unity. Thus each reaction locus contains very few propagating centres, and opportunities for mutual termination are much less than in a bulk polymerisation at a similar overall radical concentration, because mutual termination is bi-molecular with respect to propagating radicals. In a sense, the individual propagating centres have to a large extent been isolated from one another, each one or two having been provided with their own minute reaction vessel. This is the conventional explanation for the combination of high rate of polymerisation and high molecular weight of product which is characteristic of emulsion polymerisation reactions, and to which reference has been made in the preceding section.

There is, however, an alternative explanation for the combination of high rate of polymerisation and high product molecular weight which characterises emulsion polymerisation reactions. According to this alternative view, the cause is to be found in the very high viscosity of the interior of the reaction loci, especially in the later stages of the reaction. The high viscosity is due to the high ratio of polymer to monomer present in the loci at all times, but especially at high conversions. It is known that the effect of high viscosity in a free-radical polymerisation reaction is to retard mutual termination relative to propagation and this in turn leads to an enhancement of both the rate of polymerisation and the molecular weight of the product. There seems little doubt that in reality both causes contribute to the combination of high rate of polymerisation and high product molecular weight which emulsion polymerisation provides.

4.2.5 Anionic Addition Polymerisation

As has been stated in Section 4.2.1, the term *anionic addition polymerisation* is used to denote those addition polymerisation reactions involving the opening of carbon–carbon double bonds in which the propagating reactive centre is a carbanion. In order to emphasise the nature of the propagating centre, the term *carbanionic polymerisation* is also used to describe such polymerisation reactions. The initiation reaction in this type of polymerisation comprises those reactions in which are formed carbanionic centres which are then capable of propagating. The substances which are able to initiate the anionic polymerisation of ethylenic and diene monomers can be conveniently classified under two broad headings:

1. Ionic or ionogenic substances of the type $X^{\oplus}Y^{\ominus}$ or $X^{\delta\oplus}\ldots Y^{\delta\ominus}$, where the actual or potential anion $Y^{\ominus}$ is able to add to the carbon–carbon double bond to form a carbanion which can then propagate. The initiation reaction can be represented generally as

$$X^{\oplus}Y^{\ominus} + \overset{|}{\underset{|}{C}}{=}\overset{|}{\underset{|}{C}} \longrightarrow Y{-}\overset{|}{\underset{|}{C}}{-}\overset{|}{\underset{|}{C}}{}^{\ominus}\cdots X^{\oplus}$$

or

$$X^{\delta\oplus}\cdots Y^{\delta\ominus} + \overset{|}{\underset{|}{C}}{=}\overset{|}{\underset{|}{C}} \longrightarrow Y{-}\overset{|}{\underset{|}{C}}{-}\overset{|}{\underset{|}{C}}{}^{\delta\ominus}\cdots X^{\delta\ominus}$$

depending upon whether the initiator is fully ionised or merely strongly polarised. Even in cases where the initiator and the propagating centre are fully ionised, the extent to which the counterion is independent of the carbanionic centre can vary, being dependent upon such factors as the dielectric constant of the reaction medium and the presence of other compounds which can bond coordinatively to the counterion; thus low dielectric constant favours association, whilst coordination to the counterion favours dissociation. In principle, the counterion $X^{\oplus}$ may be inorganic or organic. In practice, it is usually a cation derived from an alkali or alkaline-earth metal. Examples of this class of anionic initiator are the lithium alkyls, which are more appropriately regarded as ionogenic rather than as ionic, and the alkali-metal amides and alkoxides, which are truly ionic in nature.

2.   Free metals which are able to donate an electron to the monomer, with the consequent formation of a radical ion and a metal cation. The initial reaction can be represented as follows:

$$X^{\circ} + \overset{|}{C}{=}\overset{|}{C} \longrightarrow \cdot\overset{|}{C}{-}\overset{|}{C}^{\ominus} \cdots X^{\oplus}$$

where X° denotes the free metal. Usually X° has to be an alkali metal in order for it to be sufficiently electropositive to be able to transfer an electron to the carbon–carbon double bond. This initial reaction can be followed by one of two possibilities. The first of these possibilities is that the radical–carbanion initially produced receives a further electron from a second metal atom, thereby becoming a di-carbanion:

$$X^{\circ} + \cdot\overset{|}{C}{-}\overset{|}{C}^{\ominus} \cdots X^{\ominus} \longrightarrow X^{\oplus} \cdots {}^{\ominus}\overset{|}{C}{-}\overset{|}{C}^{\ominus} \cdots X^{\ominus}$$

The second possibility is that the radical–carbanion dimerises, thereby also becoming a di-carbanion, but of different structure:

$$2 \cdot\overset{|}{C}{-}\overset{|}{C}^{\ominus} \cdots X^{\oplus} \longrightarrow X^{\oplus} \cdots {}^{\ominus}\overset{|}{C}{-}\overset{|}{C}{-}\overset{|}{C}{-}\overset{|}{C}^{\ominus} \cdots X^{\oplus}$$

Propagation proceeds by a sequence of reactions in which the carbanion attacks and adds to a carbon–carbon double bond, thereby causing one further monomer unit to be added to the growing chain and producing a species which is still a carbanion. If the initiating species is a di-carbanion, then the polymer chain propagates simultaneously from both ends. An important aspect of anionic addition polymerisations is that the extent to which the counterion is associated with the propagating carbanion can have an important effect upon the outcome of the propagation reaction in respect of the stereochemical configuration which the enchained monomer units adopt. This matter will be given further consideration in Chapter 5, where it will be discussed in relation to the industrially-important anionic polymerisations of butadiene initiated by lithium metal and lithium alkyls.

Termination reactions, analogous to the bimolecular mutual termination of free-radical addition polymerisations, are generally absent from anionic addition polymerisations. It is clearly extremely unlikely that two propagating anionic species will interact to destroy their respective anionic centres. Furthermore, there is in general no reaction whereby an ion pair can be eliminated from the end of a propagating anion leaving an unreactive polymer molecule. Termination of propagating chains generally occurs by a proton-transfer reaction with an active-hydrogen compound, H—Z, as follows:

$$\sim\!\!C\!-\!\!C^{\ominus}\cdots X^{\oplus} + H\!-\!Z \longrightarrow \sim\!\!C\!-\!\!C\!-\!H + X^{\oplus}\cdots Z^{\ominus}$$

To the extent that the species $X^{\oplus}\ldots Z^{\ominus}$ is unable to re-initiate the polymerisation, then the reaction is one of termination. Active-hydrogen compounds may be present in the reaction system as adventitious impurities, in which case they bring about an unwanted retardation or inhibition of polymerisation. They may also be added deliberately to the reaction system when it is desired to stop further polymerisation. Typical active-hydrogen compounds which are able to terminate many carbanionic polymerisations include water, alcohols, phenols and mercaptans. If the species $X^{\oplus}\ldots Z^{\ominus}$ is able to re-initiate polymerisation, then the reaction is one of transfer rather than termination. Transfer of activity to a monomer molecules can also occur, the reaction usually being one of hydride-ion transfer:

$$\sim\!\!C\!-\!\!C^{\ominus}\cdots X^{\oplus} + C\!=\!C \longrightarrow \sim\!\!C\!=\!C + H\!-\!C\!-\!C^{\ominus}\cdots X^{\ominus}$$
$$\underset{H}{\vert}$$

Termination can also be brought about by reaction with any substance which produces an ionic species which is insufficiently reactive to continue propagation by attack of a further carbon–carbon double bond, even although the ionic centre remains attached to the polymer molecule. A common example of such a reaction is interaction with carbon dioxide to produce a relatively unreactive carboxylate ion:

$$\sim\!\!\overset{|}{\underset{|}{C}}\!-\!\overset{|}{\underset{|}{C}}{}^{\ominus}\cdots X^{\oplus} + CO_2 \longrightarrow \sim\!\!\overset{|}{\underset{|}{C}}\!-\!\overset{|}{\underset{|}{C}}\!-\!CO_2^{\ominus}\cdots X^{\oplus}$$

This reaction, which in effect results in the insertion of a carbon dioxide molecule into the ionic bond $C^{\ominus}\ldots X^{\oplus}$, occurs by way of attack of the propagating carbanion on the carbon–oxygen double bond of carbon dioxide as an alternative to attack of the carbon–carbon double bond of a monomer molecule.

One consequence of the absence of an 'inherent' termination reaction is that, unless chain-terminating substances are either adventitiously present or deliberately added, a given active centre is able to propagate for as long as monomer is available. In cases where all the active centres are available for propagation virtually from the commencement of the polymerisation, this feature leads to the formation of a polymer which has a very narrow distribution of molecular weights. It is to be expected that all the active centre will be available for propagation from the commencement of the polymerisation when initiation occurs by way of reaction between the monomer and an ionic or ionogenic substance. On the other hand, if initiation occurs by reaction between an alkali metal and the monomer, then the propagating centres will form more gradually and, in consequence, a polymer having a rather broader distribution of molecular weights will be formed. Nevertheless, the distribution of molecular weights formed in even this latter type of reaction is narrow compared to that produced by free-radical addition polymerisation.

The susceptibility of an ethylenic or diene monomer to addition polymerisation by the carbanionic mechanism is affected by the nature of any substituents which may be present in the monomer molecule. In general, the presence of an electron-withdrawing substituent such as a cyano group, a nitro group, or a carboxylic-ester group will favour polymerisation by the carbanionic mechanism, because such substituents will favour nucleophilic attack of the carbon–carbon double bond on the part of the carbanion by reducing the electron density there. The mechanism by which the rate of propagation is enhanced can be represented as follows:

$$\sim\!\!\overset{|}{\underset{|}{C}}\!-\!\overset{|}{\underset{|}{C}}{}^{\ominus}\cdots X^{\oplus} + {}^{\delta\oplus}\overset{|}{C}\!=\!\overset{|}{\underset{|}{C}} \longrightarrow \sim\!\!\overset{|}{\underset{|}{C}}\!-\!\overset{|}{\underset{|}{C}}\!-\!\overset{|}{\underset{|}{C}}\!-\!\overset{|}{\underset{|}{C}}{}^{\ominus}\cdots X^{\oplus}$$
$$\underset{Z^{\delta\ominus}}{} \qquad\qquad \underset{Z}{}$$

The initiation reaction will be similarly favoured. In an analogous manner, the presence of an electron-releasing substituent, such as an alkyl group, will discourage polymerisation by the carbanionic mechanism, because such substituents tend to increase the density of electrons in the vicinity of the carbon–carbon bond and so discourage attack by a nucleophilic reagent such as a carbanion.

4.2.6 Cationic Addition Polymerisation

By analogy with anionic addition polymerisation, the term *cationic addition polymerisation* is used to denote those addition polymerisation reactions involving the opening of carbon–carbon double bonds in which the propagating reactive centre is a carbonium ion. The term *carbonium-ion polymerisation* is also used to describe polymerisations of this type, in order to emphasise the nature of the propagating centre. In contrast to polymerisation by the carbanionic mechanism, the susceptibility of an ethylenic or diene monomer to carbonium-ion polymerisation is enhanced by the presence of electron-releasing substituents and diminished by the presence of electron-withdrawing substituents. The mechanism by which the rate of polymerisation is enhanced by electron-releasing substituents can be represented as follows:

Addition polymerisations which proceed by way of carbonium-ion active centres often occur very rapidly at low temperatures. The rate of polymerisation can often be such that the reaction occurs almost instantaneously. Thus, for example, isobutene can be polymerised under the influence of suitable carbonium-ion initiators at $-100\,^{\circ}\text{C}$ within a fraction of a second. The product includes polymer molecules of high molecular weight (containing up to 10^5 monomer units) which have clearly formed extremely quickly. In practice, it may be convenient to carry out the reaction in a liquid, such as propane or butane, which has a low boiling point. This liquid functions as an internal refrigerant; by allowing the heat of polymerisation to become dissipated through vaporisation, excessive temperature rises are prevented.

The observations of the preceding paragraph provoke the question as to why carbonium-ion polymerisations proceed so much more rapidly, and at much lower temperatures, than either carbanion or free-radical addition polymerisations. The reason is that the initiation and propagation steps of this type of polymerisation involve interaction between a strongly electrophilic species (a carbonium ion) and a species which has a surplus of electrons (the monomer). It is therefore hardly surprising that the reaction is facilitated. Similarly, carbanionic polymerisations occur relatively slowly because both initiation and propagation involve interaction between a strongly nucleophilic species (a carbanion) and a species which has a surplus of electrons (the monomer). Free-radical addition polymerisations occupy an intermediate position between these extremes, but in respect of reaction rate resemble carbanionic addition polymerisations more closely than carbonium-ion addition polymerisations.

The substances which are able to initiate the carbonium-ion polymerisation of ethylenic and diene monomers are all strong Lewis acids, that is, they are all powerful electron acceptors. They can be conveniently classified as follows:

1.   protonic acids, such as sulphuric and perchloric acid;
2.   substances known collectively as 'Friedel–Crafts' catalysts, such as boron trifluoride, aluminium chloride and stannic chloride;
3.   carbonium salts, such as benzoyl perchlorate; and
4.   cationogenic substances, such as triphenylmethyl chloride.

Of these, the Friedel–Crafts catalysts are the most important.

The mechanism by which substances in the first, third and fourth of these classes initiate carbonium-ion polymerisation can be very simply represented as follows:

$$X^{\ominus} \cdots Y^{\oplus} + \overset{|}{\underset{|}{C}} = \overset{|}{\underset{|}{C}} \longrightarrow Y - \overset{|}{\underset{|}{C}} - \overset{|}{\underset{|}{C}}{}^{\oplus} \cdots X^{\ominus}$$

where for substances in the first of the above classes $Y^{\oplus}$ is a proton. The mechanism by which substances in the second of these classes initiate carbonium-ion polymerisation is more complex. They are believed to require the presence of a *co-catalyst* in order to be effective. The compounds which function as co-catalysts for this purpose are generally active-hydrogen compounds. Denoting the Friedel–Crafts catalyst by MA_n, where M is a metal or metalloid atom and A is a halogen atom, and the co-catalyst by HB, the formation of the effective catalyst can be represented as follows:

$$MA_n + HB \longrightarrow [MA_n B]^{\ominus} H^{\oplus}$$

Thus in effect the co-catalyst converts the Friedel–Crafts catalyst into a protonic acid.

To illustrate in more detail the mechanism of a carbonium-ion addition polymerisation, it is convenient to take as an example the cationic polymerisation of isobutene initiated by aluminium chloride. As will be described in Chapter 7, an important family of synthetic rubbers is made from isobutene by this type of reaction. (In fact, as has been pointed out in Chapter 2, this is the only important example of the use of cationic polymerisation in the manufacture of synthetic rubbers.) In the case of the polymerisation of isobutene using aluminium chloride as initiator, it is not certain whether the co-catalyst is water or hydrogen chloride. Assuming the latter to be the case, the initiation and propagation steps of the polymerisation can be formulated as follows:

$$AlCl_3 + HCl \rightleftharpoons H^{\oplus}AlCl_4^{\ominus}$$

$$CH_2{=}\underset{\underset{CH_3}{|}}{\overset{\overset{CH_3}{|}}{C}} + H^{\oplus}AlCl_4^{\ominus} \longrightarrow CH_3{-}\underset{\underset{CH_3}{|}}{\overset{\overset{CH_3}{|}}{C}}^{\oplus} \cdots AlCl_4^{\ominus}$$

$$CH_3{-}\underset{\underset{CH_3}{|}}{\overset{\overset{CH_3}{|}}{C}}^{\oplus} \cdots AlCl_4^{\ominus} + CH_2{=}\underset{\underset{CH_3}{|}}{\overset{\overset{CH_3}{|}}{C}} \longrightarrow CH_3{-}\underset{\underset{CH_3}{|}}{\overset{\overset{CH_3}{|}}{C}}{-}CH_2{-}\underset{\underset{CH_3}{|}}{\overset{\overset{CH_3}{|}}{C}}^{\oplus} \cdots AlCl_4^{\ominus}$$

and so on. The termination step may involve elimination of $HAlCl_4$ from the propagating centre as follows:

$$\text{\textasciitilde}CH_2{-}\underset{\underset{CH_3}{|}}{\overset{\overset{CH_3}{|}}{C}}^{\oplus} \cdots AlCl_4^{\ominus} \longrightarrow \text{\textasciitilde}CH_2{-}\underset{\underset{CH_3}{|}}{\overset{\overset{CH_2}{\|}}{C}} + H^{\oplus}AlCl_4^{\ominus}$$

Alternatively, termination may occur by reaction with an adventitious impurity containing active hydrogen, such as water, e.g.:

$$\text{\textasciitilde}CH_2{-}\underset{\underset{CH_3}{|}}{\overset{\overset{CH_3}{|}}{C}}^{\oplus} \cdots AlCl_4^{\ominus} + H_2O \longrightarrow \text{\textasciitilde}CH_2{-}\underset{\underset{CH_3}{|}}{\overset{\overset{CH_3}{|}}{C}}{-}OH + H^{\oplus}AlCl_4^{\ominus}$$

4.2.7 Addition Polymerisation Using Ziegler–Natta Catalysts

It has been noted in Section 4.2.1 that there is an important class of addition polymerisation reaction in which the propagating centre is not a free, independent entity, but is part of a coordination complex. The classification of addition polymerisations according to the extent of independence of the propagating centre is not a clear-cut matter, because all graduations of behaviour, from complete independence to complete binding to some other species, are possible. Indeed, as has already been noted, in conventional ionic polymerisations of the types considered in Sections 4.2.5 and 4.2.6, the counterion of the propagating centre can be more or less permanently associated with the propagating centre if the reaction is carried out in a medium of low dielectric constant. An important example is the polymerisation of butadiene initiated by lithium alkyls; the association between the propagating carbanion and the lithium cation has important implications for the structure of the polybutadiene which is produced, as will be seen in Chapter 5. There are, however, catalyst systems known for the addition polymerisation of ethylenic and diene monomers in which the propagating centre is very firmly bonded to some other species, and it is with an important group of such catalysts that this section is concerned.

In principle, the propagating centre which is firmly bonded to some other species can be an anion, a cation, or a free-radical. In practice, the most important catalysts of this type are those which promote addition polymerisation by way of a propagating anionic centre which is restrained by coordinative bonding to some other species. The industrial importance of this type of polymerisation catalyst arises from the ability of the resulting propagating centre to exercise tight control over the stereochemical configuration of each monomer unit which enters the polymer.

In terms of chemical composition, an important class of coordination catalysts for addition polymerisation reactions are the so-called *Ziegler–Natta catalysts*. These catalysts normally effect polymerisation by way of a coordinated anionic centre. Ziegler–Natta catalysts are formed from mixtures of an organo-metallic compound and a transition-metal compound. The organo-metallic compound is derived from a metal in Subgroups IA–IIIA of the Periodic Table; the most commonly-used organo-metallic compounds are those of aluminium, although compounds of zinc, cadmium and lithium have also been used. The transition-metal compound is derived from elements in Groups IV–VI of the Periodic Table. Most of these catalysts have the form of finely-divided precipitates of ill-defined composition and complex structure, in which the transition metal is present in a lower valence state than in the initial compound. In general, neither the precipitate nor the supernatant liquid are active polymerisation catalysts by themselves, but if the separated precipitate and liquid phases are brought together again, an active catalyst is recreated. It should also be noted that, although the formation of a precipitate in Ziegler–Natta catalyst systems is usual, it is not an essential feature.

The details of the mechanism by which Ziegler–Natta catalysts bring about the stereospecific polymerisation of a monomer is still a matter of controversy. Two aspects of the reaction mechanism do, however, appear to be established with reasonable certainty:

1. Interaction between the two components of the catalyst system results in the formation of an alkyl derivative of the transition metal, and this derivative is capable of coordinating to itself unsaturated hydrocarbon molecules.
2. Propagation of the polymer chain takes place by repeated insertion of a monomer molecule into a bond between (a) initially the transition metal atom and one carbon atom of the alkyl group, and (b) subsequently the transition metal atom and one carbon atom of the growing polymer chain.

The exact role of the organo-metallic compound is not clear. Some maintain that its function is limited to alkylating the transition metal. Others believe that it not only alkylates the transition metal, but that it also participates in the formation of the active centre at which the propagation reaction occurs.

The reactions between metal alkyls and transition-metal compounds are very complex and often have not been elucidated. It appears that, in many cases, the initial reaction is alkylation of the metal salt to form an unstable species, which then decomposes to liberate an alkyl radical and leave the metal in a reduced form. Taking a trialkylaluminium and titanium tetrachloride as an example, the reaction between them is sometimes represented as follows:

$$AlR_3 + TiCl_4 \longrightarrow AlR_2Cl + RTiCl_3$$
$$RTiCl_3 \longrightarrow R\cdot + TiCl_3$$

The titanium species may then be further alkylated. Polymerisation is envisaged as taking place by way of coordination of the monomer at the missing site in the octahedral configuration of the transition metal atom caused by the missing ligand. According to this view, the successive addition of monomer units is pictured as taking place by a sequence of reactions of the following type:

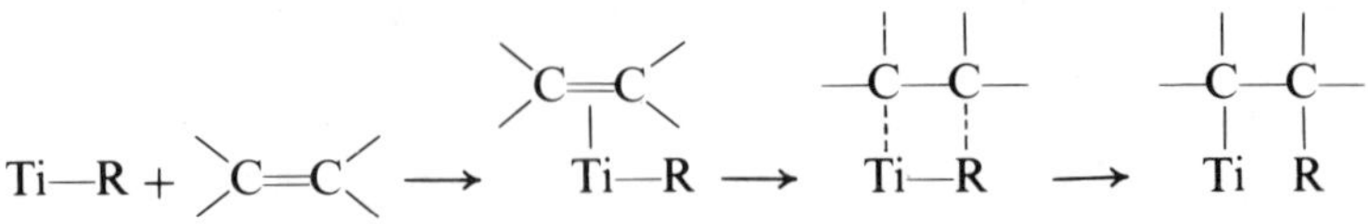

$$\text{Ti—R} + \,\,{>}\!\!\text{C}\!\!=\!\!\text{C}\!\!<\,\, \longrightarrow \quad {>}\!\!\text{C}\!\!=\!\!\text{C}\!\!< \atop \text{Ti—R} \quad \longrightarrow \quad \begin{matrix}\text{—C—C—}\\\text{Ti—R}\end{matrix} \quad \longrightarrow \quad \begin{matrix}\text{—C—C—}\\\text{Ti R}\end{matrix}$$

According to an alternative view, the monomer-insertion reaction takes place at an active centre which is formed by an organo-bridge linking the titanium and aluminium atoms:

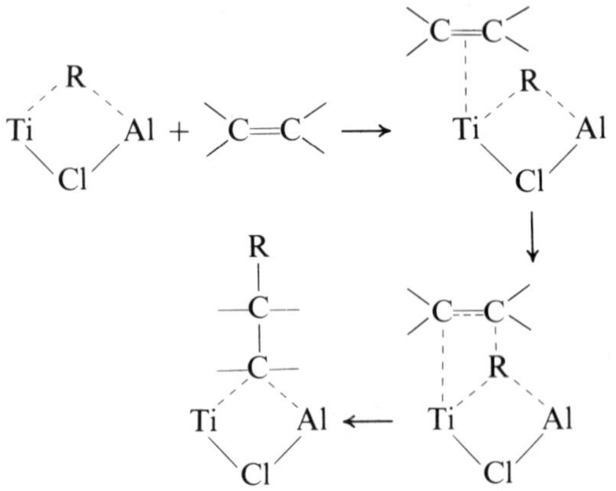

4.3 CONDENSATION POLYMERISATION

Although condensation polymerisation reactions are not nearly as important for the manufacture of synthetic rubbers as are addition polymerisation reactions, nevertheless some synthetic rubbers are made by this type of reaction. It is therefore necessary to outline the main features of this type of polymerisation. The most important types of synthetic rubber which are made by condensation polymerisation are the polysulphide rubbers, the urethane rubbers and the silicone rubbers. It may also be noted that, although all the synthetic rubbers intended as replacements for natural rubber are made by addition polymerisation, natural rubber itself is believed to be formed by a condensation polymerisation process.

It has already been pointed out in Chapter 1 that condensation polymerisations are distinguished from addition polymerisations in that a small molecule is eliminated at each step of the polymer-building reaction. A typical example of a condensation polymerisation involving a bifunctional monomer is the process whereby a polyester is formed from an $\alpha\omega$-hydroxy carboxylic acid, $HORCO_2H$, by elimination of a molecule of water at each step of the reaction:

$$HO(RCOO)_iRCO_2H + HORCO_2H \longrightarrow HO(RCOO)_{i+1}RCO_2H + H_2O$$

This reaction illustrates several of the features which are common to all condensation polymerisations involving bifunctional monomers. The more important of these features are as follows:

1. The intermediates produced in the course of the reaction are essentially no more (and no less) reactive than the monomer, because the reactive groups of the intermediates are of the same type as those of the monomer (hydroxyl and carboxyl groups in the above example). The contrast with addition polymerisation in this respect is particularly marked.

2. The polymer molecules form by a continuous 'step growth' reaction in which all the polymer molecules are growing simultaneously, and any individual polymer molecule can in principle grow for the whole of the reaction period. The latter feature is a concomitant of the fact that the polymer molecules never lose their reactive functional groups.

3. Notwithstanding the features outlined in (2), the monomer is *always* the most numerous individual molecular species present in the reaction system, however far the reaction has been taken. Similarly, the dimer is always the next most numerous individual molecular species, and so on.

4. Monofunctional impurities having the same type of functional group as the monomer have a drastic effect in reducing the maximum polymer molecular weight which can be obtained if the reaction is prolonged indefinitely. In the absence of such impurities, the maximum attainable molecular weight is infinite. For the above example, a fatty acid such as acetic acid, or a hydroxy-compound such as ethyl alcohol, would behave as a monofunctional chain terminator. An important consequence of this feature of condensation polymerisations involving bifunctional monomers is that high purity is a necessary prerequisite if it is to be possible to produce polymer of high molecular weight. Furthermore, if a condensation polymer is being made by reaction

between two bifunctional monomers (e.g., a glycol and an $\alpha\omega$-dicarboxylic acid), then a necessary prerequisite for the formation of a high-molecular-weight polymer is that the two monomers should be present in amounts which are strictly stoichiometrically equivalent.

It is a simple matter to derive expressions for the molecular-size distribution produced by the condensation polymerisation of a single bifunctional monomer in the complete absence of monofunctional impurities. The results are

$$n_i = p^{i-1}(1 - p)$$

and

$$w_i = ip^{i-1}(1 - p)^2$$

where n_i is the number fraction of polymer molecules containing exactly i repeat units (conveniently called i-mers) in the reaction mixture, w_i is the weight fraction of i-mers in the reaction mixture, and p is the *extent of polymerisation*. The latter quantity is defined as the fraction of the functional groups of either kind which have reacted. n_i and w_i are shown graphically as functions of i for various values of p in Figs. 4.2 and 4.3, respectively. (Strictly speaking, these graphs should take the form of bar diagrams, because i is a discrete variable which can take only the values 1, 2, 3, ..., etc. But for convenience of presentation and ease of making comparisons between results for different values of p, it is customary to join the tops of the bars together to form smooth curves.) The above expression for n_i shows that always $n_{i+1} < n_i$, because always $p < 1$, and this feature is clearly evident from the curves of Fig. 4.2. It implies that at all stages of the reaction $n_1 > n_2 > n_3 > ...$, thus confirming the statement which was made above concerning the relative frequencies of the various types of molecule which are present in the reaction system.

The expression for w_i as a function of i shows that, provided p exceeds a very low value, w_i passes through a maximum as i increases. The value of i at which the maximum occurs depends upon p, being approximately $-1/\log_e p$. If $p \cong 1$, this can be further approximated to $1/(1 - p)$, as is clearly evident from the curves of Fig. 4.3.

It should be apparent from the above description of the condensation polymerisation of bifunctional monomers that there is no mechanism whereby branching and crosslinking can occur in such reaction systems. For branching and crosslinking to be possible, it is necessary that the

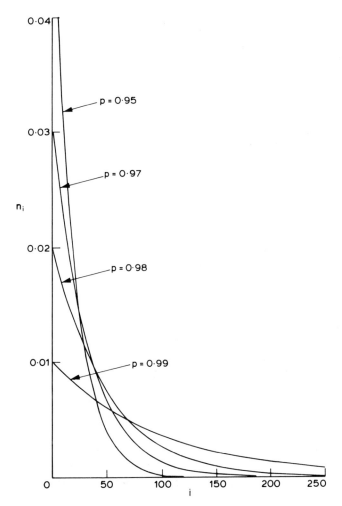

FIG. 4.2. Number fraction of i-mer, n_i, as function of i for various extents of reaction, p, for condensation polymerisation of a bifunctional monomer.

reaction system should contain molecules of functionality higher than two (it being understood that 'functionality' means here 'functionality in the context of the condensation polymerisation being considered'). Two examples of monomers of higher functionality which can be used to cause branching and crosslinking in polymerisations which proceed by the esterification of carboxylic-acid groups and hydroxy compounds are

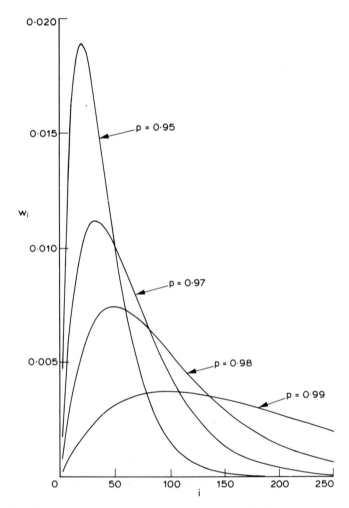

FIG. 4.3.   Weight fraction of i-mer, w_i, as function of i for various extents of reaction, p, for condensation polymerisation of a bifunctional monomer.

glycerol (XXVI) and pentaerythritol (XXVII); the functionalities of these compounds in such reactions are three and four respectively.

An important matter which arises in connection with the condensation polymerisation of monomer mixtures which contain molecules of functionality greater than two is that of *gelation*. A reaction system is said to have gelled if it contains molecules which are indefinitely large in size. The stage

$$\begin{array}{c} CH_2OH \\ | \\ CHOH \\ | \\ CH_2OH \\ (XXVI) \end{array} \qquad \begin{array}{c} CH_2OH \\ | \\ HOCH_2-C-CH_2OH \\ | \\ CH_2OH \\ (XXVII) \end{array}$$

in the reaction at which molecules of indefinite size first become present is known as the *gel point*. In order that molecules of indefinitely large size should be able to form, it is not merely sufficient that there should be opportunities for branching as polymerisation proceeds. The molecules which form by polymerisation should also be able to combine with one another, and, furthermore, the capacity to so combine should increase with the size and complexity of the polymer molecules. These conditions can clearly be fulfilled in a reaction system which comprises, say, a glycol, a dicarboxylic acid, and a trihydric alcohol. An important physical consequence of the presence of such indefinitely large molecules is that flow is no longer possible when the mixture is subjected to a shear stress. (It will be clear from this that the vulcanisation of rubbers is, in fact, a kind of gelation phenomenon.)

It is a comparatively simple matter to predict theoretically the stage of the reaction at which a mixture of two bifunctional monomers, ARA and BR'B, and a monomer of functionality f, $R''A_f$, first gels, if two assumptions are made:

1. that all the A- and B-type functional groups have equal reactivity towards each other; and
2. that no functional groups are 'wasted' by participation in the formation of cyclic structures.

The prediction is that the quantity α (called the *branching coefficient*) must have the critical value

$$\alpha^* = \frac{1}{f-1}$$

at the gel point. The quantity α is related to the extent of reaction and the stoichiometry of the reaction mixture by the following equation:

$$\alpha = \frac{rp_A^2\rho}{1 - rp_A^2(1-\rho)}$$

In this equation, p_A is the fraction of all the A-groups which have reacted, ρ is the ratio of the number of A-groups (reacted and unreacted) which are present as $R''A_f$ units to the total number of A-groups (reacted and unreacted) in the reaction mixture, and r is the ratio of the number of A-groups initially present to the number of B-groups initially present. Knowing the value of α^*, it is then a simple matter to calculate the critical value of p_A, denoted by p_A^*, corresponding to the gel point, provided that the initial stoichiometry of the reaction mixture is known. Predicted values for p_A^* generally agree fairly closely with observed values. The observed values are usually a little higher than the calculated values; this is attributed to 'wastage' of some of the functional groups by participation in the formation of cyclic structures (see assumption 2 above). It is important to note that at the gel point (and, indeed, for a certain range of extent of polymerisation beyond it) not all the polymer units are present as molecules of indefinitely large size. Only a proportion, known collectively as the *gel fraction*, are present in this form. The remainder are present as molecules of finite sizes. These latter are said to comprise the *sol fraction* of the polymer, because they are in principle soluble if a suitable solvent can be found.

GENERAL BIBLIOGRAPHY

Flory, P. J. (1953). *Principles of Polymer Chemistry*, Cornell University Press, New York. (Especially Chapters II, III, IV, V, VIII and IX.)

Smith, D. A. (Ed.) (1968). *Addition Polymers: Formation and Characterization*, Butterworths, London, Chapters 1, 2, 3 and 4.

Blackley, D. C. (1975). *Emulsion Polymerisation: Theory and Practice*, Applied Science Publishers Ltd, London.

Duck, E. W. (1966). Emulsion polymerization. In: *Encyclopedia of Polymer Science and Technology*, Vol. 5, John Wiley and Sons, New York, pp. 801f.

Ugelstad, J. and Hansen, F. K. (1976). Kinetics and mechanism of emulsion polymerization, *Rubb. Chem. Technol.*, **49**, 536.

Dunn, A. S. (1979). Emulsion polymerisation. In: *Developments in Polymerisation—2*, Haward, R. N. (Ed.), Applied Science Publishers Ltd, London.

Richards, D. H. (1977). The polymerization and copolymerization of butadiene, *Chem. Soc. Rev.*, **6**, 235.

Morton, M. and Fetters, L. J. (1975). Anionic polymerization of vinyl monomers, *Rubb. Chem. Technol.*, **48**, 359.

Hsieh, H. L. and Glaze, W. H. (1970). Kinetics of alkyllithium initiated polymerizations, *Rubb. Chem. Technol.*, **43**, 22.

Natta, G. and Giannini, U. (1966). Coordinate polymerization. In: *Encyclopedia of Polymer Science and Technology*, Vol. 4, John Wiley and Sons, New York, pp. 137f.

Chapter 5

Butadiene Polymers and Copolymers

5.1 INTRODUCTION

This chapter is concerned with rubbery polymers in which the major proportion of the repeat units is derived from butadiene, and which are intended to be vulcanised by conventional means before product application. We commence with a study of the type which over the years has comprised by far the largest proportion of the general-purpose synthetic rubbers which have been produced, namely, the emulsion-polymerised styrene–butadiene rubbers. We proceed then to consider the emulsion-polymerised acrylonitrile–butadiene rubbers, and rubbery terpolymers of styrene, vinyl pyridines and butadiene produced by emulsion polymerisation. We then consider the rubbery polybutadienes and styrene–butadiene copolymers which are produced by solution polymerisation. Consideration of solution-polymerised styrene–butadiene block copolymers intended for application as thermoplastic rubbers will be deferred until a later chapter (Chapter 10). This chapter concludes with a brief description of an unusual and interesting type of polybutadiene known as an 'Alfin' rubber.

The principal repeat units which are present in the styrene–butadiene rubbers are those derived from butadiene, whose nominal structure is shown below (XXVIII), and those derived from styrene (XXIX). Although conventionally represented as shown (XXVIII), the actual structures which

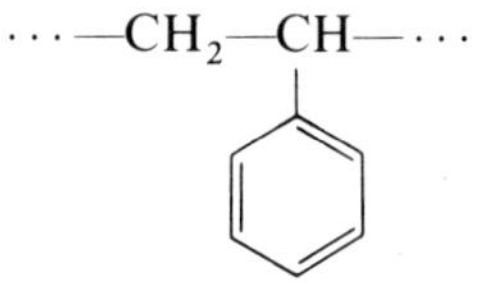

$$\cdots -CH_2-CH- \cdots$$

$$\cdots -CH_2-CH=CH-CH_2- \cdots$$

(XXVIII)

a butadiene unit can adopt are rather more diverse than is suggested by this single structure. Reference has already been made to this matter in the previous chapter, and some consideration of its implications will be given subsequently in this chapter.

5.2   EMULSION-POLYMERISED STYRENE–BUTADIENE RUBBERS (SBR†)

5.2.1 Outline of Production of Emulsion-polymerised SBR

Formerly known briefly as GR–S, and now known as SBR, emulsion-polymerised styrene–butadiene rubbers can be produced either by batch processes or by continuous processes. Most, if not all, of the large-tonnage grades of SBR are produced by continuous emulsion polymerisation, the advantages being greater efficiency and a more consistent product. However, batch processes are still appropriate for the production of low-tonnage speciality grades, and are used, for instance, for the production of carboxylated SBR grades for application as latices to carpets and to paper.

A schematic flowsheet for the production of SBR by a batch emulsion polymerisation process is shown in Fig. 5.1. The details of the so-called 'mutual' GR–S recipe for 'hot' polymerisation at $c.$ 50 °C are given in Table 5.1. In order to polymerise this recipe in a batch reactor, the aqueous phase is run into the reactor, purged with nitrogen, and then the monomers and mercaptan modifier run in. Reaction is started by heating the mixture to $c.$ 50 °C with stirring. The pressure in the reaction vessel is considerably above atmospheric, being in the range 45–60 lb in^{-2} depending upon the ratio of butadiene to styrene in the recipe. The reaction system is kept at $c.$ 50 °C for about 10 h, by which time the conversion is $c.$ 70 %. At this stage, the polymerisation is terminated (*shortstopped*) for reasons which will become apparent below. Shortstopping is effected by the addition of a small amount of a substance such as hydroquinone or a water-soluble dithiocarbamate. Unreacted butadiene is removed first by flash distillation

---

† Groups of initials such as 'SBR' are widely used, at least in the English-speaking Western world, to denote generically the members of various classes of synthetic rubbers. Often, but not always, the initials are derived from the generic chemical name for the particular class of synthetic rubbers. The purpose of introducing these abbreviations has been 'to provide a standardisation of terms for use in industry, commerce, and government...' (ANSI/ASTM D 1418-77). These abbreviations are not intended to conflict with, but rather to act as a supplement to, existing trade names and trade marks.

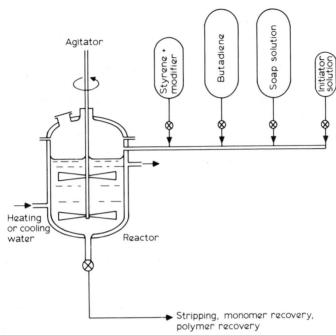

Fɪɢ. 5.1.   Schematic flowsheet for production of styrene–butadiene rubber by
batch emulsion polymerisation.

at atmospheric pressure and then by application of reduced pressure.
Unreacted styrene is removed by steam-stripping in a column. A dispersion
of a suitable antioxidant (which might be staining or non-staining) is added
in amount equivalent to *c*. 1·25 pphr in order subsequently to protect the
rubber against oxidative degradation. A common staining antioxidant

TABLE 5.1

'MUTUAL' GR–S ʀᴇᴄɪᴘᴇ ꜰᴏʀ ᴘʀᴏᴅᴜᴄᴛɪᴏɴ ᴏꜰ ꜱᴛʏʀᴇɴᴇ–
ʙᴜᴛᴀᴅɪᴇɴᴇ ʀᴜʙʙᴇʀ ʙʏ ᴇᴍᴜʟꜱɪᴏɴ ᴄᴏᴘᴏʟʏᴍᴇʀɪꜱᴀᴛɪᴏɴ ᴀᴛ
50 °C

	Parts by weight
Styrene	25
Butadiene	75
Water	180
Fatty-acid soap	5
Lauryl mercaptan	0·50
Potassium persulphate	0·30

which has been used is phenyl-β-naphthylamine. The latex can then be creamed and partially coagulated by the addition of brine, and fully coagulated by the addition of dilute sulphuric acid or alum solution. The crumb which forms on coagulation is washed, dried and baled. Modern processes for the production of carboxylated SBR latices are similar except that (a) the polymerisation is usually taken almost to completion by employing longer reaction times and slightly higher polymerisation temperatures (especially towards the end of the reaction), (b) no anti-oxidant is added, and (c) the latex is neither creamed nor coagulated but is usually subjected to a concentration process.

Figure 5.2 shows a schematic flowsheet for the production of SBR by continuous emulsion polymerisation. The various recipe ingredients are pumped continuously from storage tanks at appropriate rates into and through a series of agitated reactors which are maintained at a suitable temperature. The overall rate of throughput of material is such that the mean residence time in the reactor train allows the desired degree of conversion to be achieved by the time the reaction mixture effluxes from the final reactor. By means of a by-passing arrangement, it is possible to make relatively fine adjustments to the mean residence time in order to make allowances for small fluctuations in rate of polymerisation. Polymerisation is terminated by the addition of shortstopper to the latex which effluxes from the final reactor. The latex is then heated by the addition of steam, the unreacted butadiene stripped off, and the unreacted styrene removed by steam-stripping. The latex is then creamed, coagulated and filtered, and the resultant crumb dried and baled. If oils or carbon blacks are to be blended with the rubber at the production stage, they are added to the latex as aqueous emulsions or dispersions prior to the creaming and coagulation step.

From the foregoing outline of the production methods for emulsion-polymerised styrene–butadiene rubber, it will be apparent that the producer has at his disposal a large number of production variables, each of which will, in principle at least, have an effect upon the properties of the rubber produced. Some consideration will subsequently be given to the more important of these variables and to their effects upon the product. However, before turning to these, a fundamental question will be discussed which arises in connection with the production of many types of synthetic rubber and which can be conveniently dealt with here because it is very clearly exemplified by the production of styrene–butadiene rubber. This is the matter of why synthetic rubbers are frequently produced from two or more monomers by copolymerisation rather than by blending the corre-

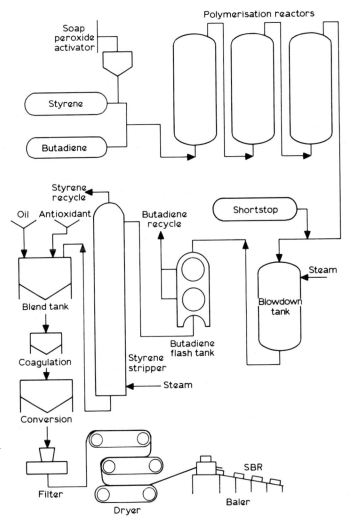

FIG. 5.2. Schematic flowsheet for production of styrene–butadiene rubber by continuous emulsion polymerisation.[1]

sponding homopolymers. In the case of styrene–butadiene rubber, for instance, it would be possible to polymerise the styrene and the butadiene separately to polystyrene and polybutadiene respectively, and then to blend the homopolymers in appropriate ratios either at the plant where the polymers were produced or in the factory where the rubber is used. One obvious advantage which would accrue from this alternative procedure would be increased versatility in respect of the styrene–butadiene ratio in the rubber; in particular, it would be possible for the rubber compounder to vary the styrene–butadiene ratio at will by varying the ratio of the two homopolymers in his mix.

5.2.2 Why Copolymerise the Styrene and Butadiene Rather than Blend the Homopolymers?

The answer to this question is to be found in the inherent incompatibility of polystyrene and polybutadiene at the molecular level. It is now widely recognised that complete compatibility between homopolymers of different types is seldom achieved. Although apparently homogeneous mixtures can often be prepared, for example, by casting from mixed solutions in a mutual solvent, close examination of the product shows a tendency for the repeat units of the one polymer to separate from the repeat units of the other. The consequence is that, at the molecular level, the mixture tends to comprise regions (or *domains*) which are relatively rich in the units of the one polymer, intimately mixed with regions which are relatively rich in the units of the other polymer. The properties of such homopolymer mixtures are often rather different from those of the more random mixture of the same overall numbers of monomer units which may result from copolymeris-ation. Whilst good use is now being made of the distinctive properties offered by polymer systems comprising domains of one polymer type embedded in domains of another polymer type, it is the case that, for rubbers intended for processing by conventional methods, copolymers offer a much more useful balance of properties than do homopolymer blends.

Results published by Shundo et al.[2] show the effects of copolymerisation and blending for styrene–butadiene polymers. Figures 5.3 and 5.4 show their results for the variation with styrene content of hardness and elongation at break respectively for styrene–butadiene copolymers, poly-styrene–polybutadiene blends made by latex blending, and polystyrene–polybutadiene blends made by roll-mixing. The mechanical properties of the blends are generally consistent with the view that the blends are inhomogeneous two-phase systems. Thus, for instance, the tensile strengths of physical blends of high overall styrene content can actually be rather higher than those of copolymers of similar overall composition, because of the reinforcing effect of the separate polystyrene domains.

A further very important advantage which, in the case of styrene and butadiene, copolymerisation has over physical blending is that it allows the whole polymer subsequently to be bound into a molecular network during

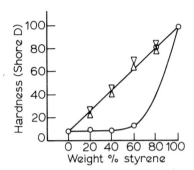

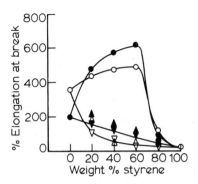

FIG. 5.3. Hardness as function of styrene content for sulphur vulcanisates from styrene–butadiene copolymers and polystyrene–polybutadiene blends.[2] Compound: polymer 100, stearic acid 1·5, zinc oxide 5, sulphur 1·5, N-cyclohexyl-2-benzthiazyl sulphenamide 1·5. Roll blending carried out at 110°C. Vulcanisation at 150°C. ○ copolymers, △ latex blends, ▽ roll blends.

FIG. 5.4. Elongation at break as function of styrene content for sulphur and peroxide vulcanisates from styrene–butadiene copolymers and polystyrene–polybutadiene blends.[2] Compound for sulphur vulcanisates as given in caption to Fig. 5.3. Compound for peroxide vulcanisates: polymer 100, stearic acid 0·5, zinc oxide 5, sulphur 0·4, dicumyl peroxide 4·0. Roll blending carried out at 110°C. Vulcanisations at 150°C.

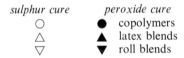

sulphur cure	peroxide cure	
○	●	copolymers
△	▲	latex blends
▽	▼	roll blends

the vulcanisation reaction. This feature is of great significance in relation to certain mechanical properties of the vulcanisate such as stress relaxation, creep, and compression set. Results reported by Shundo et al.[2] for the extraction of vulcanised copolymers and polymer blends with hot benzene illustrate the differences between vulcanised copolymers and vulcanised homopolymer blends in this respect. These results are summarised in Fig. 5.5. From the copolymers, an almost constant (and small) quantity of extract (sol fraction) was obtained, regardless of the styrene content of the copolymer. By contrast, the amount of extract obtained from the blends increased almost linearly with the styrene content; in fact, for the polymer blends, the proportion of polymer which can be extracted by hot benzene is virtually the same as the polystyrene content of the blend. The molecular interpretation of these observations is, of course, that crosslinking during vulcanisation occurs exclusively through the polymer units derived from

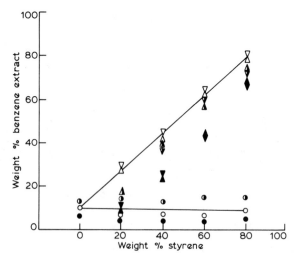

FIG. 5.5. Effect of extraction in benzene upon sulphur, peroxide and radiation vulcanisates from styrene–butadiene copolymers and polystyrene–polybutadiene blends.[2] Compounds for sulphur and peroxide vulcanisates as given in captions to Figs. 5.3 and 5.4. Compound for radiation vulcanisation: polymer 100, calcium carbonate 100. Roll blending carried out at 110 °C. Sulphur and peroxide vulcanisations carried out at 150 °C. Radiation vulcanisation: 10^8 rads from ^{60}Co source in air at room temperature.

○   Copolymers, sulphur-cured.
●   Copolymers, radiation-cured.
◐   Copolymers, peroxide-cured.
△   Latex blends, sulphur-cured.
▲   Latex blends, radiation-cured.
▲   Latex blends, peroxide-cured.
▽   Roll blends, sulphur-cured.
▼   Roll blends, radiation-cured.
▼   Roll blends, peroxide-cured.

butadiene (because they alone contain residual unsaturation after polymerisation). Only if the styrene units are chemically combined with butadiene units is it possible for the styrene units to be bound into the molecular network which forms during vulcanisation. It is certainly worth noting that, whereas an 80/20 polystyrene–polybutadiene blend loses approximately 80 % of its polymer when extracted by hot benzene, the amount lost by a copolymer of this composition is only approximately 10 %. It appears that the crosslinking which takes place through the 20 % of butadiene units is sufficient to bind all the polymer units into a molecular network.

5.2.3 Effects of Varying the Styrene/Butadiene Ratio

The ratio of styrene to butadiene is clearly one of the most obvious variables which the formulator of a recipe for the emulsion copolymerisation of styrene and butadiene has at his disposal, especially since the two monomers can be copolymerised in all proportions. Some information concerning the effects of varying this ratio has already been given in the preceding section incidentally to consideration of the question of why the two monomers are copolymerised. More systematically, the effects of varying the monomer ratio can be considered under three headings:

1. effects upon polymerisation behaviour;
2. effects upon the properties of the raw polymer; and
3. effects upon the properties of the compounded and vulcanised rubber.

1. *Effects upon polymerisation behaviour:* Storey and Williams[3] have summarised the effects of increasing the styrene/butadiene ratio upon polymerisation behaviour as (a) an increase in the overall rate of polymerisation, (b) a decrease in the rate of consumption of mercaptan modifier, (c) a decrease in the amount of monomer which is required to attain a polymer of given plasticity, (d) a tendency towards more uniform molecular weight distribution, and (e) a decreased tendency to form polymer gel. Of these, (c), (d) and (e) are directly attributable to the reduced unsaturation of the polymer which is produced. Opportunities for branching and cross-linking are thereby diminished, because, as has been seen in Chapter 4, these processes occur either by reactions involving carbon–carbon double bonds in the polymer, or by reaction at sites adjacent to carbon–carbon double bonds. The effect of monomer composition upon the rate at which styrene and butadiene copolymerise in an emulsion system is illustrated in Fig. 5.6. Figure 5.6(a) shows a series of conversion–time curves for reaction systems which are similar except that the ratio of the two monomers has been varied. Figure 5.6(b) shows the steady rate of polymerisation as a function of monomer composition.

2. *Effects upon the properties of the raw polymer:* The homopolymer of styrene is a rigid, glassy material having a glass-transition temperature in the region of $+90\,^{\circ}\mathrm{C}$. The polybutadiene which is produced by free-radical polymerisation is a rubbery material with a glass-transition temperature of $c.\,-90\,^{\circ}\mathrm{C}$. Random copolymers

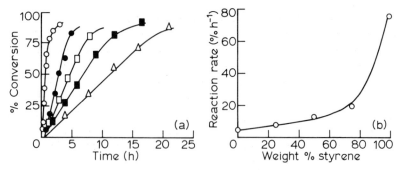

FIG. 5.6.   (a) Effect of monomer feed composition upon conversion–time curve for emulsion copolymerisation of styrene and butadiene.[4] ○ styrene, ● butadiene/ styrene 25/75, □ butadiene/styrene 50/50, ■ butadiene/styrene 75/25, △ bu- tadiene. (b) Effect of monomer feed composition upon steady rate of polymerisa- tion for emulsion copolymerisation of styrene and butadiene.[4] Reaction system: monomers 100, water 180, soap 6, potassium persulphate 0·3, t-hexadecyl mercaptan variable. Polymerisation temperature 50 °C.

of styrene and butadiene have glass-transition temperatures inter- mediate between these extreme values, as is illustrated by the results summarised in Fig. 5.7. It is seen that the variation of glass- transition temperature with styrene content is regular, in the sense that the glass-transition temperature always increases with increas- ing styrene content. Thus the effect of increasing the styrene content is to produce a copolymer which is less rubbery at room temperature and which is rather stiffer, and which becomes rigid at a higher temperature as the temperature is lowered. Figure 5.7 also shows that the nature of the variation of glass-transition tempera- ture with styrene content is such that increasing styrene content has less effect when the styrene content is low than when it is high. This has the important technological consequence that usefully rubbery copolymers can be produced over a rather wide range of styrene contents.

It is important to note that the glass-transition temperatures of random styrene–butadiene copolymers are in general very different from those of physical blends having the same overall composition. This fact provides important evidence for the belief that the molecular structure of the blends differs significantly from that of the random copolymers (see Section 5.2.2). In the case of the random copolymers, a single glass-transition temperature is ob- served which is intermediate in value between that for polybutadiene

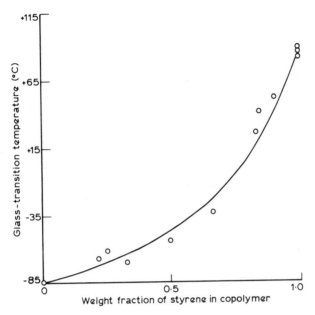

FIG. 5.7.   Glass-transition temperatures of styrene–butadiene copolymers as function of styrene content.[5]

and that for polystyrene. By contrast, in the case of the blends, *two* transitions are generally observed, one at approximately the temperature for the transition in polybutadiene and the other at approximately the temperature for the transition in polystyrene. This latter observation clearly accords with the view that the blend consists of domains which are very rich in styrene units admixed with domains which are very rich in butadiene units.

One practical consequence of the increase in glass-transition temperature of the copolymers with increasing styrene content is that the ease with which the particles of a styrene–butadiene latex coalesce at room temperature decreases as the styrene content increases. Thus as long as the styrene/butadiene ratio is below about 70/30 by weight, the latices will dry down at room temperature to give coherent deposits in which the particles have at least partially fused together. However, if the ratio exceeds 70/30, the latices dry down to give powdery deposits in which the particles remain essentially discrete from one another.

3. *Effects upon the properties of the compounded and vulcanised rubber:* Data for the effect of styrene/butadiene ratio upon the properties of vulcanisates obtained from styrene–butadiene rubbers present a somewhat confusing picture. This is no doubt because the comonomer ratio is merely one of the variables which affects the vulcanisate properties. Other important variables include the type of vulcanising system which has been used, and the level and type of reinforcing filler which is present in the vulcanisate. Another important matter is whether the various copolymers have been compared at the same initial plasticity, that is, whether or not when preparing the copolymers the modifier level was adjusted in such a way that the copolymers all had similar plasticities before they were compounded and vulcanised. Storey and Williams[3] have summarised the results of their investigation (for which a sulphur-cured vulcanisate containing a reinforcing carbon black was used) as follows: 'As the styrene content is increased the tensile strength passes through a maximum at about 50/50 ratio and diminishes thereafter slightly. The modulus increases with increasing styrene content.' These investigators also found that the rebound resilience decreased with increasing styrene content. It is also appropriate to refer under this heading to the effect of styrene/butadiene ratio upon processing behaviour. In general, as the styrene content is increased, so the elastic character of the copolymer diminishes and its plastic character increases. This tendency is evidenced by, for instance, increased extrusion rate, reduced die swell, and reduced mill shrinkage, as the styrene content is increased.

Having considered in some detail the effects which accompany variation of the styrene/butadiene ratio in random styrene–butadiene copolymers, it remains to note the ratios which are actually used in practice. The styrene/butadiene weight ratio for most copolymers intended to be used as general-purpose synthetic rubbers is in the region of 25/75. This corresponds to a molecular ratio of approximately six butadiene units to each styrene unit. A similar ratio is used in non-carboxylated latices which are intended as competitors for natural rubber latex in products such as latex foam rubber. However, in carboxylated styrene–butadiene rubber latices which are intended for applications such as carpet backing and paper coating, a higher styrene/butadiene weight ratio is commonly used, e.g., 50/50 or 60/40. From what has been said above, it will be observed that copolymers of this composition are still rubbery at room temperature and that the latices of such copolymers still integrate readily to coherent films when they are dried down. There are, however, styrene–butadiene copolymers of even higher styrene content which are produced industrially by emulsion copolymerisation. These are resinous, non-rubbery copolymers which are intended primarily for use as organic stiffening or reinforcing fillers for rubbers. A range of styrene/butadiene ratios can be used for this application, typical weight ratios being 85/15 and 90/10. Latices of these so-called *high-styrene resins* are particularly well adapted to use as reinforcing agents for rubbery polymers in latex form, because mixing can be achieved very simply by latex blending. It may be noted that, although numerous attempts have been made to use styrene homopolymers as a stiffening filler for rubbers, these attempts have been unsuccessful, principally because of the limited compatibility of polystyrene and most rubbers on the molecular scale.

5.2.4 Effects of Varying the Overall Conversion of Monomers to Copolymer

Economic considerations dictate that the conversion of monomers to copolymer should be as complete as possible, consistent with a satisfactory product being obtained from the reaction. The separation, recovery and recycling of unreacted monomers is an expensive, troublesome and time-consuming process. Indeed, if possible (as in fact is commonly the case in the production of carboxylated styrene–butadiene copolymer latices for application as latices) the reaction is taken to virtually 100 % conversion, there being left at the end of the reaction only minor amounts of unreacted monomers to be removed by stripping.

However, in the case of styrene–butadiene copolymers intended for application as general-purpose rubbers, there are two factors which make it impracticable to take the conversion beyond *c.* 70 %. These factors are:

1. the effect of conversion upon the composition of the copolymer which is produced; and
2. the effect of conversion upon the formation of polymer gel.

1. *Effect of conversion upon copolymer composition:* The relative reactivities of styrene and butadiene in free-radical copolymeris-ation are such that the butadiene tends to copolymerise preferen-tially to styrene. The copolymer which is formed initially is therefore richer in butadiene than was the monomer feed mixture from which it was formed. The feed mixture therefore tends to become progressively more rich in styrene, and the copolymer

molecules which form towards the end of the reaction are relatively rich in styrene. These effects are illustrated in Fig. 5.8, which shows both the overall composition of the copolymer formed and the instantaneous (or incremental) composition of the copolymer formed in a styrene–butadiene emulsion copolymerisation as a function of conversion. Curves are given for three different initial weight ratios of styrene to butadiene, namely, 10/90, 30/70 and 50/50. These curves show clearly that the composition of the copolymers formed at the various stages of the reaction is not uniform, and that the degree of non-uniformity increases with the proportion of styrene in the initial feed. The copolymers which form at low conversions are reasonably uniform as regards comonomer content, but above about 60 % conversion the heterogeneity increases rapidly with conversion. The technological significance of these observations will become clear when it is recalled (see Section 5.2.2) that butadiene and styrene units are not

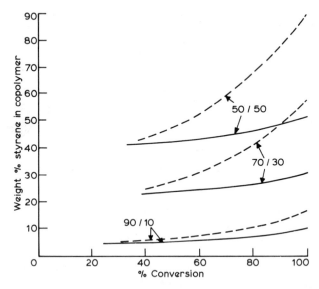

FIG. 5.8.   Effect of conversion upon overall copolymer composition and instantaneous incremental copolymer composition for emulsion copolymerisation of styrene and butadiene.[6] Reaction system: monomers 100, water 180, soap 5, potassium persulphate 0·3, mercaptan variable (to give Mooney viscosity 50 ± 3). Polymerisation temperature 50 °C. Numbers appended to curves indicate weight ratio styrene/butadiene. ——, overall composition; – – –, incremental composition.

inherently miscible at the molecular level. The styrene-rich co-polymer molecules which form in the later stages of a styrene–butadiene polymerisation tend not to be completely compatible with the butadiene-rich molecules which formed during the earlier stages. This incompatibility tends to have a detrimental effect upon the properties of the vulcanisates which are obtained from such copolymers.

2. *Effect of conversion upon gel content:* The main reason why the copolymerisation is stopped short of full conversion has, however, little to do with the effect of extent of conversion upon the heterogeneity of the copolymer produced. It has instead to do with the fact that highly branched and crosslinked polymer (polymer gel) forms during the later stages of the reaction, and this gel has an adverse effect upon polymer processing and upon the properties of the vulcanisate which is subsequently produced. In particular, raw rubbers containing a high proportion of tight gel are very difficult to mill satisfactorily; they tend to be rather tough and 'nervy', that is to say their behaviour on the mill is elastic rather than plastic. The formation of polymer gel during a styrene–butadiene emulsion copolymerisation is illustrated in Fig. 5.9(a). The inherent viscosity of the soluble portion of the polymer as a function of conversion is shown in Fig. 5.9(b); it will be seen that the viscosity (and therefore molecular weight) of the soluble fraction passes through a maxi-mum at the conversion at which gel first begins to form, and thereafter falls sharply with further conversion. The main reason why gel forms increasingly as the conversion increases is that gel formation arises from reactions involving carbon–carbon double bonds in polymer molecules which have already been formed (see Section 4.2.2 of Chapter 4). The rate of formation of gel will therefore increase as the number of polymer molecules in the reaction system increases, that is, as the conversion of monomer to polymer increases. A secondary reason why gel tends to form during the later stages of the reaction is that the modifier (which, it will be recalled, is added specifically to prevent the formation of polymer molecules of too high a molecular weight) tends to be consumed more rapidly than are the monomers, so that the modifier/monomer ratio falls as the reaction progresses. Some mitigation of this latter effect can be achieved by withholding some of the modifier from the initial reaction mixture and adding it incrementally during the later stages of the reaction; however, the

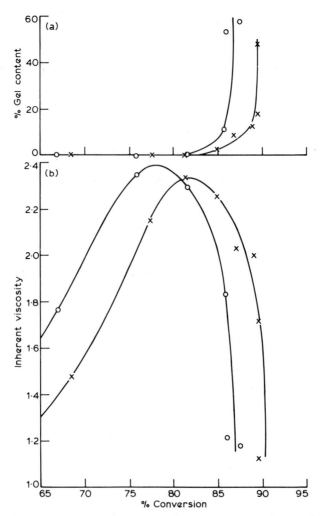

FIG. 5.9. (a) Formation of polymer gel during emulsion copolymerisation of styrene and butadiene.[7] (b) Inherent viscosity of sol fraction as function of conversion for emulsion copolymerisation of styrene and butadiene.[7] ○, 0·30 part mercaptan; ×, 0·35 part mercaptan.

principal way of overcoming the problem of untoward gel formation in practice has been to shortstop the reaction well before complete conversion of monomers to copolymers, even although this incurs the consequent problems of the separation, recovery and recycling of unreacted monomers.

5.2.5 Effects of Varying the Modifier Content

The effects of increasing the amount of mercaptan modifier in the polymerisation recipe are summarised in Fig. 5.10. The effects upon polymer plasticity and gel content are largely as expected; both fall sharply as the modifier level is increased; indeed, if the modifier content of the polymerisation recipe is sufficiently great, the product is a very viscous liquid rather than an elastomer.

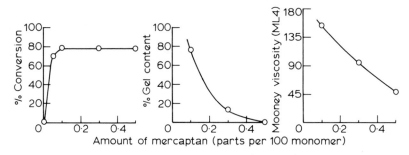

FIG. 5.10.   Effect of increasing level of mercaptan in emulsion copolymerisation of styrene and butadiene.[8] Reaction system: styrene 25, butadiene 75, water 180, soap 5, potassium persulphate 0·3, dodecyl mercaptan variable. Polymerisation conditions: 12 h at 50 °C.

5.2.6 Effect of Polymerisation Temperature

From the earliest days of the synthetic rubber industry, interest has been shown in the effect of polymerisation temperature upon the properties of synthetic rubbers obtained from diene monomers. As far back as World War I, German scientists and technologists were aware that the rubber obtained by the bulk polymerisation of dimethylbutadiene had better physical properties when it was manufactured at room temperature than when it was manufactured at elevated temperature. The problem was that the reaction rate was so slow at room temperature that it took several months to produce the rubber. The commercial production of styrene–butadiene rubber by low-temperature polymerisation was not possible until shortly after the end of World War II. By this time, very active 'redox' initiation systems† had been developed, enabling emulsion polymerisation to be carried out rapidly at low temperatures.

Significant changes in polymer structure result from reducing the temperature at which the emulsion copolymerisation of styrene and

† These initiation systems depend for their effectiveness upon interaction between a reducing agent and an oxidising agent; hence their name.

butadiene is effected. These changes have important consequences for the properties of the vulcanisates which are obtained from the raw rubber. This is illustrated in Fig. 5.11, which shows the effect of polymerisation temperature upon the tensile strength of styrene–butadiene rubber vulcanisates containing 50 pphr of carbon black. Other mechanical vulcanisate properties which are improved significantly by reducing the temperature at which the raw rubber is produced include resistance to flex cracking and resistance to abrasion. The latter characteristics make low-temperature-polymerised rubbers of great interest for the manufacture of tyres. Furthermore, as will be described subsequently, low-temperature-polymerised styrene–butadiene rubber is able to accept substantial quantities of oil without serious deterioration of vulcanisate properties. There are, however, certain advantages associated with styrene–butadiene rubber which has been made by polymerisation at *c.* 50 °C. Compared to low-temperature-polymerised rubbers, the filler acceptance is superior, processing is easier, and smoother calenderings and extrusions can be obtained.

Reference to Fig. 5.11 shows that the tensile strength of a styrene–butadiene rubber vulcanisate increases approximately linearly as the polymerisation temperature is reduced, at least for the temperature range covered by the results summarised in Fig. 5.11. It therefore appears that the lower the polymerisation temperature, the more satisfactory are the

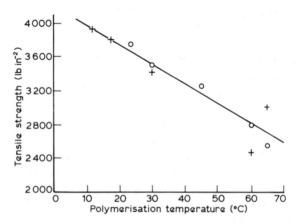

FIG. 5.11.   Effect of polymerisation temperature upon tensile strength of styrene–butadiene rubber vulcanisates containing 50 pphr of carbon black.[9] Different styles of point refer to data for vulcanisates prepared using rubbers from different types of polymerisation recipe.

mechanical properties of the vulcanisate likely to be. However, a lower limit to the polymerisation temperature is effectively set by the freezing point of water and the economics of the process. Although some interest has been shown in the emulsion copolymerisation of styrene and butadiene at sub-zero temperatures (in which case it is necessary to include anti-freeze additives in the polymerisation recipe), most low-temperature copolymerisations of styrene and butadiene are effected at *c*. 5 °C.

The well-known 'custom' recipe for the production of styrene–butadiene rubber by 'cold' polymerisation is given in Table 5.2. Important features of this recipe are as follows:

1. The initiation system is a 'redox' couple comprising an organic hydroperoxide (the oxidising component) and an iron(II) salt (the reducing component).
2. The iron(II) ions are complexed with pyrophosphate anions before being added to the reaction system, in order to ensure that they are present in the aqueous phase in an appropriately low concentration.
3. A reducing sugar (dextrose) is present to reduce the iron(III) ions, produced during initiation, back to iron(II) ions. Direct interaction between the reducing sugar and the hydroperoxide is not thought to contribute significantly to initiation.
4. An inorganic electrolyte is present to ensure that the reaction system remains fluid throughout the polymerisation.

TABLE 5.2

'CUSTOM' RECIPE FOR PRODUCTION OF STYRENE–BUTADIENE RUBBER BY EMULSION COPOLYMERISATION AT 5 °C

	Parts by weight
Styrene	28
Butadiene	72
Water	180
Potassium soap of disproportionated rosin acid	4·7
Mixed tertiary mercaptans	0·24
Cumene hydroperoxide	0·10
Dextrose	1·00
Ferrous sulphate heptahydrate	0·14
Potassium pyrophosphate	0·177
Potassium chloride	0·50
Potassium hydroxide	0·10

5.   Although the soap is still of the carboxylate type, it is derived from
a rosin acid and not a fatty acid. The principal reason is that rosin-
acid soaps are much more soluble in water at low temperatures than
are fatty-acid soaps.

The rate of polymerisation for recipes of this type at 5 °C is such that 60 %
conversion of monomer to copolymer is achieved after 12–15 h reaction.

As in the case of 'hot' styrene–butadiene emulsion copolymerisations,
the overall composition of the copolymer produced in 'cold' polymerisations depends upon the extent of conversion of monomers to copolymer.
The nature of the variation is, however, somewhat different, as is illustrated
in Fig. 5.12, for copolymerisations at 50 °C and −20 °C. Whereas in 'hot'
polymerisations the styrene content of the copolymer increases uniformly
with conversion, in 'cold' polymerisations the styrene content passes
through a shallow minimum in the region of 60 % conversion. Although

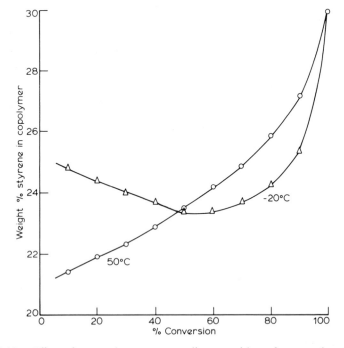

FIG. 5.12.   Effect of conversion upon overall composition of styrene–butadiene
copolymers produced by emulsion copolymerisation at 50 °C, ⊙, and at −20 °C.
△.[10] Initial charge ratio: styrene/butadiene 30/70.

well-established experimentally, the origin of this minimum is obscure. It has been suggested that it arises from differences in the rates of dissolution of the two monomers in the growing polymer particle.

Molecular explanations for the effect of polymerisation temperature upon the properties of styrene–butadiene rubbers, and of vulcanisates obtained from them, are sought in terms of the effects of polymerisation temperature upon branching, crosslinking, the stereochemical configuration of the enchained butadiene units, and the molecular-weight distribution of the polymer. In general, the lower the polymerisation temperature, the less branched and cross-linked is the rubber and the lower is its gel content. The effect of polymerisation temperature upon the microstructure of emulsion-polymerised polybutadiene is summarised in Fig. 5.13. The proportion of vinyl groups in the polymer (from 1,2 addition) varies little over the range of polymerisation temperatures from $-20\,°C$ to $100\,°C$, being about 20%. By contrast, the balance between cis-1,4 and trans-1,4

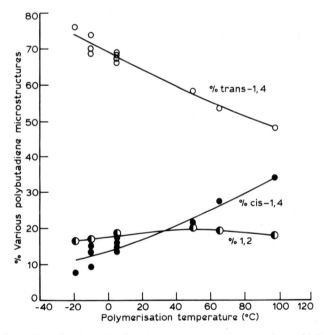

FIG. 5.13.   Effect of polymerisation temperature upon proportions of 1,2, ◑, cis-1,4, ●, and trans-1,4, ○, microstructures in polybutadienes prepared by emulsion polymerisation at various temperatures.[11]

structures is markedly affected by polymerisation temperature, the ratio of *cis* and *trans* falling as the polymerisation temperature is decreased. Although the results shown in Fig. 5.13 refer to the homopolymerisation of butadiene, the presence of styrene has little effect upon the microstructure of the enchained butadiene units, at least in the amounts conventionally used in synthetic rubber manufacture. The improved mechanical properties of vulcanisates from 'cold' styrene–butadiene rubbers have been attributed, in part at least, to the greater regularity of structure occasioned by the increase in the content of *trans*-1,4 units, as well as to the formation of less branched and crosslinked polymer molecules. However, whilst improved structural regularity is no doubt important, the major effect is thought to arise from reduction in the low-molecular-weight fraction of the rubber when polymerised at low temperature, and from the fact that it is possible by low-temperature polymerisation to obtain a processable rubber at higher average molecular weight.

5.2.7 Oil-extended Styrene–Butadiene Rubber

The possibility of extending and cheapening rubbers by the addition of plasticising oils has been recognised since the earliest days of rubber technology. It first became feasible commercially in the early 1950s, following the successful production of processable high-molecular-weight styrene–butadiene rubbers by low-temperature polymerisation. Oil-extended styrene–butadiene rubbers are produced by blending a latex of a rubber of high average molecular weight (Mooney viscosity 110–130) with an emulsion of a suitable petroleum oil, and then co-coagulating the mixture. The oil content is in the range 25–50 parts by weight per 100 parts by weight of polymer. In effect, the added oil provides the low-molecular-weight 'tail' of the molecular-weight distribution of the composite material more cheaply than is possible by polymerisation. The oils used are of the naphthenic, aromatic or highly-aromatic types, but not of the paraffinic type because of problems of limited compatibility with the rubber. The miscibility of oils with styrene–butadiene rubbers diminishes in the order aromatic > naphthenic > paraffinic. The paraffinic oils have such limited miscibility that they tend to 'sweat out' of the rubber if an attempt is made to blend them to any appreciable concentration. Oil-extended rubbers intended for the manufacture of lightly-coloured products are extended with lightly-coloured oils, usually of the naphthenic type. Ester-type plasticisers are not used for this application to any significant extent, principally because of their cost.

There are two common methods of blending the oil emulsion with the latex:

1. a batch process, in which the two dispersions are blended in the appropriate quantities in a tank prior to coagulation; and
2. a continuous process, in which the oil emulsion and latex are pumped through flow meters at appropriate rates and allowed to mix before coagulation.

Coagulation is effected in a similar way to that for non-oil-extended styrene–butadiene rubbers, e.g. by the use of a brine/acid coagulant. During the subsequent filtration, drying and baling of the crumb, allowance has to be made for the possibility that oil-extended crumb may be more subject to plastic flow at elevated temperatures than, say, non-oil-extended 'cold' styrene–butadiene rubber.

The compounding principles for oil-extended styrene–butadiene rubbers are similar to those for non-oil-extended rubbers, except that the compounder has the choice of whether to base the weights of the various ingredients upon 100 parts of rubber or upon 100 parts of rubber plus oil. It is recommended by some that the additions of all ingredients excluding the curatives should be made on the basis of the rubber plus oil, and that the levels of the curatives only should be based upon the rubber alone. Others have maintained that, because of the diluting effect of the extending oil, the curatives also should be added on the basis of the rubber plus oil.

5.2.8 Carbon Black Masterbatches

Interest in the possibility of preparing rubber–carbon black masterbatches by the blending and subsequent co-coagulation of latex and a carbon black dispersion dates from the early 1920s. Masterbatching of carbon black with natural rubber in latex form has never achieved successful commercial development for the following reasons:

1. Natural rubber and carbon black are produced in widely-separated geographical regions, so that masterbatch production would necessitate transportation of either carbon black or water (in the latex) over large distances.
2. Natural rubber latex is rather variable both as regards rubber content and the plasticity of the contained rubber.
3. Adsorption of certain of the non-rubber constituents of the latex by the carbon black can cause accelerated degradation of the masterbatch in storage, because some of these non-rubber constituents otherwise act as rubber antioxidants.

None of these disadvantages applies to carbon black masterbatches based upon styrene–butadiene rubber. The rubber and the carbon black are produced from the same raw material, often in close proximity to one another. The rubber content and plasticity of the contained rubber can be controlled quite closely. The amounts of non-rubber constituents in the latex (other than soap) are low. For these reasons, carbon black master-batching is largely confined to synthetic rubbers, in particular, to those of the styrene–butadiene copolymer type produced by emulsion polym-erisation.

The production of carbon black masterbatches by latex blending comprises four essential steps:

1. production of an aqueous dispersion of the carbon black;
2. mixing of the latex and the carbon black dispersion;
3. co-coagulation of the blend to give a crumb; and
4. filtration, drying and baling of the crumb.

The carbon black is first subjected to dry micropulverisation in order to break down large aggregates. It may then be dispersed by mechanical agitation in water which contains a dispersing agent such as an alkylnaph-thalene sulphonate or a lignin sulphonate. Some alkali is also added in order to raise the pH. The carbon black content of the dispersion may be as high as 20 % by weight. It has been shown that in well-stabilised carbon black dispersions, the size of the aggregates has been reduced to a similar size to that of the rubber particles in a styrene–butadiene rubber latex. Blending of the latex and the carbon black dispersion presents little difficulty if the dispersing agent selected for the latter is compatible with the stabiliser system of the former.

The coagulation step is critical if a homogeneous masterbatch is to be obtained. For instance, if the blend is coagulated by the slow addition of acid, then the material initially precipitated can contain a very high concentration of carbon black, and may give rise to beads of carbon black which are very difficult to disperse subsequently. One suitable technique is that of 'shock' coagulation in which an excess of salt and sulphuric acid is added rapidly to the latex–carbon black mixture under high-speed stirring.

In the process described above, a surface-active substance is used to aid the dispersion of the carbon black in water. The use of a surface-active substance for this purpose gives rise to the problem that it is adsorbed on the surface of the carbon black (which must happen if it is to function as a dispersing agent), and remains so adsorbed when the carbon black becomes incorporated in the rubber. The effect is to deactivate part of the surface of the carbon black and so interfere with the ability of the carbon black to function as a reinforcing agent in the vulcanisate. Because of this, 'dispersant-free' processes have been developed in which, for example, the carbon black is dispersed by a high-energy agitation process or by blending the carbon black with wet-coagulated rubber crumb in an internal mixer.

It is also possible to produce carbon black masterbatches which contain added oil as well. In this case, the latex is blended with a suitable oil emulsion as well as with a carbon black dispersion prior to coagulation.

The principal advantages of using a carbon black–rubber masterbatch as compared with using the carbon black and rubber separately are as follows:

1. Economies of storage, handling and weighing can be effected.
2. The power required for mixing can be reduced.
3. The mixing process is cleaner.
4. The dispersion of the carbon black in the rubber can be improved, and this can lead to easier processing and better product performance, e.g. improved abrasion resistance in a tyre tread.

The principal disadvantages are as follows:

1. The masterbatch may have unexpected effects upon the rate of vulcanisation of the mix; these effects are associated principally with the additives which were used to aid the dispersion of the carbon black in water.
2. The compounder is restricted to the use of those grades of rubber and carbon black which the producer has seen fit to incorporate in his masterbatches.

It is probably the lack of versatility implied by the second of these disadvantages which has retarded the application of carbon black masterbatches.

It should be noted that the reinforcing properties of the carbon black in a carbon–rubber masterbatch produced by latex blending do not become apparent until after the rubber has been subjected to a mastication/mixing process.

5.2.9 Other Modifications of Emulsion-polymerised Styrene–Butadiene Rubber

Styrene–butadiene rubbers do not process as satisfactorily as does natural rubber, nor do the unvulcanised compounds have the same degree of 'green strength'. In an attempt to improve the processing behaviour of emulsion-polymerised styrene–butadiene rubbers, small amounts of divinylbenzene

(XXX) have been included in the polymerisation recipe. The effect of this addition is to cause a few permanent crosslinks to form within the individual rubber latex particles. Rather paradoxically, it appears that such crosslinks have a beneficial effect upon processing behaviour, provided that

$$CH\!=\!CH_2$$

(XXX)

they are confined within the individual latex particles which subsequently coalesce together to form the bulk rubber.

A more recent approach has been to introduce a small concentration of ionic crosslinks into the rubber by copolymerising with the styrene and butadiene a third monomer which contains t-amine groups, and then quaternising the nitrogen atoms of these amine groups at the latex stage by reaction with an organic dihalide. The chemistry is similar to that for the quaternisation of styrene–vinylpyridine–butadiene terpolymers, to be described in Section 5.4 of this chapter. The effect of introducing these labile ionic crosslinks is said to be to improve both green strength and processing behaviour of the rubber.

5.2.10 Types of Emulsion-polymerised Styrene–Butadiene Rubber

From the foregoing summary of production methods for styrene–butadiene rubbers, it will come as no surprise that a bewildering array of types of this rubber is available commercially. The principal classes into which the available grades fall are summarised in Table 5.3, which also gives the code numbers which have been assigned to the various classes by the International Institute of Synthetic Rubber Producers.

5.2.11 Compounding Principles

Vulcanisation

Styrene–butadiene rubbers can be vulcanised by most of the methods which are used for natural rubber, although there are differences of detail. It has been rightly said that the differences between styrene–butadiene rubber and natural rubber in practical vulcanisation behaviour are matters of degree rather than of kind. Sulphur is the most widely-used vulcanising agent. It is used in conjunction with conventional accelerators of sulphur vulcanisation (notably the thiazoles and the thiurams), and with an activator such as zinc oxide. Rather less sulphur is required than for natural rubber (presumably because of the reduced overall unsaturation in the polymer), and somewhat higher accelerator/sulphur ratios. For a given vulcanising system, styrene–butadiene rubber vulcanises more slowly than does natural rubber; it is therefore good practice to 'boost' the rate of crosslink insertion by the addition of small amounts of a very active accelerator such as tetramethyl thiuram disulphide. As in the case of natural rubber, the rate of vulcanisation and the scorchiness of the compound is greatly affected by the presence of other compounding ingredients, notably carbon blacks. Styrene–butadiene rubbers can also be vulcanised by various non-sulphur systems, notably organic peroxides such as dicumyl peroxide.

As with natural rubber, the physical properties of a styrene–butadiene rubber vulcanisate change progressively with degree of vulcanisation, but again there are some differences when the behaviours of the two rubbers are compared in detail. As with natural rubber, the tensile strength and the tear strength increase to a maximum and then decrease as vulcanisation progresses. Unlike natural rubber, the modulus, resilience and abrasion resistance generally increase progressively; this is the phenomenon of the so-called 'marching modulus'. Properties such as elongation at break, permanent set, swelling in solvents and flex resistance decrease progressively with increasing degree of vulcanisation.

Reinforcement
The tensile strength of styrene–butadiene rubber 'gum' (that is, unfilled) vulcanisates is very low (of the order of $300\,lb\,in^{-2}$). This is in marked contrast to the high tensile strength of natural rubber gum vulcanisates. The difference between the two rubbers in this respect is attributed to the absence of crystallisation on stretching in the case of styrene–butadiene rubbers. A practical consequence is that, if high tensile strength is required in a styrene–butadiene rubber vulcanisate, then it is essential to incorporate a reinforcing filler. If the colour of the vulcanisate is of no consequence, then carbon blacks are normally used. For white or light coloured vulcanisates, silica or metal silicates (calcium or aluminium) are used. Cheap materials such as precipitated calcium carbonate are used as non-reinforcing extending and cheapening fillers.

The physical properties of styrene–butadiene rubber vulcanisates reinforced with carbon black depend upon both the type of carbon black used

TABLE 5.3

PRINCIPAL CLASSES OF EMULSION-POLYMERISED STYRENE–BUTADIENE RUBBERS, TOGETHER WITH CODE NUMBERS ASSIGNED BY THE INTERNATIONAL INSTITUTE OF SYNTHETIC RUBBER PRODUCERS (I.I.S.R.P.)

I.I.S.R.P. code no.	% Bound styrene	Mooney viscosity (ML 1 + 4 100°C)	Type of soap in recipe	Coagulant	Product stain[a]	Other comments
(a) Hot-polymerised non-pigmented rubbers: 1000 series						
1 000	23·5	48	FA	A or SA	st	
1 001	23·5	48	FA	A or SA	slst	
1 002	23·5	54	RA	A or SA	st	
1 004	23·5	50	FA	AL	st	
1 006	23·5	50	FA	A or SA	nst	
1 009	23·5	97[b]	FA	A or SA	nst	(contains divinylbenzene as crosslinker)
1 010	23·5	30	FA	AL	nst	
1 011	23·5	54	RA	A or SA	nst	
1 012	23·5	105	FA	A or SA	nst	
1 013	43·0	45	FA	AL	nst	
1 014	40·0	70	RA	S/AL	slst	
1 018	23·5	125[b]	FA	GA	nst	
1 019	23·5	50	FA	GA	nst	
1 027	23·5	50	FA	A or SA	nst	
1 028	48·0	58	FA	A or SA	nst	
(b) Cold-polymerised non-pigmented rubbers: 1500 series						
1 500	23·5	52	RA	A or SA	st	
1 502	23·5	52	FA/RA	A or SA	nst	
1 503	23·5	52	FA	GA	nst	
1 505	9·5	40	RA	A or SA	nst	

1506	23·5	25	FA/RA	AL	nst
1507	23·5	35	FA/RA	A or SA	nst
1509	23·5	34	FA/RA	AL	nst
1510	23·5	32	FA	A or SA	nst
1512	29·0	52	FA	GA	nst
1513	40·0	36	FA/RA	AL	nst
1515	28·0	52	FA	A or SA	nst
1516	40·0	40	FA/RA	A or SA	nst
1518	23·5	100	FA	GA	nst
1519	23·5	22	FA	A or SA	nst
1551	23·5	52	RA	A or SA	nst
1570	23·5	117	FA/RA	A or SA	nst
1573	5·0	22	FA	A or SA	nst

(c) Cold-polymerised carbon black–oil–rubber masterbatches containing 14 pphr or less of oil: 1600 series

1601	23·5	68[b]	FA/RA	A	st	50 pphr HAF
1605	23·5	62[b]	FA	A	nst	50 pphr FEF
1606	23·5	56[b]	RA	A	st	52 pphr HAF + 10 pphr ar
1608	23·5	58[b]	RA	A	st	52 pphr ISAF + 12·5 pphr ar
1609	23·5	61[b]	RA	A	st	40 pphr SAF + 5 pphr ar
1610	23·5	64[b]	RA	A	st	52 pphr ISAF + 10 pphr ar
1611	23·5	62[b]	RA	A	slst	62·5 pphr HAF + 12 pphr hp
1613	23·5	48[b]	RA	A	st	40 pphr SAF + 10 pphr ar
1614	23·5	48[b]	FA	A	nst	50 pphr EPF
1615	23·5	59[b]	RA	A	st	52 pphr ISAF + 12·5 pphr ar
1616	23·5	56[b]	RA	A	st	52 pphr HAF + 10 pphr ar
1617	28·0	53[b]	FA	A	nst	50 pphr HAF
1621	23·5	50[b]	FA/RA	A	nst	50 pphr ISAF + 12·5 pphr naph
1623	23·5	45[b]	RA	A	st	52 pphr HAF + 12·5 pphr ar$_\gamma$

(continued)

TABLE 5.3—contd.

I.I.S.R.P. code no.	% Bound styrene	Mooney viscosity (ML 1 + 4 100°C)	Type of soap in recipe	Coagulant	Product stain[a]	Other comments
(d) Cold-polymerised oil–rubber masterbatches: 1700 series						
1703	23·5	60	FA/RA	A or SA	nst	25 pphr naph
1707	23·5	55	RA	A or SA	nst	37·5 pphr naph
1708	23·5	60	FA	GA	nst	37·5 pphr naph
1710	23·5	50	FA/RA	A or SA	st	37·5 pphr ar
1712	23·5	55	FA/RA	A or SA	st	37·5 pphr ar
1713	23·5	52	FA/RA	A or SA	nst	50 pphr naph
1714	23·5	52	FA/RA	A or SA	st	50 pphr ar
1715	23·5	52	FA	GA	nst	50 pphr naph
1718	16·0	40	FA/RA	A or SA	st	37·5 pphr ar
1773	23·5	60	FA/RA	A or SA	nst	25 pphr naph
1778	23·5	55	FA/RA	A or SA	nst	37·5 pphr naph
1779	23·5	85	FA/RA	A or SA	nst	15 pphr naph
(e) Cold-polymerised carbon black–oil–rubber masterbatches containing more than 14 pphr of oil: 1800 series						
1801	23·5	60[b]	FA/RA	A	st	50 pphr HAF + 25 pphr naph
1805	23·5	58[b]	FA/RA	A	nst	75 pphr HAF + 37·5 pphr naph
1808	23·5	48[b]	FA/RA	A	st	75 pphr HAF + 50 pphr ar
1811	23·5	46[b]	RA	A	st	75 pphr SRF + 17·5 pphr ar
1813	23·5	62[b]	FA/RA	A	st	60 pphr ISAF + 37·5 pphr ar
1814	23·5	60[b]	FA/RA	A	st	75 pphr ISAF + 50 pphr ar

1815	23·5	45^b	FA/RA	A	nst	75 pphr HAF + 50 pphr naph
1820	23·5	55^b	FA/RA	A	nst	69 pphr FEF + 37·5 pphr naph
1821	23·5	58^b	FA/RA	A	nst	80 pphr FEF + 37·5 pphr naph
1823	23·5	52^b	FA/RA	A	st	82·5 pphr HAF + 62·5 pphr ar
1824	23·5	52^b	FA/RA	A	st	82·5 pphr ISAF + 62·5 pphr ar
1828	23·5	52^b	FA/RA	A	nst	75 pphr SRF + 17·5 pphr naph
1829	23·5	45^b	RA	A	nst	75 pphr SRF + 17·5 pphr naph
1830	23·5	48^b	FA/RA	A	nst	69 pphr SRF + 37·5 pphr naph
1831	23·5	50^b	FA/RA	A	st	82·5 pphr ISAF + 62·5 pphr ar
1832	23·5	46^b	FA/RA	A	nst	75 pphr HAF + 50 pphr naph
1833	23·5	45^b	FA/RA	A	st	82·5 pphr HAF + 62·5 pphr ar
1836	23·5	48^b	FA/RA	A	st	75 pphr HAF + 50 pphr ar
1837	23·5	46^b	FA/RA	A	st	69 pphr SRF + 37·5 pphr ar
1839	23·5	55^b	FA/RA	A	st	75 pphr ISAF + 50 pphr ar
1845	23·5	54^b	FA	A	nst	75 pphr HAF + 50 pphr naph
1846	23·5	45^b	FA/RA	A	st	90 pphr SRF + 68 pphr ar
1847	23·5	45^b	FA/RA	A	st	75 pphr HAF + 50 pphr ar
1848	23·5	45^b	FA/RA	A	st	82·5 pphr HAF + 62·5 pphr ar
1850	23·5	42^b	FA/RA	A	st	75 pphr GPF + 50 pphr ar

[a] Staining determined principally by antioxidant present.

[b] Compound viscosity.

Abbreviations: RA rosin acid, FA fatty acid, A acid, AL alum, GA glue-acid, SA salt-acid, st staining, slst slightly staining, nst non-staining, FEF fast-extrusion furnace black, SAF super-abrasion furnace black, ISAF intermediate super-abrasion furnace black, HAF high-abrasion furnace black, EPF easy-processing furnace black, SRF, semi-reinforcing furnace black, GPF general-purpose furnace black, naph naphthenic oil, ar aromatic oil, hp heavy process oil.

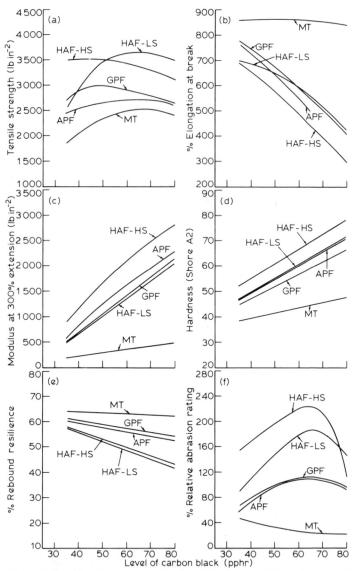

FIG. 5.14. Effect of various types of carbon black upon (a) tensile strength, (b) elongation at break, (c) modulus, (d) hardness, (e) resilience and (f) abrasion resistance of vulcanisates from a cold-polymerised non-pigmented styrene–butadiene rubber.[12] Compound: styrene–butadiene rubber type 1500 100, stearic acid 1·5, zinc oxide 3, naphthenic process oil 12, diphenylamine–acetone condensate antioxidant 2, N-t-butyl-2-benzthiazyl sulphenamide 1, sulphur 1·75, carbon black variable. Significance of abbreviations for types of carbon black: MT = medium thermal; GPF = general-purpose furnace; APF = all-purpose furnace; HAF–LS = high-abrasion furnace—low structure; HAF–HS = high-abrasion furnace—high structure.

and the loading. Figure 5.14 summarises results for (a) tensile strength, (b) elongation at break, (c) modulus at 300% extension, (d) hardness, (e) rebound resilience and (f) abrasion resistance for various loadings of five carbon blacks of widely different reinforcing characteristics. As a broad generalisation, it may be said that, as with natural rubber, the reinforcing ability of a carbon black increases as the particle size decreases.

Protection against ageing
In order to achieve maximum protection against the various deteriorative influences, it is necessary to add an antioxidant (and possibly also an antiozonant) to the compound. The additives used are similar to those used for natural rubber, being mainly aromatic phenols or amines, and are added in similar amounts (*c*. 1 pphr).

5.2.12 Processing Behaviour
Mastication and mixing
Styrene–butadiene rubber differs significantly from natural rubber in that it does not soften appreciably when masticated in air. It is therefore necessary to ensure that the grade selected has approximately the correct plasticity for the subsequent processing, making due allowance for any plasticising oils which may be added during compounding; little change in plasticity can be achieved by mastication. The mixing procedure is broadly similar to that for natural rubber; allowance should be made for the fact that sulphur is less soluble in styrene–butadiene rubber at room temperature than it is in natural rubber, and is rather more difficult to disperse. The mixing of styrene–butadiene rubber requires more power and generates more heat than does that of natural rubber. Some difficulty is experienced in ensuring coherence of the mix on an open mill. Hot compounded styrene–butadiene is rather weaker than is an equivalent natural rubber compound, and so can create handling difficulties.

Calendering and extrusion
Most styrene–butadiene rubber compounds extrude at a slower speed than do similar natural rubber compounds. Higher extrusion temperatures are required. Extruded and calendered stocks tend to be rougher in the case of styrene–butadiene rubber than are those from natural rubber, and the die swell is greater. Most of these differences in processing behaviour are consistent with styrene–butadiene rubber stocks containing a higher proportion of polymer gel than do similar natural rubber stocks. This is in

turn a consequence of styrene–butadiene rubber being reluctant to undergo plasticisation during the mastication step.

Self-adhesive tack

An important deficiency of styrene–butadiene rubber in relation to natural rubber is that it lacks the marked tendency to self-adhesion which is characteristic of natural rubber stocks. Self-adhesive tack is a most important characteristic in connection with the building of rubber articles such as tyres, wrapped hoses, etc. The self-adhesive tack of a styrene–butadiene rubber is somewhat improved if the rubber contains a rosin acid (derived from the rosin-acid soap used during the emulsion polymerisation), but is still poor in comparison with that of natural rubber. Significant improvements can be made by adding 'tackifying' resins.

Moulding

The behaviour of styrene–butadiene stocks in moulding processes is generally similar to that of natural rubber stocks. However, the hot tensile strength of styrene–butadiene rubber vulcanisates is inferior to that of natural rubber vulcanisates, and this means that styrene–butadiene rubber mouldings tear more readily when being removed from a hot mould.

5.3   EMULSION-POLYMERISED ACRYLONITRILE–BUTADIENE RUBBERS (NBR)

The principal repeat units which are present in the acrylonitrile–butadiene rubbers are those derived from butadiene (their nominal structure being as shown at (XXVIII) and those derived from acrylonitrile. Most, if not all, of these latter repeat units have the structure shown at (XXXI).

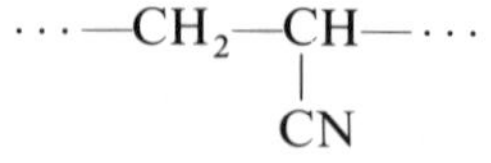

$$\cdots -CH_2-CH- \cdots$$
$$|$$
$$CN$$

(XXXI)

5.3.1 Outline of Production of Emulsion-polymerised NBR

The principles which underly the production of acrylonitrile–butadiene rubber by emulsion polymerisation are broadly similar to those which underly the production of styrene–butadiene rubber by emulsion polymerisation. Detailed discussion is not therefore appropriate. In outline, the polymerisation may be carried out in batch reactors or in continuous reactors. It may be carried out at high temperature (c. 50 °C) or at low temperature (c. 5 °C), or, indeed, at intermediate temperatures. Table 5.4 gives details of two older German recipes for the production of acrylonitrile—butadiene rubbers at 29 °C and 23 °C respectively. The rate of polymerisation obtained using a given initiating system at a given reaction temperature increases with the acrylonitrile/butadiene ratio in the recipe. The composition of the copolymer which is obtained from an acrylonitrile–butadiene emulsion polymerisation system depends, of course, upon the ratio of the two monomers in the reaction system.

The polymerisation is usually stopped short of complete conversion (at least when a dry rubber is being produced) for the same reasons that styrene–butadiene emulsion copolymerisation is shortstopped. Seventy to eighty per cent conversion is the usual range for shortstopping. Removal of unreacted monomers from the latex is very important because of the toxicity of acrylonitrile and the fact that monomer dimers formed in the course of the polymerisation give rise to strong odour in the finished product. Coagulation of the latex is achieved by mixing with electrolytes, such as sodium chloride, and acid. It is important to adjust the conditions of coagulation so that the coagulum is coarse and flocculant, containing neither lumps nor milky suspensions. The texture of the coagulum is important for the subsequent filtering, washing and drying operations.

TABLE 5.4

TWO OLD GERMAN RECIPES FOR PRODUCTION OF ACRYLONITRILE–BUTADIENE RUBBER BY EMULSION COPOLYMERISATION[13]

| | Parts by weight | |
	I ('Perbunan')	II ('Perbunan Extra')
Acrylonitrile	27	40
Butadiene	73	60
Water	150	200
Sodium dibutylnaphthalenesulphonate	3·6	3·6
Potassium persulphate	0·2	0·2
Sodium pyrophosphate	0·3	0·3
Sodium hydroxide	0·08	0·1
Diisopropylxanthogen disulphide	0·3	0·27
Phenyl-2-naphthylamine	3·0	3·0
Polymerisation temperature	c. 29 °C	c. 23 °C

As has been indicated above, acrylonitrile–butadiene rubbers can be produced by low-temperature as well as by high-temperature emulsion copolymerisation. The advantages which accrue from reducing the polymerisation temperature are similar to those which pertain to styrene–butadiene rubber produced at low temperature, namely, improved processing in respect of mixing, calendering and extrusion, and improved vulcanisate properties. The molecular interpretations of these changes are similar to those for styrene–butadiene rubber.

5.3.2 Effect of Varying the Acrylonitrile Butadiene Ratio

The comonomer ratio is the single most important recipe variable for the production of acrylonitrile–butadiene rubbers. Table 5.5 summarises those properties of the rubber which are strongly dependent upon the comonomer ratio, and gives a qualitative indication of the direction in which the various properties change. Of the properties listed in Table 5.5, the two most important are oil resistance and low-temperature flexibility. This is because (a) the major use for acrylonitrile–butadiene rubbers is in applications where a strongly oil-resistant rubber is required, and (b) as the oil resistance of a rubber increases, so its ability to withstand stiffening at

TABLE 5.5

QUALITATIVE SUMMARY OF EFFECT OF ACRYLONITRILE/BUTADIENE RATIO UPON PROPERTIES OF ACRYLONITRILE–BUTADIENE RUBBERS[13]

Acrylonitrile rubber with low acrylonitrile content		*Acrylonitrile rubber with high acrylonitrile content*
——————————— Acrylonitrrile content ———————————→		
————————— Density —————————→		
————————— Processability —————————→		
————————— Rate of vulcanisation —————————→		
————————— Permanent set —————————→		
————————— Oil resistance —————————→		
——————— Compatibility with polar polymers ———————→		
————————— Hysteresis loss —————————→		
←————————— Resilience —————————		
←————————— Low-temperature flexibility —————————		
←————————— Solubility in aromatic liquids —————————		
←————————— Gas permeability —————————		

Arrows show directions of increase.

low temperatures is reduced. What this means in practice is that oil resistance and low-temperature flexibility are mutually incompatible requirements in a rubber: the greater is the ability to withstand the swelling action of oils and solvent, the higher is the temperature at which the vulcanisate loses its elastomeric properties when it is cooled. These trends are illustrated in Fig. 5.15 for acrylonitrile–butadiene rubbers which do not contain a plasticiser. Acrylonitrile–butadiene rubbers are commonly

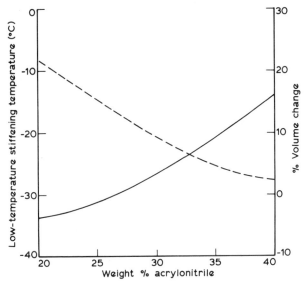

Fig. 5.15. Low-temperature stiffening temperature and swelling in mineral oil as functions of acrylonitrile content for vulcanisates from acrylonitrile–butadiene rubbers.[14] ——, low-temperature stiffening temperature; – – –, volume change.

classified very broadly according to acrylonitrile content as low (10–25 %), medium (25–35 %), and high (35–50 %). The manner in which the glass-transition temperature of an acrylonitrile–butadiene rubber varies with the acrylonitrile content is shown in Fig. 5.16.

Acrylonitrile–butadiene rubber vulcanisates are considerably less elastic at room temperature than are similar vulcanisates of either natural rubber or styrene–butadiene rubber. Other things being equal, the rebound resilience of an acrylonitrile–butadiene rubber vulcanisate depends very much upon the acrylonitrile content of the rubber, falling with increasing acrylonitrile content in the manner indicated in Fig. 5.17. This tendency is

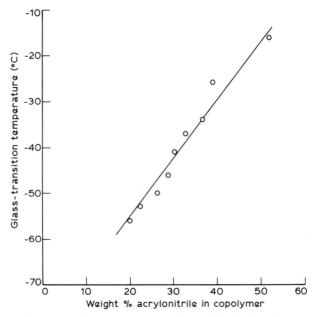

F<small>IG.</small> 5.16.   Glass-transition temperatures of acrylonitrile–butadiene copolymers as function of acrylonitrile content.[13]

really a consequence of the effect of acrylonitrile content upon the glass-transition temperature of the rubber: the higher is the glass-transition temperature of the rubber, the nearer is the rubber to its glass transition at, say, ambient temperatures. A further consequence of these considerations is that the elasticity of acrylonitrile–butadiene rubber vulcanisates is more sensitive to change of temperature than is that of, say, similar natural rubber vulcanisates. Thus, for instance, the rebound resilience of acrylo-nitrile—butadiene rubber vulcanisates increases more sharply with increasing temperature than does that of natural rubber vulcanisates.

5.3.3 Compounding Principles

Vulcanisation

The normal vulcanising agent for acrylonitrile–butadiene rubbers is sulphur in conjunction with conventional accelerators and zinc oxide as an activator. Other curatives, notably organic peroxides, are also available, but these are used in very specialised applications only.

The principles which underly the vulcanisation of acrylonitrile–butadiene rubbers by sulphur and accelerators are very similar to those for

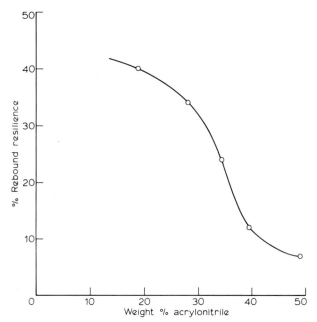

FIG. 5.17.   Effect of acrylonitrile content upon rebound resilience of acrylonitrile–
butadiene rubber vulcanisates.[13] Compound: acrylonitrile–butadiene rubber 100,
plasticiser 5, stearic acid 2, zinc oxide 5, HAF carbon black 40, sulphur 1·5, N-
cyclohexyl-2-benzthiazyl sulphenamide 0·8.

styrene–butadiene rubbers. The amount of sulphur required to effect
satisfactory vulcanisation is less than that required for natural rubber, and
decreases with increasing acrylonitrile content of the rubber. The accel-
erator/sulphur ratio is usually higher than that used for natural rubber.
Increasing the amount of sulphur in an acrylonitrile–butadiene rubber
vulcanisate tends to lower properties such as tear resistance, elongation at
break and heat resistance. Di-2-benzthiazyl disulphide/sulphur and
sulphenamide/sulphur combinations can be used where a good all-round
balance of vulcanisate properties is required. Typical combinations would
be di-2-benzthiazyl disulphide/sulphur = 1·0/1·5 pphr and N-cyclohexyl-2-
benzthiazylsulphenamide/sulphur = 1·5/0·75 pphr. Where good compression set combined with freedom from scorch is required, a combination
of a thiuram monosulphide and sulphur is recommended, e.g., tetramethyl
thiuram monosulphide/sulphur = 1·0/1·0 pphr. Where superior resistance to
ageing at high temperatures is required, a thiuram disulphide at a level of
3–4 pphr may be used in the absence of added sulphur.

Plasticisation

Plasticising additives may be incorporated into acrylonitrile–butadiene rubber compounds for three distinct reasons:

1. to improve processing behaviour;
2. to impart self-adhesive tack; and
3. to improve low-temperature flexibility.

Of these, reason (3) is probably the most important technologically, because of the inverse relationship between oil resistance and low-temperature flexibility which exists within the family of acrylonitrile–butadiene rubbers.

Included amongst the processing plasticisers are additives such as stearic acid, coal tar oils, factice, alkyd resins, aldol resins, and aromatic petroleum plasticisers. Amongst the 'tackifying' additives are materials such as coumarone resins and xylene–formaldehyde resins.

The plasticisers which are used to improve the low-temperature flexibility of the vulcanisate are usually esters, either of carboxylic acids or of phosphoric acid. The esters may be either monomeric or polymeric. Dioctyl phthalate is a typical example of a low-cost general-purpose monomeric ester plasticiser. A more expensive monomeric ester plasticiser is triethylene glycol dicaprylate. Tritolyl phosphate is a typical phosphate ester; it has the advantage of reducing the flammability of any vulcanisate in which it is incorporated. Monomeric plasticisers have the disadvantage that they tend to be extracted when the vulcanisate comes into contact with oils. Although this effect may be superficially attractive in that the degree of swell which occurs when the vulcanisate is immersed in the oil may be very small (in some cases actually negative) because of the volume lost through the plasticiser being extracted, it is an undesirable effect. This is because (a) the extracted plasticiser will contaminate the oil into which it is extracted, and (b) the resistance of the vulcanisate to low-temperature stiffening will be reduced, especially if the swelling oil is allowed subsequently to dry out of the vulcanisate. For this reason, polymeric ester plasticisers have been developed; low-molecular-weight polypropylene sebacate is an example. Although not so efficient as plasticisers as are the monomeric esters, they show considerable resistance to extraction by oils. They are also considerably more expensive than are the monomeric ester plasticisers. Polymeric polyether plasticisers have also been developed, such as dibutyl carbitol formal. The low-temperature flexibility of acrylonitrile—butadiene rubber vulcanisates can also be improved by blending with other rubbers such as natural rubber, styrene–butadiene rubber, polybutadiene, or an acrylonitrile–butadiene rubber of lower acrylonitrile content.

The amount of plasticiser which is added to an acrylonitrile–butadiene rubber compound is determined by the balance of several factors, of which the more important are (a) the level of low-temperature flexibility which is required, and (b) the degree of deterioration of other properties, such as tensile strength, tear strength, compression set and stress relaxation, which can be tolerated. The effects of increasing amounts of a typical plasticiser (a mixed dialkyl phthalate) upon some of the properties of an acrylonitrile–butadiene rubber vulcanisate are summarised in Fig. 5.18.

Reinforcement
Like styrene–butadiene rubber, acrylonitrile–butadiene rubber does not crystallise on stretching. The tensile strength of gum vulcanisates is therefore very low, and some degree of reinforcement is necessary if adequate strength is to be achieved in the vulcanisate. Where colour is of no consequence, carbon blacks are the preferred fillers. The mechanical properties of the resultant vulcanisate depend upon both the level and the type of carbon black. Some typical results are summarised in Fig. 5.19. The effects of various non-black fillers are summarised in Fig. 5.20.

Blending with resinous polymers
Acrylonitrile–butadiene rubbers are readily compatible with certain resinous, polar polymers. Insofar as the rubber is the major component, the resinous polymer acts as a kind of organic reinforcing agent, and, as such, might well have been considered under the heading 'Reinforcement' above. Included amongst such resinous polymers are (a) incompletely-reacted phenol–formaldehyde resins, and (b) polyvinyl chloride.

The compatibility of phenol–formaldehyde resins with butadiene–acrylonitrile rubbers increases with the acrylonitrile content of the rubber. The addition of such resins brings about several significant changes to the rubber, including improvement of processing behaviour and tack at processing temperatures, improvement of adhesion to metals, and improvement of tensile strength and tear resistance. The compounds containing these resins are, however, very hard at room temperature. These effects, like the compatibility of the resin and the rubber, increase with the acrylonitrile content of the rubber.

The effect of blending acrylonitrile–butadiene rubber with polyvinyl chloride is to increase the tensile strength and tear resistance of the vulcanisate, and also its resistance to abrasion, to solvents, to deteriorative

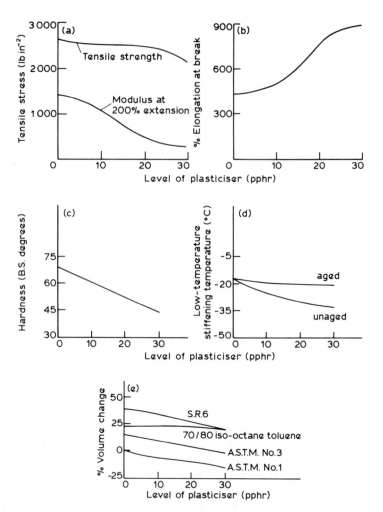

FIG. 5.18.   Effect of a typical plasticiser (a mixed dialkyl phthalate) upon properties of vulcanisates from a medium-acrylonitrile-content low-temperature-polymerised acrylonitrile–butadiene rubber.[14] Compound: acrylonitrile–butadiene
rubber 100, stearic acid 1, zinc oxide 5, SRF carbon black 60, di-2-benzthiazyl
disulphide 1, sulphur 1·5, plasticiser variable. Vulcanisation: 30 min at 153 °C.
'Aged' in (d) refers to immersion in 70/30 isoöctane/toluene mixture for 24 h at 40 °C
    followed by drying at 100 °C. Labels at (e) denote various swelling liquids.

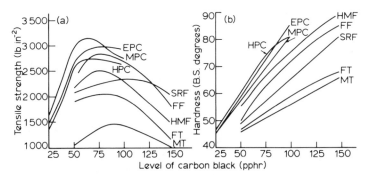

FIG. 5.19. Effects of various carbon blacks upon (a) tensile strength and (b) hardness of acrylonitrile–butadiene rubber vulcanisates.[14] Significance of abbreviations for types of carbon black: FT = fine thermal; MT = medium thermal; FF = fine furnace; SRF = semi-reinforcing furnace; HMF = high-modulus furnace; EPC = easy-processing channel; MPC = medium-processing channel; HPC = hard-processing channel.

influences such as ozone, and to fire. The proportion of polyvinyl chloride used is usually within the range 20–80 pphr. Again, both the compatibility and the reinforcing effect increase with the acrylonitrile content of the rubber. Soft rubbers produced by low-temperature polymerisation are preferred for the preparation of such blends, because they give the best compromise between compatibility with the polyvinyl chloride on the one hand and low-temperature flexibility on the other. The blends extrude and calender excellently at temperatures between 100 °C and 120 °C. In order to

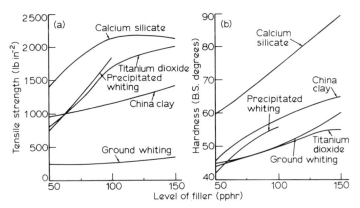

FIG. 5.20. Effects of various non-reinforcing fillers upon (a) tensile strength and (b) hardness of acrylonitrile–butadiene rubber vulcanisates.[14]

achieve the maximum advantage of blending polyvinyl chloride with an acrylonitrile–butadiene rubber, it is necessary to flux the blend at a temperature in the range 150–160 °C. However, there are blends produced commercially in such a way that they are already fully fused before they leave the manufacturer. It is therefore not necessary to flux them during processing in order to achieve the maximum enhancement of desirable properties.

It must also be pointed out that, since acrylonitrile–butadiene rubber and polyvinyl chloride are compatible with each other in all proportions, it is possible to prepare blends in which the major proportion is polyvinyl chloride. In such blends, the rubber in effect acts as a polymeric 'elasticising' plasticiser for the polyvinyl chloride which does not migrate and which improves properties such as permanent set, compression set, cold flow and low-temperature flexibility.

5.3.4 Processing Behaviour

Mastication and mixing

Like styrene–butadiene rubbers, the acrylonitrile–butadiene rubbers are not so susceptible to softening during mastication as is natural rubber. Higher proportions of plasticisers and processing aids are normally required. The expenditure of energy and heat build-up during mixing is higher than with natural rubber.

The principal mixing problem which calls for comment is that of the incorporation of sulphur. Sulphur is much less soluble in acrylonitrile–butadiene rubber than it is in natural rubber; a problem of sulphur dispersion may therefore arise. The consequences can be serious, because poorly-dispersed sulphur can lead to localised over-curing and under-curing. There are several approaches to the problem of sulphur dispersion. Thus special grades of sulphur, such as a magnesium-carbonate-coated grade, can be used; the sulphur can be added as a pre-mixed masterbatch; and the sulphur can be added early in the mixing cycle, in particular, before the fillers are added.

Calendering and extrusion

The calendering behaviour of acrylonitrile–butadiene rubber stocks is similar to that of natural rubber stocks, but somewhat lower temperatures are used. The higher the polymer content of the stock, the higher is the bowl temperature required. The range of calendering temperatures is 30–80 °C. Acrylonitrile–butadiene rubber stocks tend to be more difficult to extrude than do natural rubber stocks, unless the compound has been carefully designed and prepared. For both calendering and extrusion, it is necessary to control the rheological behaviour of the stock by way of such variables as (a) choice of base polymer, (b) selection of level and type of plasticiser, and (c) selection of level and type of filler. Very little control is possible by way of softening during mastication.

5.3.5 Swelling of Acrylonitrile–Butadiene Rubber Vulcanisates

Since the major use for acrylonitrile–butadiene rubbers is in applications where an oil-resistant elastomer is required, it is necessary to give brief consideration to the factors which influence the swelling of these rubbers in oils and solvents. The two most important factors are the nature of the solvent and the acrylonitrile content of the rubber. The data summarised in Table 5.6 illustrate both of these effects. Clearly the extent of swelling of a given vulcanisate varies enormously with the nature of the fluid. Concerning the effect of the acrylonitrile content of the rubber, a broad generalisation is that for relatively non-polar liquids, such as an isoöctane–toluene mixture, the degree of swelling *decreases* with increasing acrylonitrile content, but with relatively polar liquids, such as acetone, the degree of swelling *increases* with increasing acrylonitrile content. Table 5.6 also illustrates the effect of compounding with polyvinyl chloride upon resistance to swelling.

Further information concerning the effects of immersion of acrylonitrile–butadiene rubber vulcanisates in hydrocarbon liquids is summarised in Fig. 5.21. This shows the effect of immersion in hydrocarbon liquids of varying aromatic content upon the extent of swelling, and also upon the retention of some of the mechanical properties of the vulcanisate. The extent of swelling rises sharply with increasing aromatic content in the hydrocarbon, and falls sharply with increasing acrylonitrile content of the rubber. As might have been anticipated, the extent to which the mechanical properties are reduced by immersion in the hydrocarbon largely parallels the extent to which the rubber swells.

Other factors which influence the extent to which an acrylonitrile–butadiene rubber vulcanisate swells in a given fluid are (a) the density of crosslinks in the rubber phase, (b) the filler content, and (c) the level of extractable plasticiser in the vulcanisate. The crosslink density has only a minor effect upon the swelling of these rubbers, the reason being that the origin of the resistance to swelling is principally the attraction between the dipoles of the $C \equiv N$ groups of the acrylonitrile units. The principal effect of fillers is to dilute the rubber, thus leaving less material which can imbibe fluid. However, the nature of the filler also has an effect; as a generalisation,

TABLE 5.6

EFFECT OF NATURE OF SOLVENT, ACRYLONITRILE CONTENT OF RUBBER, AND COMPOUND-
ING WITH POLYVINYL CHLORIDE UPON SWELLING OF ACRYLONITRILE–BUTADIENE
RUBBER VULCANISATES[13]

	Weight % of acrylonitrile in base rubber				
	28	33	38	55	28
Compounds	Parts by weight				
Acrylonitrile–butadiene rubber	100	100	100	100	70
Polyvinyl chloride	0	0	0	0	30
Diphenylthiourea	0	0	0	0	0·2
Stearic acid	1	1	1	1	1
Zinc oxide	5	5	5	5	5
Phenyl-1-naphthylamine	1·5	1·5	1·5	1·5	1·5
FEF carbon black	40	40	40	40	40
Sulphur	1·8	1·8	1·8	1·8	1·8
N-cyclohexyl-2-benzthiazyl sulphenamide	1·2	1·2	1·2	1·2	1·2
Cure: minutes at 151 °C	15	20	20	20	20
Swelling	% Weight increase after 20 days immersion in solvent indicated				
Isoöctane at 20 °C	4·3	1·6	0·5	0·0	2·3
Isoöctane at 50 °C	5·7	4·7	2·0	0·1	3·8
70/30 Isoöctane/toluene at 20 °C	29·0	23·3	18·5	11·2	22·3
70/30 Isoöctane/toluene at 50 °C	30·2	24·0	18·6	11·3	22·8
50/50 Isoöctane/toluene at 20 °C	43·8	35·2	30·7	18·3	39·3
50/50 Isoöctane/toluene at 50 °C	50·8	40·7	31·3	20·1	38·5
Methanol at 50 °C	14·3	13·5	13·3	10·3	6·5
n-Propanol at 50 °C	18·5	16·2	15·2	14·2	11·1
n-Butanol at 50 °C	16·7	14·6	11·3	10·5	11·0
n-Amyl alcohol at 50 °C	16·6	15·7	13·1	8·5	10·8
Glycerol at 50 °C	0·4	0·3	0·1	0·0	0·1
Diethylene glycol at 50 °C	2·5	3·3	2·4	2·5	1·1
1,4 Dioxane at 20 °C	169·5	158·0	143·7	120·3	145·3
Tetrahydrofuran at 20 °C	208·9	192·1	172·9	167·9	186·6
Methyl glycol acetate at 50 °C	138·7	129·6	123·7	112·7	114·1
Acetone at 20 °C	91·1	104·7	110·7	145·9	84·6
Methyl ethyl ketone at 20 °C	129·5	128·4	129·3	148·1	117·0
Hexachlorobutadiene at 20 °C	69·3	36·0	11·8	1·2	48·3
Ethylene dichloride at 20 °C	306·0	308·5	303·0	306·5	249·0
Perchloroethylene at 20 °C	102·5	69·2	46·5	23·6	77·8

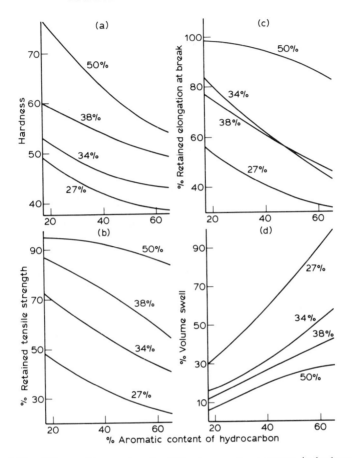

Fig. 5.21. Effects of immersion for 70 h at room temperature in hydrocarbon liquids of various aromatic contents upon properties of acrylonitrile–butadiene rubber vulcanisates.[15] Compound: acrylonitrile–butadiene rubber 100, stearic acid 0·5, zinc oxide 5, SRF carbon black 75, dioctyl phthalate 5, tetramethylthiuram disulphide 0·5, sulphur 1·25. Vulcanisation: 10 min at 166 °C. Figures appended to curves indicate acrylonitrile content of rubber.

it can be said that the more reinforcing the filler, the greater will be its tendency to inhibit swelling in addition to any diluting effect which it has. As has been noted earlier, the effect of the presence of an extractable plasticiser is to reduce the volume change which accompanies the imbibition of the swelling fluid, simply because material is lost from the rubber by extraction.

It is appropriate to conclude this discussion of the swelling behaviour of acrylonitrile–butadiene rubbers by referring to a useful quantitative index which is available for predicting the tendency of a rubber to swell when immersed in a fluid. This index is the *cohesive energy density* (γ), or, what amounts to the same thing, the *solubility parameter* (δ). These two quantities are very simply related, since $\delta = \sqrt{\gamma}$. The cohesive energy density of a liquid is defined as

$$\gamma = \frac{\Lambda - RT}{V}$$

where Λ is the molar latent heat of evaporation of the liquid, V is its molar volume, R is the molar gas constant, and T is the absolute temperature. The general rule is that a rubber vulcanisate will swell most in those fluids which have similar solubility parameters to the rubber itself; indeed, solubility parameters for rubber vulcanisates are determined by observing the way in which the equilibrium swelling of the vulcanisate depends upon the solubility parameter of the fluid in which it is immersed. Figure 5.22 shows the variation of the solubility parameter of acrylonitrile–butadiene rubbers with acrylonitrile content. Also noted on this diagram are the positions of various fluids on the solubility parameter scale. Taken in conjunction with

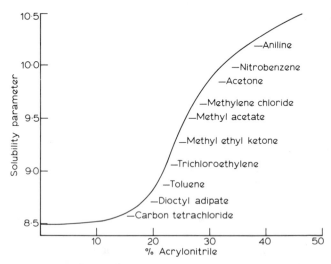

FIG. 5.22.   Effect of acrylonitrile content upon solubility parameter of acrylonitrile–butadiene rubbers. Also indicated are the solubility parameters of various organic liquids.[15]

the curve, these indications show which acrylonitrile–butadiene rubber will have *minimum* resistance to swelling in a given fluid, that is, maximum swelling; thus, for example, of the entire range of acrylonitrile–butadiene rubber vulcanisates, it is one containing about 24% of acrylonitrile and 76% of butadiene (by weight) which is predicted to show the least resistance to swelling in methyl ethyl ketone.

5.4   EMULSION-POLYMERISED RUBBERY TERPOLYMERS OF STYRENE, VINYLPYRIDINE AND BUTADIENE (PSBR)

Although they have found very limited application as dry rubbers in the past, rubbery terpolymers of styrene, vinylpyridines and butadiene are used exclusively as latices at the present time. These latices are used almost entirely in a single application, namely, for the treatment of textile fibres (such as tyre cords) in order to improve adhesion to the matrix of vulcanised rubber in which they will eventually be embedded. Some interest has been shown in recent years in the possibility of reviving the application of these terpolymers as solid rubbers, but large rises in the price of vinylpyridines following the oil crisis of 1974 have effectively inhibited serious consideration of this possibility.

The vinylpyridine which has been most commonly used for this application is 2-vinylpyridine (XXXII). Others include 3-vinylpyridine (XXXIII), 4-vinylpyridine (XXXIV), 6-methyl-2-vinylpyridine (XXXV), 6-methyl-3-vinylpyridine (XXXVI) and 5-ethyl-2-vinylpyridine (XXXVII).

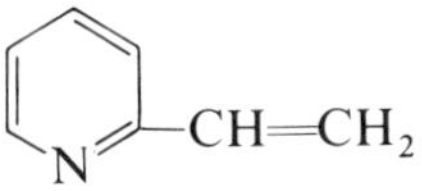

XXXII 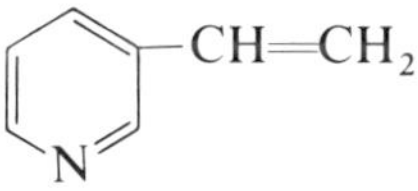   XXXIII 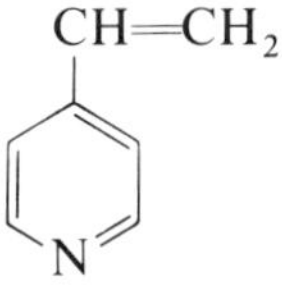   XXXIV

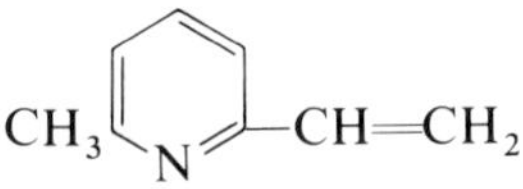

XXXV 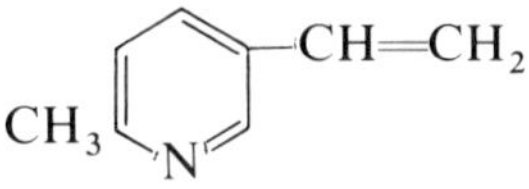   XXXVI

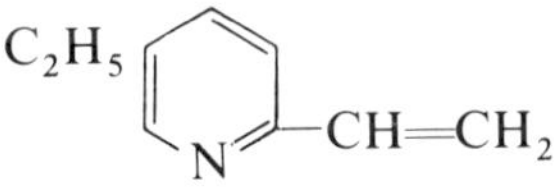

XXXVII

TABLE 5.7

RECIPE FOR PRODUCTION OF STYRENE–VINYLPYRIDINE–BUTADIENE TERPOLYMERS BY EMULSION COPOLYMERISATION AT 60 °C

	Parts by weight
Styrene	15
2-Vinylpyridine	15
Butadiene	70
Water	138
Potassium soap of disproportionated rosin acid	6
Sodium alkylnaphthalenesulphonates	0·6
t-Dodecyl mercaptan	0·7
Ammonium persulphate	0·27

The weight ratio of monomers is typically approximately 15/15/70 styrene/ vinylpyridine/butadiene. Thus these materials are in effect 30/70 styrene– butadiene rubbers in which about half of the styrene by weight has been replaced by a vinylpyridine.

A recipe for the preparation of these terpolymers by emulsion polymerisation is given in Table 5.7. Mixtures of styrene, vinylpyridine and butadiene polymerise considerably more rapidly than do the corresponding mixtures of styrene and butadiene. This is attributed to the inductive effect of the electronegative nitrogen atom of the pyridine nucleus making the vinyl group more susceptible to attack from radicals. The rate of polymerisation depends upon the particular vinylpyridine which is used, as is illustrated by the results summarised in Table 5.8. 4-Vinylpyridine gives a particularly high rate of polymerisation, the rates for the three isomeric vinylpyridines being in the order 4 > 2 > 3. This order is consistent with the explanation given above for the enhancement of the rate of polymerisation when styrene is replaced by vinylpyridine. Alkyl substitution of vinylpyridine reduces its activity in an emulsion polymerisation system.

Concerning the application of these terpolymers *as solid rubbers*, there are three matters which call for comment:

1.   their self-adhesive tack;
2.   their vulcanisation behaviour; and
3.   their ability to undergo quaternisation reactions.

1.   *Self-adhesive tack:* The progressive replacement of the styrene in a styrene–butadiene rubber by a vinylpyridine results in a progressive enhancement of the ability of the vulcanised rubber to adhere

TABLE 5.8

COMPARATIVE RATES OF EMULSION POLYMERISATION OF BUTADIENE WITH VARIOUS VINYLPYRIDINES AT $50\,^{\circ}\text{C}$[16]

Monomer composition (parts by weight)	% Conversion after 8 h reaction
Butadiene 75, 4-vinylpyridine 25	91·0
Butadiene 75, 2-vinylpyridine 25	82·0
Butadiene 75, 2-vinylpyridine 12·5, 4-vinylpyridine 12·5	78·0
Butadiene 75, 2-methyl-6-vinylpyridine 25	78·5
Butadiene 75, 2-vinylpyridine 12·5, 3-vinylpyridine 12·5	78·0
Butadiene 75, 3-vinylpyridine 25	74·2
Butadiene 75, 2-vinylpyridine 12·5, 2-vinyl-5-ethylpyridine 12·5	73·0
Butadiene 75, 5-vinyl-2-methylpyridine 25	69·5
Butadiene 75, 2-vinyl-5-ethylpyridine 25	66·5
Butadiene 75, styrene 25	48·8

to itself. This property would be of considerable technological value, were it not for the high price of the vinylpyridine relative to styrene.

2. *Vulcanisation behaviour:* Styrene–vinylpyridine–butadiene terpolymers are vulcanised with sulphur, accelerators and zinc oxide in a manner similar to natural rubber and styrene–butadiene rubber. However, relative to styrene–butadiene rubber, the rate of crosslink insertion is much faster. Only a very small replacement of styrene by vinylpyridine is necessary in order for this effect to become apparent. The reason for the enhancement of rate of vulcanisation is obscure. It might be thought that the copolymerised vinylpyridine units are functioning as amine-type accelerators of vulcanisation, but this does not appear to be the case.

3. *Quaternisation of vinylpyridine rubbers:* Tertiary amines are known to react with alkyl halides to give quaternary ammonium salts:

$$R_3N + R'X \longrightarrow R_3R'N^{\oplus}X^{\ominus}$$

Rubbers which contain vinylpyridines undergo significant changes in physical properties when they are blended with organic halogen compounds and heated to vulcanising temperatures. These changes are attributed primarily to salt formation at the nitrogen atom of the pyridine nucleus of the copolymerised vinylpyridine units, the salt-forming reaction being analogous to that given above for simple amines. The process is therefore described as 'quaternisation' of the vinylpyridine rubber. Amongst the many organic halogen compounds which can be used for this purpose are benzyl chloride (XXXVIII), α,α',α''-trichlorotoluene (XXXIX) and chloranil (tetrachloro-p-benzoquinone) (XL). The principal

XXXVIII   XXXIX   XL

advantageous change which quaternisation is said to bring about in a vinylpyridine rubber is an improvement in resistance to swelling by organic liquids. In the case of quaternisation by a compound such as benzyl chloride, this is attributed exclusively to the increased polarity of the polymer chain brought about by ionisation of some of the nitrogen atoms and the presence of the associated counterions. In the case of quaternisation by poly-halogen organic compounds such as chloranil, there are several other possibilities, including the formation of ionic crosslinks arising from quaternisation, by the same molecule, of nitrogen atoms in different polymer chains. There is, however, some doubt as to whether quaternisation is always accompanied by an increase in resistance to swelling.

5.5 SOLUTION-POLYMERISED BUTADIENE RUBBERS (BR)

5.5.1 Introduction
Polybutadiene produced by emulsion polymerisation is of little interest as a dry rubber, although it has found some application in the manufacture of high-impact polystyrene, and as a latex, where it has been used in the form of a concentrate as a partial or complete replacement for natural rubber latex. The principles which underly the manufacture of emulsion-polymerised butadiene rubber are identical with those which underly the manufacture of emulsion-polymerised styrene–butadiene rubber, and so

do not need to be considered here. The structure of emulsion-polymerised polybutadiene is almost identical with that of the butadiene part of a styrene–butadiene rubber produced under similar conditions.

The butadiene rubbers which are currently available commercially are manufactured by two different types of process. In both of these processes, the polymerisation is carried out in the strict absence of water and, indeed, of all active-hydrogen compounds. Both processes can be loosely described as 'solution' polymerisations, and in both processes the polymerisation proceeds by an ionic mechanism. The one type of process uses lithium metal or an alkyllithium as the polymerisation initiator. The other type of process uses a Ziegler–Natta catalyst, that is a complex prepared from an organo-metallic compound (such as triethylaluminium) and a transition-metal compound (such as titanium trichloride) (see Section 4.2.7 of Chapter 4). Polymerisations initiated by lithium metal and alkyllithiums are properly described as 'solution' polymerisations, because the propagation step of the polymerisation takes place in solution in the monomer and hydrocarbon solvents. Although it is convenient to apply the adjective 'solution' to polymerisations effected by Ziegler–Natta catalysts, these are not usually true solution polymerisations, since the species responsible for the catalysis are frequently not soluble in the monomer. The polymerisation often takes place at the surface of the catalyst precipitate, and, indeed, it is probably the case that these catalysts owe their unusual behaviour to their essentially heterogeneous nature.

5.5.2 Production of Butadiene Rubber by Polymerisation Initiated by Lithium Metal and by Alkyllithiums

The polymerisation of butadiene can be initiated by all the alkali metals from lithium to caesium. In all cases, the polymerisation proceeds by a mechanism which is essentially anionic in nature. An outline of the mechanism of anionic polymerisation involving the opening of carbon–carbon double bonds has been given in Chapter 4 (Section 4.2.5). In what follows here, further details will be given of the way in which the anionic mechanism operates in the case of butadiene polymerisation initiated by lithium metal and by alkyllithiums. In particular, an attempt will be made to explain the more important features of these polymerisations, which have made them so useful for the production of butadiene rubbers.

The microstructure of the polybutadiene obtained by alkali-metal catalysis varies with the alkali metal, as shown in Table 5.9 (which also includes data for emulsion-polymerised polybutadiene for purposes of

TABLE 5.9

MICROSTRUCTURES OF POLYBUTADIENES OBTAINED USING VARIOUS ALKALI-METAL INITIATORS IN PENTANE AT $0\,°C^{17}$ AND OF POLYBUTADIENE OBTAINED BY EMULSION POLYMERISATION AT $5\,°C^{11}$

Initiator	Polybutadiene microstructure		
	% 1,2	% cis-1,4	% trans-1,4
Lithium	13	35	52
Sodium	65	10	25
Potassium	45	15	40
Rubidium	62	7	31
Caesium	59	6	35
Free radical in emulsion	18·3	14·4	67·3

comparison). Whilst each of the alkali metals gives a polybutadiene having a different microstructure from that given by the others, the greatest difference is between the behaviour of lithium and the behaviour of the other alkali metals. Initiation by lithium gives a polymer having a much lower content of 1,2 microstructures, and in consequence, a much higher content of 1,4 microstructures, than does initiation by the other alkali metals. That lithium should behave rather differently from the other alkali metals in respect of the polymerisation of butadiene is not surprising; the first member of the elements in a given Group of the Periodic Table frequently differs somewhat in behaviour from the others, and nowhere is this more clearly demonstrated than in the case of the chemistry of lithium compared to that of the other alkali metals. The unusual behaviour of lithium is attributed to the small size of the lithium atom and, especially, of the lithium cation. One consequence of the small size of the cation is that the strength of the electric field which surrounds the cation is greater than for the other alkali-metal cations, because the charge carried by the ion is in effect spread out over a smaller area of cation surface. In relation to the initiation of butadiene polymerisation, it is believed that the small size of the lithium cation means that the lithium–carbon bonds which form during the polymerisation (see below) do not ionise or dissociate to the same extent as do the metal–carbon bonds which are formed by the other alkali metals. The lithium–carbon bonds formed in the course of the polymerisation are therefore more covalent in character than are the bonds formed between carbon and the other alkali metals. It appears that ionisation of the metal–carbon bond, with consequent increase in the separation between the metal cation and the carbanion, favours the enchainment of butadiene units by

1,2 addition. A practical consequence of the differences noted in Table 5.9 is that, of the alkali-metal-polymerised polybutadienes, it is only those produced by initiation with lithium which have proved useful as rubbers in the Western world, although sodium initiation is said to be used for the production of low-molecular-weight liquid polybutadienes intended for use as resins. It is understood that polybutadiene produced by sodium initiation is still a major industrial rubber in the U.S.S.R. At the present time, the higher alkali metals appear to be quite unsuitable for the industrial initiation of butadiene polymerisation for synthetic rubber production. There are at least three reasons for this: (a) the molecular weight of the polymer formed decreases as the alkali metal becomes more electropositive, (b) the higher alkali metals are more costly, and (c) the higher alkali metals are more hazardous to use.

The microstructure of the polybutadiene obtained using an alkali-metal alkyl as initiator is similar to that which is obtained using the corresponding alkali metal. However, the polymer formed by alkali-metal alkyl initiation has a narrower distribution of molecular weights than does that produced by initiation with the corresponding alkali metal. In the case of initiation by the alkyllithiums, the polymer microstructure is independent of the initiator concentration, the monomer concentration, the extent of conversion, and the polymerisation temperature over the range 4–80 °C. However, as will be described subsequently, the microstructure is very much affected by the presence of small quantities of polar additives such as ethers.

A consequence of the small size of the lithium cation is that the alkyllithiums tend to be associated in non-polar solvents to give aggregates. The degree of association (that is, the average number of alkyllithium molecules per aggregate) depends upon the alkyl group, the n-alkyls being more highly associated than are those whose alkyl groups are branched at the first or second carbon atom from the lithium. For n-butyllithium, for example, the degree of association is approximately six. The first step in the initiation of polymerisation by alkyllithiums is thought to be the dissociation of the alkyllithium aggregate into the corresponding monomeric species:

$$(LiR)_x \rightleftharpoons (LiR)_{x-1} + LiR$$

As indicated, this dissociation is a reversible process. Initiation is believed to involve the addition of a butadiene molecule to a monomeric alkyllithium molecule by a reaction of the following type:

$$LiR + CH_2{=}CH{-}CH{=}CH_2 \longrightarrow R{-}CH_2{-}CH{=}CH{-}CH_2^{\ominus} \cdots Li^{\oplus}$$

In the structure which is formed in this reaction, $-CH_2^{\ominus} \cdots Li^{\oplus}$ represents a partially-ionised, partially-dissociated lithium–carbon bond. The lithium atom is believed to be essentially permanently associated with the carbon atom, and because of this is able to control the way in which the subsequent butadiene units become enchained during the consecutive steps of the propagation reaction. The typical propagation step is as follows:

$$\sim CH_2^{\ominus} \cdots Li^{\oplus} + CH_2{=}CH{-}CH{=}CH_2 \longrightarrow$$

$$\sim CH_2{-}CH_2{-}CH{=}CH{-}CH_2^{\ominus} \cdots Li^{\oplus}$$

An interesting question is why catalysis by alkyllithiums favours the formation of 1,4-polybutadienes, whereas this is not so in the case of alkyls derived from the other alkali metals. The most plausible explanation is that, because of its small size, the lithium cation encourages the formation of a six-membered ring with the incoming butadiene molecule. Having become part of a six-membered ring, the butadiene unit is then more likely to become enchained as a 1,4 unit than as a 1,2 unit. According to this view, the detailed mechanism of each propagation step is according to a scheme such as the following:

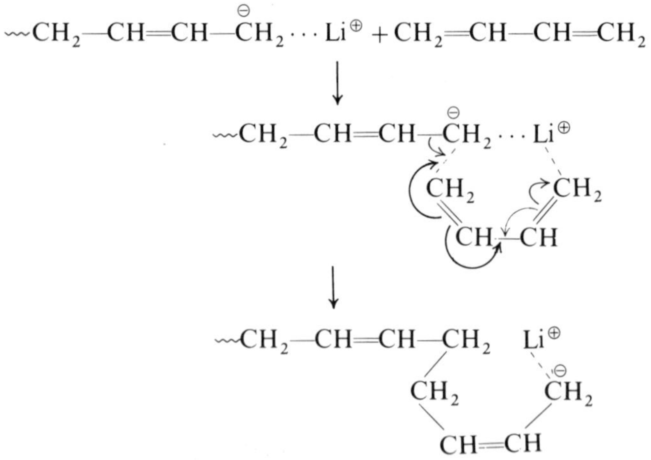

The other alkali-metal alkyls show less tendency to promote 1,4 addition because they do not display that propensity towards divalency which is necessary for the formation of the six-membered rings shown above.

An important feature of polymerisation initiated by alkyllithiums is that they do not in themselves have any inherent termination mechanism. It has already been noted in Chapter 4 that in this respect, as in several others, they contrast markedly with free-radical polymerisations. Such termination of propagating chains as does occur in an alkyllithium-initiated polymerisation is brought about by adventitious impurities in the reaction system (e.g., active-hydrogen compounds, carbon dioxide, etc.). Otherwise, a given active centre propagates for as long as monomer is available. This feature, together with the fact that all the active centres are available from the commencement of the polymerisation, leads to the formation of a polymer which has a very narrow molecular-weight distribution. Indeed, in respect of processing behaviour, a narrow distribution of molecular weights can be a disadvantage. For this reason, interest has been shown in 'modifying agents' which can be added to alkyllithium-initiated butadiene polymerisation systems in order to obtain rubbers of broadened molecular-weight distribution which band better during roll milling, mix faster and better in an internal mixer, extrude faster and more smoothly, and have reduced cold flow. One such group of additives which has been developed is the lithium alkenoxides of general structure (XLI), in which each R is hydrogen or a hydrocarbon moiety. A typical example is lithium allyloxide.

$$CH_2\!\!=\!\!\underset{\underset{R}{|}}{\overset{\overset{R}{|}}{C}}\!\!-\!\!\underset{\underset{R}{|}}{\overset{\overset{R}{|}}{C}}\!\!-\!\!O^{\ominus}Li^{\oplus}$$

(XLI)

The preferred range for the mole ratio lithium alkenoxide/alkyllithium initiator is 2:1 up to 5:1.

The kinetics of butadiene polymerisation initiated by alkyllithiums in non-polar solvents are relatively simple. The rate of polymerisation is first-order with respect to both monomer and initiator concentrations. The rate at which polymerisation is initiated depends upon the structure of the alkyl group of the alkyllithium. For the butyllithiums, the order of decreasing reactivity is as follows:

$$CH_3CH_2\underset{\underset{CH_3}{|}}{CH}\!\!-\; > \;CH_3\overset{\overset{CH_3}{\diagdown}}{\underset{\diagup}{C}}\!\!-\; >$$

sec-butyl   t-butyl

$$\overset{CH_3}{\diagdown}\underset{\diagup}{\underset{CH_3}{}}CHCH_2\!\!-\; > \;CH_3CH_2CH_2CH_2\!\!-$$

iso-butyl   n-butyl

n-Butyllithium is probably the alkyllithium which has been most widely used for the production of rubbery polybutadienes. Unlike the other butyllithiums, it is indefinitely stable in hydrocarbon solution at room temperature. Although it is advantageous to use sec-butyllithium because it gives a higher rate of polymerisation, it is less stable than n-butyllithium and so presents problems of storage and transport.

The mechanism of butadiene polymerisation initiated by lithium metal is similar to that initiated by the alkyllithiums. The main difference is that, whereas an organo-metallic compound is already present in the reaction system when an alkyllithium is used, the organo-metallic compound has to be formed by reaction between the butadiene and the lithium when lithium metal is used as the initiator. This leads to a slower rate of polymerisation and to a broader distribution of molecular weights in the polymer which is formed. Both of these features arise because the active centres for the polymerisation are not now all present at the beginning of the reaction, but are formed gradually at the same time as the polymerisation itself is proceeding by way of the active centres which were formed earlier in the reaction. The first step of initiation by lithium metal is probably as follows:

$$CH_2{=}CH{-}CH{=}CH_2 + Li \longrightarrow \cdot CH_2{-}CH{=}CH{-}CH_2^\ominus \cdots Li^\oplus$$

This reaction involves the transfer of the single outer electron of the lithium atom to a butadiene molecule. The latter acquires a negative charge in consequence, and so becomes a radical anion whose most probable structure is that shown above. The reaction may then proceed in one of two different ways, both of which lead to the formation of a di-anionic species. In the first of these, reaction occurs with a further atom of lithium:

$$Li + \cdot CH_2{-}CH{=}CH{-}CH_2^\ominus \cdots Li^\oplus \longrightarrow$$

$$Li^\oplus \cdots {}^\ominus CH_2{-}CH{=}CH{-}CH_2^\ominus \cdots Li^\oplus$$

This reaction is likely to occur whilst the initial radical anion is still in the vicinity of the surface of the lithium metal. The second possibility is dimerisation of the initial radical anion:

$$Li^\oplus \cdots {}^\ominus CH_2{-}CH{=}CH{-}CH_2 \cdot + \cdot CH_2{-}CH{=}CH{-}CH_2^\ominus \cdots Li^\oplus$$

$$\downarrow$$

$$Li^\oplus \cdots {}^\ominus CH_2{-}CH{=}CH{-}CH_2{-}CH_2{-}CH{=}CH{-}CH_2^\ominus \cdots Li^\oplus$$

Once an organo-metallic compound has been formed, propagation takes place from all the anionic centres simultaneously in a manner similar to that which has been described for polymerisation by alkyllithiums.

Reference has been made to the fact that the microstructure of polybutadienes obtained by alkyllithium or lithium-metal-initiated polymerisation is markedly affected by the presence of small quantities of polar additives such as ethers. This effect is illustrated in Fig. 5.23, from which it is apparent that the presence of these substances can increase markedly the proportion of 1,2 microstructures which is formed in the course of the reaction. Indeed, Fig. 5.23 shows that, by introducing small quantities of diethylene glycol dimethyl ether or tetramethyl ethylene diamine into a butyllithium-initiated polymerisation in a non-polar hydrocarbon solvent, it is possible to obtain polybutadienes of almost any desired content of 1,2 microstructures. The reason why polar additives can have such a large effect upon the structure of the polybutadiene produced by lithium catalysis is that they can coordinate with the lithium (via, for example, a nitrogen or oxygen atom in the additive), thereby increasing the effective size of the cation and increasing the degree of dissociation and ionic character of the $—CH_2^{\ominus} \cdots Li^{\oplus}$ bond. Catalysis by organo-lithium initiators thus becomes similar to catalysis by other alkali-metal compounds. A polybutadiene of very high 1,2 content (c. 96%) can be made by polymerising butadiene at 0 °C in an ether solvent (tetrahydrofuran) using as initiator the electron-transfer complex which forms between lithium and naphthalene. It is interesting that the other alkali-metal–naphthalene complexes also give polybutadienes of predominantly 1,2 microstructure

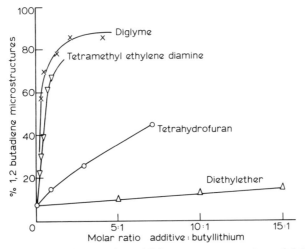

FIG. 5.23. Effect of various polar additives upon proportion of 1,2 microstructures formed during butyllithium-initiated polymerisation of butadiene.[18]

under similar conditions; thus the distinctive character of lithium-initiated polymerisation is again lost. However, the proportion of 1,2 micro-structures falls as the electronegativity of the alkali metal increases. A plausible mechanism for the enchainment of butadiene in both 1,2 and 1,4 modes by lithium-ion catalysis is as follows:

$$\sim CH_2-CH=CH-CH_2^{\ominus}\cdots Li^{\oplus} \xrightarrow{\;CH_2=CH-CH=CH_2\;} \text{1,4 addition}$$

$$\Updownarrow$$

penultimate unit enchained
in 1,2 mode

1,4 addition

penultimate unit enchained
in 1,2 mode

5.5.3 Production of Butadiene Rubber by Polymerisation Initiated by Ziegler–Natta Catalysts

The bewildering array of Ziegler–Natta catalyst combinations which is available for the polymerisation of butadiene is matched by a similar diversity in the microstructure of the polybutadiene which is obtained. Table 5.10 summarises the microstructures which are produced by a small selection of the catalysts which are available. At the present time, notwithstanding the enormous amount of information which exists concerning the subject, our knowledge of the correlations between the catalyst composition and the microstructure of the polybutadiene produced is essentially empirical; it is not possible to predict from theoretical considerations which catalysts will produce a given type of microstructure. Some systematic dependencies of polymer microstructure upon catalyst composition are observed. One example is given in Fig. 5.24; this shows the

TABLE 5.10

MICROSTRUCTURES OF POLYBUTADIENES OBTAINED USING VARIOUS ZIEGLER–NATTA CATALYST SYSTEMS[1]

Catalyst system	Polybutadiene microstructure		
	% 1,2	% cis-1,4	% trans-1,4
$R_3Al + VCl_4$	2–3		97–98
$R_3Al + TiCl_4$	2	49	49
$R_3Al + TiCl_4$	5	70	25
$R_3Al + TiI_4$ (5 °C)	2	95	3
$R_3Al + TiI_4$ (50 °C)	5	92	4
$R_3Al + TiBr_4$	9	80–88	3–11
$R_3Al + Ti(OBu)_4$	90	0	0–10
$R_2AlCl + CoCl_2$	3–4	93–94	3
$R_2AlCl + CoBr_2$	1–2	95–97	2–3
$R_2AlCl + CoI_2$	2–4	94–95	3–5
$R_2AlCl + Co(naphthenate)_2$	1–3	94–97	2–3
$R_2AlCl + Co(acetylacetonate)_3$	2	95	3
$LiAlH_4 + TiI_4$	4	9	86
$R_3Al + VOCl_3$	1		99
$R_3Al + VCl_3$	2–3		97–98
$R_2Mg + TiI_4$		92	
$RMgI + TiCl_4$		92	
$R_2AlCl + NiCl_2$		90	
$R_2AlH + CrCl_3$			high
$R_3Al + ZrI_4$			high
$Et_3Al + Cr(acetylacetonate)_3$	70	12	18

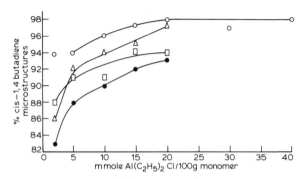

FIG. 5.24.   Effect of diethylaluminium chloride concentration upon proportion of cis-1,4 microstructures formed during polymerisation of butadiene initiated by diethylaluminium chloride and a cobalt(II) chloride–pyridine complex.[1] Concentrations of cobalt(II) chloride–pyridine complex in mmole per 100 g of monomer: ● 0·05, □ 0·04, △ 0·03, ○ 0·02.

effect of catalyst composition upon the cis-1,4 content of polybutadiene obtained using a catalyst prepared from diethylaluminium chloride and a cobalt(II)–pyridine complex.

A schematic flow diagram for the production of high-cis-polybutadiene by solution polymerisation using an aluminium–cobalt Ziegler–Natta catalyst combination is given in Fig. 5.25. The catalyst components are dissolved or slurried in dried hydrocarbon solvent. The butadiene and solvent are mixed and dried by passing them through alumina, silica or molecular sieves. The catalyst components, monomer and solvent are continuously pumped into agitated polymerisation vessels. Usually, several reactors in series are used, each being designed to agitate and pump solutions of increasing viscosity (as increasing polymerisation occurs). The liquid which effluxes from the final reactor contains 7–25% solids. It is shortstopped by the addition of an antioxidant (which is usually rather conveniently an active-hydrogen compound). The solvent and unreacted monomer are removed either by flash or steam distillation, and are then recycled. The polymer is then steam-stripped and dewatered. Drying and baling of the rubber is accomplished as for emulsion-polymerised styrene–butadiene rubber.

5.5.4 Effect of Polymer Microstructure upon Properties of Butadiene Rubbers

As shown in Fig. 5.26, the glass-transition temperature of a butadiene rubber increases almost linearly with the proportion of butadiene units

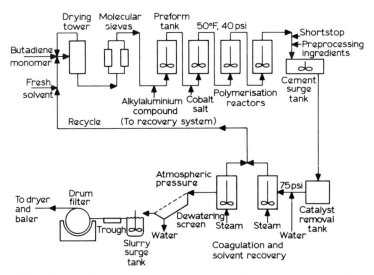

FIG. 5.25. Schematic flowsheet for production of high-*cis*-1,4-polybutadiene by solution polymerisation using an aluminium–cobalt Ziegler–Natta catalyst.[1]

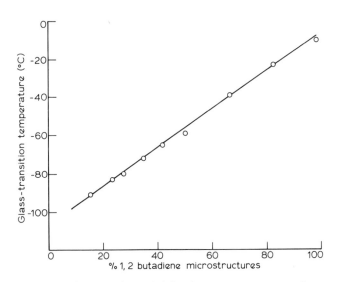

FIG. 5.26. Effect of proportion of 1,2 microstructures upon glass-transition temperature of polybutadiene.[19]

which are enchained by 1,2 addition. This is of particular interest in connection with the application of the rubber in tyre-tread compounds, because the skid resistance of tyre treads is known to improve as the glass-transition temperature of the rubber upon which they are based is increased. The ability to vary the glass-transition temperature of a general-purpose hydrocarbon rubber systematically, as implied by Fig. 5.26, offers the prospect of being able to optimise the balance of tyre properties such as good abrasion resistance and good road-holding ability in a single elastomer (as opposed to using a blend of elastomers).

A few generalisations concerning the effect of polymer microstructure upon the mechanical properties of butadiene rubber vulcanisates can be noted. The ratio of cis-1,4 to trans-1,4 microstructures has little effect in the range 25–80 % cis-1,4 content. Vulcanisates of rubbers having high cis-1,4 content are characterised by high resilience and low hysteresis. The very high cis-1,4 rubbers (> 96 %) crystallise on stretching, and so, like natural rubber, give gum stocks having high tensile strength. As might be expected, the microstructure of a butadiene rubber has a pronounced effect upon its tendency to crystallise, and this has important consequences for the mechanical properties. Crystallinity has been observed in high cis-1,4-, high trans-1,4- and high 1,2-polybutadienes. The melting point of 100 % cis-1,4 polybutadiene is c. 4 °C; the melting point decreases with decreasing cis-1,4 content. Thus most high cis-1,4-polybutadienes do not crystallise on stretching at room temperature, and so pure gum vulcanisates do not show high tensile strength. Polymers of high trans-1,4 content are crystalline, give gum vulcanisates of high tensile strength, high hysteresis and low resilience. Amorphous rubbers of high 1,2 content (c. 95 %) have moderate tensile strength.

5.5.5 Types of Butadiene Rubber Produced Commercially

The types of butadiene rubber which are currently available commercially can be classified broadly under three headings:

1.   low cis-1,4-polybutadienes (i.e., approx. 40 % cis-1,4) produced by lithium-initiated polymerisation;

2.   medium–high cis-1,4-polybutadienes (i.e., approx. 92 % cis-1,4) produced by Ziegler–Natta catalysis using combinations such as a trialkylaluminium and titanium tetraiodide; and

3.   high cis-1,4-polybutadienes (i.e., approx. 97 % cis-1,4) produced by Ziegler–Natta catalysis using combinations such as a dialkyl-aluminium chloride and cobalt(II) bromide.

The distribution of molecular weights in these rubbers depends primarily upon the catalyst system or initiator used in their preparation. Those prepared by alkyllithium initiation have the narrowest distribution of molecular weights. The order of increasing molecular-size heterogeneity is then: polymers prepared by titanium-based Ziegler–Natta catalysts, polymers prepared by cobalt-based Ziegler–Natta catalysts, and, finally, emulsion-polymerised polybutadienes.

A problem which has been encountered with low *cis*-1,4 butadiene rubbers is that of 'cold flow'. The rubbers tend to behave as very viscous liquids at low shear rates, and this can cause problems during storage. Some grades of rubber produced by alkyllithium catalysis—the so-called 'non-flow grades'—have a small amount of crosslinking induced during polymerisation. This can be achieved by including a small amount of crosslinking monomer (such as divinylbenzene) in the monomer feed. The amount of crosslinking induced is sufficient to inhibit cold flow, but is not sufficient to induce significant gel formation in the polymer, nor to affect the processability significantly.

A further point to note concerning the various types of commercially available butadiene rubber is that those produced by Ziegler–Natta catalysis may contain residues of transition-metal compounds. Some of these can be very powerful catalysts of oxidative degradation. It is therefore essential to ensure the presence of sufficient antioxidant in the vulcanisate to afford adequate protection.

Solution-polymerised butadiene rubbers can be extended with oils. Aromatic oils are preferred for this purpose.

5.5.6 Technology of Butadiene Rubbers

The technology of butadiene rubber is generally similar to that of styrene–butadiene rubbers, although there are some significant differences. The vulcanisation of butadiene rubbers is achieved by heating with sulphur and conventional accelerators. The sulphur and accelerator requirements are intermediate between those of natural rubber and those of styrene–butadiene rubbers. As compared with natural rubber, rather less sulphur and more accelerator is required. The level of fatty-acid activator should be similar to that used for natural rubber; for unlike synthetic rubbers produced by emulsion polymerisation, those produced by solution polymerisation do not contain any fatty acids which have originated from the polymerisation reaction system. Unlike most of the other diene rubbers, some butadiene rubbers cannot be conveniently vulcanised by means of organic peroxides. Instead of a rubbery vulcanisate, a resinous, densely-crosslinked material tends to be produced. It appears that, instead of inducing controlled crosslinking, the decomposing peroxide induces free-radical polymerisation through the carbon–carbon double bonds of the polybutadiene chain. Furthermore, polymerisation seems to occur through the double bonds of the vinyl side-chains which arise from 1,2 addition, rather than through the main-chain double bonds which arise from 1,4 addition. Thus, whereas it is possible to obtain rubbery peroxide vulcanisate from butadiene rubbers of high 1,4 content, the tendency to glass formation increases as the 1,2 content of the butadiene rubber increases.

For general-purpose applications, butadiene rubbers are normally used as blends with other rubbers. This is partly because the processing behaviour of butadiene rubbers alone is poor, and they are difficult to handle with conventional factory equipment. Furthermore, a better balance of vulcanisate properties may be achieved by blending with other rubbers. Butadiene rubbers are very resistant to breakdown during mastication, either on an open mill or in an internal mixer. Chemical peptisers have little effect upon the polymer plasticity. Some curious transitions in milling behaviour are observed as the temperature is varied. Thus, in the case of the high *cis*-1,4 polybutadienes, the rubber bands coherently and smoothly on the mill at temperatures below about 40 °C, whereas above that temperature the rubber loses its cohesion and sags on the mill. In the case of butadiene rubbers produced by alkyllithium initiation, they process badly on a cool mill, but the processing becomes progressively better as the temperature is raised, becoming quite satisfactory at temperatures above about 65 °C. Butadiene rubbers having high Mooney viscosities are difficult to process whatever the temperature.

5.6   SOLUTION-POLYMERISED STYRENE–BUTADIENE RUBBERS

As noted in the Introduction to this chapter, we are concerned here with essentially random copolymers of styrene and butadiene produced by solution polymerisation. The block copolymers will be dealt with in a later chapter (Chapter 10). Random copolymers of styrene and butadiene can be, and indeed are commercially, produced by alkyllithium initiation. Special precautions are, however, necessary, because, if the copolymerisation is carried out in non-polar solvents, the butadiene in the mixture first polymerises to the virtual exclusion of the styrene, and then, when the butadiene has been consumed, the styrene polymerises. This phenomenon will be described and discussed in more detail subsequently (see Section 10.2.1 of Chapter 10). All that needs to be noted here is that such a reaction will clearly lead to the formation of a block copolymer and not a random copolymer. Various procedures have been adopted for overcoming this problem when it is desired to produce random copolymers. Thus the butadiene monomer may be added to the system incrementally, so that the molar ratio of unreacted styrene to unreacted butadiene is always high. Another approach is to add the monomer mixture to an initiator-containing solution at a slower rate than the rate of polymerisation, that is the polymerisation is carried out under essentially 'monomer-starved' conditions. The result of the reaction should then be an oft-repeated sequence of short blocks of butadiene and styrene units, rather than one long block of butadiene units followed by one long block of styrene units. However, the most interesting approach to this problem from the chemical point of view is that which makes use of 'randomising' additives. These can be of the ether or amine type, in which case a rubber with a relatively high content of 1,2 microstructures will be obtained (and therefore it will have a rather higher glass-transition temperature than if the 1,2 content were low). Potassium t-butoxide has also been proposed as a randomising agent, and apparently this does not cause the formation of a rubber with significantly enhanced content of 1,2 microstructures. (Of course, the random copolymers produced by simple alkyllithium initiation with adjustment of the monomer feed rate do not have an enhanced content of 1,2 microstructures either.)

The technology of the solution-polymerised styrene–butadiene rubbers is broadly similar to that of the emulsion-polymerised rubbers. However, the solution polymers have a much narrower molecular-weight distribution than do the emulsion polymers, and this has consequences for the processing behaviour. Very low-molecular-weight polymer is absent from the solution polymer, as also is polymer of very high molecular weight and polymer gel. Furthermore, the solution polymers are essentially unbranched. A further point to note is that the solution polymer does not contain fatty acids or rosin acids, and allowance should be made for this when compounding. Solution-polymerised styrene–butadiene rubbers are available as oil-extended grades; the loadings of oil are of the order of 40 pphr, the oil being aromatic.

An interesting development has been that of solution-polymerised styrene–butadiene rubbers which are susceptible to mechanical degradation in the presence of a suitable peptising agent. In one such development, three or four of the 'living' polymer molecules are coupled together by reaction with a suitable tin compound, thereby giving a star-shaped molecule which can be degraded by shearing in the presence of a reagent such as stearic acid. Degradation is believed to occur at the tin-coupled sites.

5.7   'ALFIN' RUBBERS

This chapter concludes with brief reference to an unusual type of butadiene rubber which was first prepared about 30 years ago but which, notwithstanding the interest which has been shown in it from time to time, has never been developed commercially. The complex mixtures which are used to polymerise butadiene to these rubbers are known as 'Alfin' catalysts. This name is derived by combination of parts of the two words 'alcohol' and 'olefin'; it will be seen from what follows that these are two of the components of the catalyst.

A typical Alfin catalyst is prepared by reacting amyl chloride and sodium metal in pentane at about $-15\,°C$, giving amylsodium and sodium chloride. Isopropyl alcohol is then added in sufficient quantity to react with about half of the amyl sodium and form sodium isopropoxide in a finely-divided condition. Propylene is then bubbled into the mixture to form allylsodium from the remainder of the amylsodium. The reactions which occur during the formation of the catalyst can be summarised as follows:

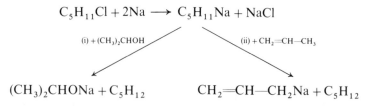

$$C_5H_{11}Cl + 2Na \longrightarrow C_5H_{11}Na + NaCl$$

(i) $+ (CH_3)_2CHOH$ (ii) $+ CH_2{=}CH{-}CH_3$

$$(CH_3)_2CHONa + C_5H_{12} \qquad CH_2{=}CH{-}CH_2Na + C_5H_{12}$$

The result of this series of reactions is a heterogeneous mixture of sodium chloride, sodium isopropoxide and allylsodium. It is capable of polymerising monomers such as butadiene, isoprene and styrene very rapidly to polymers of extremely high molecular weight. Thus the catalyst prepared as described above is capable of polymerising butadiene in pentane at $30\,°C$ to a polymer of weight-average molecular weight as high as $c.\ 10^7$ in a matter of minutes. The microstructure of the polybutadiene produced is approximately 10% cis-1,4, 70% trans-1,4 and 20% 1,2. The microstructure is thus very different from that obtained by polymerisation initiated by sodium metal. In fact, the microstructure is similar to that of polybutadienes obtained by emulsion polymerisation. This has led to the suggestion that the reaction occurs by a free-radical mechanism rather than by an ionic mechanism.

The mechanism of polymerisation by Alfin catalysts is not well understood. There are several unusual features of the reaction. In the first place, polymers of extremely high molecular weight are formed very fast, much faster in fact than polymerisations brought about by conventional sodium initiators. Then there is the question of the microstructure of the polymer which is formed; this has already been referred to in the previous paragraph. A third curious feature is that the presence of sodium chloride is essential to the functioning of the Alfin catalyst. Furthermore, the sodium chloride must be produced by reaction between sodium metal and amyl chloride. It has been proposed that the catalyst has the following type of structure:

However, this structure fails to explain the function of the sodium chloride.

Because of their very high molecular weight, polybutadienes made using Alfin catalysts are too tough for the manufacture of rubber products by conventional processes. They can be oil extended, and thereby softened, but excessive amounts of oil are required. A significant development has been the discovery that the molecular weight of Alfin polybutadienes can be greatly reduced if a small amount of a suitable dihydroaromatic compound is present during the polymerisation. Two such compounds are 1,4-dihydrobenzene (XLII) and 1,4-dihydronaphthalene (XLIII). When used

(XLII)   (XLIII)

for this purpose, such compounds are known as 'modifiers' or 'moderators', by analogy with the behaviour of mercaptans in emulsion polymerisation reactions. There is evidence to suggest that, like mercaptans, they bring about a reduction in molecular weight by acting as transfer agents. Whatever are the underlying reactions involved, it appears that the presence of 1.5% of 1,4-dihydronaphthalene in the butadiene gives a polybutadiene of molecular weight $c.\ 3 \times 10^5$. This corresponds to a Mooney viscosity of approximately 40, and brings the polymer well within the range of processability. Whether or not this development will lead eventually to the commercial exploitation of Alfin rubbers remains to be seen. The use of modifiers does not appear to influence significantly the microstructure of the polymer produced.

GENERAL BIBLIOGRAPHY

Whitby, G. S., Davis, C. C. and Dunbrook, R. F. (Eds.) (1954). *Synthetic Rubber*, John Wiley and Sons, New York, Chapters 7, 11, 14 and 23.

Saltman, W. M. (1965). Butadiene polymers. In: *Encyclopedia of Polymer Science and Technology*, Vol. 2, John Wiley and Sons, New York, pp. 678 f.

Storey, E. B. (1961). Oil extended rubbers, *Rubb. Chem. Technol.*, **34**, 1402.

Janssen, H. J. J. and Weinstock, K. V. (1961). Carbon black–latex masterbatches, *Rubb. Chem. Technol.*, **34**, 1485.

Powers, P. O. (1963). Resins used in rubber, *Rubb. Chem. Technol.*, **36**, 1542.

Hofmann, W. (1964). Nitrile rubber, *Rubb. Chem. Technol.*, **37** (2[2]), 1.

Dunn, J. R., Coulthard, D. and Pfisterer, H. A. (1978). Advances in nitrile rubber technology, *Rubb. Chem. Technol.*, **51**, 389.

Saltman, W. M. (Ed.) (1977). *The Stereo Rubbers*, John Wiley and Sons, New York, Chapters 2, 3 and 4.

REFERENCES

1. Saltman, W. M. (1965). *Encyclopedia of Polymer Science and Technology*, Vol. 2, John Wiley and Sons, New York, p. 678.
2. Shundo, M., Hidaka, T., Goto, K., Imoto, M. and Minoura, Y. (1968). *J. Appl. Polym. Sci.*, **12**, 975.
3. Storey, E. B. and Williams, H. L. (1951). *Rubber Age*, **68**, 571.
4. MacLean, D. B., Morton, M. and Nicholls, R. V. V. (1949). *Ind. Engng. Chem.*, **41**, 1622.
5. Gordon, M. and Taylor, J. S. (1952). *J. Appl. Chem.*, **2**, 493.
6. Maher, E. D. and Davies, T. L. (1946). *Rubber Age*, **59**, 557.
7. Schulze, W. A. and Crouch, W. W. (1948). *Ind. Engng. Chem.*, **40**, 151.

8. Bebb, R. L., Carr, E. L. and Wakefield, L. B. (1952). *Ind. Engng Chem.*, **44**, 724.
9. Howland, L. H., Messer, W. E., Neklutin, V. C. and Chambers, V. S. (1949). *Rubber Age*, **64**, 459.
10. Meyer, A. W. (1949). *Ind. Engng Chem.*, **41**, 1570.
11. Hampton, R. R. (1949). *Anal. Chem.*, **21**, 923.
12. Hansen, E. B. and Church, F. C. *A Practical Approach to Carbon Black Selection for Rubber Compounding*, Cabot Corporation Technical Report RG-127.
13. Hofmann, W. (1964). *Rubb. Chem. Technol.*, **37** (2[2]), 1.
14. *Breon Nitrile Rubbers*, BP Chemicals (U.K.) Ltd, London, 1967.
15. *Krynac Oil Resistant Rubbers*, Polysar Ltd, Sarnia, Canada.
16. Frank, R. L. *et al.* (1948). *Ind. Engng Chem.*, **40**, 879.
17. Tobolsky, A. V. and Rogers, C. E. (1959). *J. Polym. Sci.*, **40**, 73.
18. Duck, E. W. and Locke, J. M. (1968). *J. Instn. Rubb. Ind.*, **2**, 223; Duck, E. W. (1973). *Eur. Rubb. J.*, **155**(12), 38.
19. Duck, E. W. and Locke, J. M. (1977). *The Stereo Rubbers*, Saltman, W. M. (Ed.), John Wiley and Sons, New York, p. 139.

Chapter 6

Rubbers Obtained from Butadiene Derivatives

6.1 INTRODUCTION

This chapter deals with rubbers which are produced by the polymerisation of substituted butadienes. There are only two families of such rubbers which are of commercial importance at the present time. These are the rubbers which are obtained by the polymerisation of isoprene and of chloroprene respectively. These two types of rubber are rather disparate as regards nature, technology, and areas of application. Isoprene rubber is a general-purpose hydrocarbon rubber which is intended as a competitor primarily for natural rubber, and to a lesser extent for styrene–butadiene rubber and butadiene rubber. Chloroprene rubber, on the other hand, is a special-purpose rubber which is used in applications which call for such attributes as moderate resistance to swelling in hydrocarbon oils, resistance to heat, and resistance to the action of ozone. Nevertheless, it is convenient and appropriate to consider these two rubbers together, in order to demonstrate how dissimilar two formally-similar rubbers can be. Chloroprene can be regarded as being derived from isoprene by replacing the methyl group of the latter by a chlorine atom, an atom which has almost the same size as the methyl group. Although the one monomer is derived from the other by a very simple substitution, the two rubbers derived from these monomers are very different, as the following pages will show. The origin of these differences is to be found in the electronegativity of the chlorine atom relative to that of the methyl group.

6.2 ISOPRENE RUBBER (IR)

6.2.1 Introduction

The repeat unit of the isoprene rubber molecule is conventionally represented as shown at (XLIV). However, as has been explained in an

164

earlier chapter, the actual structure can be more complex than is indicated at (XLIV) because of the possibility of 1,2 and 3,4 addition, as well as of 1,4 addition.

$$\cdots-CH_2-\underset{\underset{CH_3}{|}}{C}=CH-CH_2-\cdots$$

(XLIV)

Isoprene can be polymerised to a rubbery polymer by emulsion polymerisation. Processes similar to those employed for the emulsion copolymerisation of styrene and butadiene can be used. Thus isoprene can be emulsion polymerised in a reaction system which comprises monomer, water, a fatty-acid soap, a persulphate initiator and a mercaptan modifier. Polymerisation can also be effected by means of redox initiators, such as a combination of diisopropylbenzene hydroperoxide and tetraethylene-pentamine. However, isoprene rubber has never been produced in significant commercial quantities by emulsion polymerisation. There are several reasons for the failure to exploit emulsion-polymerised isoprene rubber as a general-purpose synthetic rubber. The high cost of the monomer and its non-availability in large quantities in the early days of the synthetic rubber industry was an important factor. So too was the fact that the emulsion polymerisation of isoprene is a non-stereospecific reaction. It has not therefore been possible to produce a molecular structure analogous to that of natural rubber by the emulsion polymerisation of isoprene. The product of the emulsion polymerisation reaction is a polymer of the same type of mixed microstructure as that which is obtained by the emulsion polymeris-ation or copolymerisation of the cheaper and more readily-available butadiene. In short, emulsion-polymerised polyisoprene suffered from the disadvantages of butadiene polymers and copolymers produced by emul-sion polymerisation, but did not offer the advantages of the cheapness and availability of the monomer. Isoprene can also be emulsion copolymerised with styrene in a manner similar to butadiene, but the product suffers from the same disadvantages as emulsion-polymerised isoprene rubber. It has therefore never been developed commercially.

Isoprene can also be polymerised by most of the methods which are available for butadiene, in particular, by alkali metals and alkali-metal alkyls, by Ziegler–Natta catalysts, and by Alfin catalysts. Of the methods available, only two are of commercial importance at the present time. These are polymerisation initiated by lithium metal and alkyllithiums, and polymerisation using Ziegler–Natta catalysts. The principles which underlie the use of these initiators and catalysts are similar to those which underlie their application to the polymerisation of butadiene. It will therefore be unnecessary to consider the mechanistic aspects of isoprene polymerisation brought about by these means. As in the case of butadiene, these polymerisations are carried out under strictly anhydrous conditions, and are broadly described as 'solution' polymerisations, even although some of the catalysts may in fact be insoluble in the reaction medium.

6.2.2 Production of Isoprene Rubbers by Polymerisations Initiated by Lithium Metal and by Alkyllithiums

As in the case of butadiene polymerisations, of the various alkali metals and alkali-metal alkyls which are available, it is only lithium metal and the alkyls derived from it which are able to effect the stereospecific polymerisation of isoprene. This is illustrated by the data summarised in Table 6.1. (In considering the information given in this table, it should be recalled that a further distinctive microstructure is possible in polyisoprene in addition to those which are possible in polybutadiene, namely, that which results from 3,4 addition of an incoming isoprene unit.) It will be seen from Table 6.1 that initiation by lithium metal and by alkyllithiums results in *c.* 95 % 1,4 addition. In this respect, the behaviour of these initiators toward isoprene resembles that towards butadiene. However, unlike lithium-initiated butadiene polymerisation, which results in the formation of a mixture of *cis*-1,4 and *trans*-1,4 units, the lithium-initiated polymerisation of isoprene gives a polymer whose 1,4 units have almost exclusively the *cis* configuration. It is this feature which excited great interest in lithium-polymerised polyisoprenes when they were first announced. This is because natural rubber happens to be a stereoregular *cis*-1,4-polyisoprene; there-

TABLE 6.1

MICROSTRUCTURES OF POLYISOPRENES OBTAINED USING VARIOUS ALKALI-METAL AND ALKALI-METAL–ALKYL INITIATORS IN NON-POLAR MEDIA[1]

Initiator	*Solvent*	*Polyisoprene microstructure*			
		% 1,2	% 3,4	% cis-1,4	% trans-1,4
Lithium	Pentane	0	6	94	0
Butyllithium	Pentane	0	7	93	0
Sodium	Pentane	6	51	0	43
Ethylsodium	Pentane	7	45	6	42
Potassium	Bulk	8	40	0	52
Rubidium	Bulk	8	39	3	50
Caesium	Bulk	8	17	8	67

fore, rather fortuitously, the lithium-initiated polymerisation of isoprene produces a polymer having a similar microstructure to that of natural rubber. (It may be added that the mechanism by which the rubber tree produces natural rubber is very different from the lithium-initiated polymerisation of isoprene. Natural rubber is not, in fact, produced in the tree by the addition polymerisation of isoprene, but by the condensation polymerisation of isopentenyl pyrophosphate. Furthermore, the tree produces natural rubber under very mild reaction conditions, and in an aqueous environment which is completely hostile to the operation of lithium initiators.)

Table 6.1 also shows that the polyisoprenes produced using the other alkali metals and their alkyls as initiator have quite different microstructures to those obtained using lithium initiators. In these cases, the polyisoprene is non-stereospecific, consisting mainly of mixtures of *trans*-1,4 units and 3,4 units in roughly equal numbers, except in the case of caesium initiation where *trans*-1,4 units predominate.

Again, as in the case of butadiene polymerisation, the presence of a polar solvent has a marked effect upon the microstructure of the polyisoprene obtained by lithium initiation. This is illustrated in Table 6.2, which also gives information relating to the microstructure of polyisoprenes produced by sodium-metal initiation in non-polar and polar solvents. If the lithium-initiated polymerisation of isoprene is carried out in a polar solvent such as diethyl ether, then the distinctive character of the polymer produced is lost. The product now has a similar microstructure to that of sodium-polymerised polyisoprene produced in a non-polar solvent. Furthermore, the product obtained by sodium initiation in an ether solvent (tetrahydrofuran) does not differ greatly from that produced by polymerisation in a non-polar solvent.

TABLE 6.2

EFFECT OF NATURE OF SOLVENT UPON MICROSTRUCTURE OF POLYISOPRENES OBTAINED USING LITHIUM AND SODIUM AS INITIATORS[1]

Initiator	Solvent	Polyisoprene microstructure			
		% 1,2	% 3,4	% cis-1,4	% trans-1,4
Lithium	Pentane	0	6	94	0
Lithium	Diethyl ether	5	46	0	49
Sodium	Pentane	6	51	0	43
Sodium	Tetrahydrofuran	13	54	0	33

Further information concerning the effect of the solvent upon the microstructure of polyisoprene produced by alkyllithium-initiated polymerisation is given in Table 6.3. This information relates to the effect upon the proportion of isoprene units which become enchained by 3,4 addition. Amongst hydrocarbon solvents, the number of 3,4 units enchained is lowest for aliphatic solvents, rather greater for the polymerisation of isoprene in bulk (in which case the solvent is isoprene itself) and greatest for aromatic solvents. This parallels the increasing dielectric constant of the solvent. Polymerisation in ether solvents produces higher levels of 3,4 addition than does polymerisation in hydrocarbon solvents. Tetrahydrofuran is particularly effective in promoting 3,4 addition.

A further illustration of the importance of the solvent medium in determining the microstructure of polyisoprene produced by lithium-initiated polymerisation is to be found in the marked effects which very small amounts of methyl ethers have upon the microstructure of the product. Again, there are strong similarities with lithium-initiated buta-diene polymerisation. In the case of isoprene polymerisation, some ethers are able to effect significant changes in microstructure when present at the extent of only half of the molar amount of the alkyllithium initiator. Dimethoxyethane is so efficient a promoter of 3,4 addition that there is no longer any *cis*-1,4 addition taking place when it is present at a molar ratio of 100:1 dimethoxyethane:alkyllithium.

The interpretation of these observations in terms of the tendency or otherwise of the $-CH_2^{\ominus} \cdots Li^{\oplus}$ bond to remain undissociated is identical to that which has been given for butadiene polymerisation in the previous

TABLE 6.3

EFFECT OF SOLVENT UPON 3,4 CONTENT OF POLYISOPRENE PREPARED USING *n*-BUTYLLITHIUM AS INITIATOR[2]

Solvent	*% 3,4 Units in polymer*
Cyclohexane	6
Decalin	6
n-Hexane	7
Benzene	7
Isoprene (i.e., bulk polymerisation)	8
Toluene	9
Diphenyl ether	18
Diethyl ether	47
Tetrahydrofuran	54

chapter. The effect of hydrocarbon solvents upon the polymer micro-structure parallels the dielectric constant of the solvent because the dielectric constant influences the ease with which ion pairs can dissociate. Increasing the dielectric constant of the solvent increases the tendency for the $-CH_2^{\ominus} \cdots Li^{\oplus}$ bond to dissociate, and in consequence lessens the steric control which this bond can exercise over the incoming monomer unit. Polar additives and solvents such as ethers affect the microstructure of the polymer produced because they coordinate to the lithium cation, thereby increasing its effective size and again increasing the tendency for the $-CH_2^{\ominus} \cdots Li^{\oplus}$ bond to dissociate.

Other significant facts concerning the polymerisation of isoprene initiated by alkyllithiums are as follows:

1. The microstructure of the polymer formed is essentially inde-pendent of the organic moiety of the alkyllithium, and this generalisation extends to aryllithiums as well, such as phenyl-lithium, tolyllithium and naphthyllithium. Indeed, as Table 6.1 shows, almost the same microstructure is obtained whether lithium metal or an alkyllithium is used for effecting initiation.

2. The extent of monomer conversion has no effect upon the polymer microstructure.

3. Reduction of the polymerisation temperature decreases slightly the extent to which 3,4 addition occurs.

4. Excessive concentrations of initiator may lead to a decrease in the extent of cis-1,4 addition. This has been attributed to the strong tendency of lithium compounds to form association complexes.

Although the nature of the organic moiety of the alkyllithium has little effect upon the microstructure of the polymer produced, it does have an effect upon the rate of polymerisation. Amongst the family of butyl-lithiums, the same correlations between structure and rate of polymerisation hold as for butadiene polymerisation. It is understood that butyllithium is used for the initiation of isoprene polymerisation industrially, and that, of the various isomers, sec-butyllithium is preferred to n-butyllithium because it gives a much faster rate of initiation. This is important for isoprene polymerisation, because isoprene initiates and polymerises with alkyl-lithiums more slowly than does butadiene.

6.2.3 Production of Isoprene Rubber by Polymerisation Initiated by Ziegler–Natta Catalysts

A large number of Ziegler–Natta catalyst combinations is available for effecting the polymerisation of isoprene. Most of these catalysts do not produce stereoregular polyisoprenes. Because the primary commercial interest in polyisoprene has been as a direct replacement for natural rubber, the main effort has been directed towards the discovery of catalysts which produce polymers having a high content of *cis*-1,4 units. Interest has also been shown in the production of polyisoprene of high *trans*-1,4 content; this polymer is used as a replacement for balata and gutta percha which, like natural rubber, are obtained from certain trees.

Much of the literature which relates to the production of *cis*-1,4-polyisoprene by Ziegler–Natta catalysis is concerned with a catalyst system comprising a trialkylaluminium and titanium tetrachloride. It appears that catalyst combinations of this type are used for the commercial production of high-*cis*-1,4 polyisoprene. Much interest has been shown in the catalyst which is formed specifically from triisobutylaluminium and titanium tetrachloride. The ratio of aluminium to titanium in the catalyst strongly influences the microstructure of the polymer which is produced. The preferred atomic ratio is approximately 1:1. This ratio produces polymer with the highest content of *cis*-1,4 microstructure (*c.* 96 %), and it also gives a fast rate of reaction. At lower aluminium/titanium ratios, the yield of polymer decreases and the nature of the polymer changes. At ratios of less than 0·8, the product of the polymerisation is a leathery crosslinked material which has a mixed microstructure. At ratios higher than 1:1, the yield of polymer also decreases, and an increase in the amount of low-molecular-weight material is observed. The polyisoprene which is produced by aluminium/titanium catalysts contains a certain amount of gel of rather variable crosslink density. A certain content of loose gel (that is, gel which breaks down during mastication) can be advantageous in respect of processing behaviour. However, tight gel (which does not break down during mastication) is undesirable because it forms discrete domains in the vulcanisates and thereby leads to inferior vulcanisate mechanical properties.

The production of *cis*-1,4-polyisoprene by Ziegler–Natta catalysis is normally carried out in an inert hydrocarbon solvent. Low-molecular-weight aliphatic solvents, such as *n*-butane or *n*-pentane, are preferred to solvents such as heptane or benzene. Because of their greater volatility, such solvents permit closer control of the temperature at which polymerisation is occurring. Furthermore, the evolution of solvent vapour within the reactor aids agitation. The reaction mixtures produced using these solvents have a lower viscosity than those produced using aromatic solvents or aliphatic solvents of higher molecular weight; this facilitates the removal of catalyst residues which (as has been pointed out in connection with the production of butadiene rubber) are deleterious because they are able to catalyse the oxidative degradation of diene rubbers.

Trans-1,4-polyisoprene can be produced using several Ziegler–Natta catalyst systems. One example is the catalyst formed from triisobutyl-aluminium plus titanium tetrachloride plus iron (III) chloride. The optimum atomic ratio Al:Ti:Fe is 10:1:5. It is used with benzene as solvent at a solvent:monomer ratio of approximately 3:1. Polymerisation is carried out at 50 °C.

6.2.4 Types of Isoprene Rubber Produced Commercially
The types of isoprene rubber which are currently available commercially can be classified broadly under two headings:

1.   those produced by lithium-initiated polymerisation; and
2.   those produced by Ziegler–Natta catalysis.

Isoprene rubbers produced by lithium-initiated polymerisation have a *cis*-1,4 content of approximately 92 %, are essentially linear and are free of gel. Those produced by Ziegler–Natta catalysis have a higher content of *cis*-1,4 microstructures (*c*. 96 %), are branched and contain some 20 % of loose gel which breaks up during mastication. The presence of transition-metal-catalyst residues in isoprene rubber produced by Ziegler–Natta catalysis should be borne in mind when designing compounds which are to have good ageing properties.

Oil-extended grades of isoprene rubber are available. The oil used for extension is usually of the naphthenic type.

6.2.5 Technology of Isoprene Rubbers
The compounding principles for isoprene rubber are very similar to those for natural rubber. Such differences as do exist between the compounding of the two rubbers arise mainly because of the absence from synthetic isoprene rubbers of the non-rubber constituents which are found in natural rubber. These differences are most evident in relation to vulcanisation behaviour. Like natural rubber, the synthetic isoprene rubbers are normally vulcanised with sulphur and conventional accelerators. However, certain of the non-rubber constituents of natural rubber—notably the fatty acids, proteins and resinous substances—function as mild accelerators and activators of vulcanisation. These substances are, of course, absent from the synthetic isoprene rubbers as received from the manufacturer. Substances having similar effects must therefore be added if the vulcanis-ation behaviour of natural rubber is to be reproduced. Thus, whereas

fatty-acid activator is usually added to natural rubber compounds at a level in the range 1–2 pphr, a higher level (2–3 pphr) has to be added to synthetic isoprene rubber in order to ensure adequate activation of vulcanisation. Then again, rather higher levels of accelerator (approximately 10 % extra) are required in order to give similar rates of vulcanisation. It has also been found advantageous to add small amounts (c. 0·1 pphr) of an aldehyde–amine condensate to act as a secondary accelerator of vulcanisation in much the same way as the proteinaceous substances function in natural rubber.

It may be noted that synthetic polyisoprenes can be vulcanised without difficulty by heating with organic peroxides such as dicumyl peroxide. Their behaviour in this respect is again very similar to that of natural rubber and stands in marked contrast to that of butadiene rubbers, especially those of high 1,2 content (to which reference has been made in the previous chapter).

The mastication behaviour of two types of synthetic isoprene rubber as compared with natural rubber is illustrated in Fig. 6.1. The pattern of molecular weight reduction during mastication is generally similar for all three polyisoprenes. There is an initial rather rapid decrease. The rate of decrease of molecular weight with time of mastication then falls rapidly and a stage is soon reached where little further reduction in molecular weight occurs. Since the synthetic isoprene rubbers as received are rather softer than natural rubber (which contains a large and variable amount of gel), the initial reduction of molecular weight accompanying mastication is not so pronounced; indeed, it has been suggested that synthetic isoprene

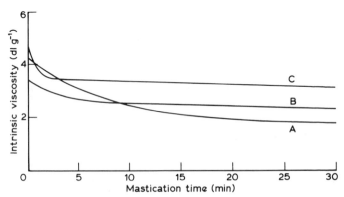

FIG. 6.1.   Mastication behaviour of synthetic isoprene rubbers compared with that of natural rubber. A, alkyllithium-polymerised polyisoprene; B. Ziegler–Natta-polymerised polyisoprene; C, natural rubber.[2]

rubber is best treated as though it were natural rubber whose plasticity has been increased by pre-mastication.

Synthetic isoprene rubber can be blended with other general-purpose diene rubbers, such as natural rubber, styrene–butadiene rubber and *cis*-1,4-polybutadiene. Few problems are encountered as a consequence of blending. In particular, since all the above-mentioned diene rubbers are usually vulcanised with sulphur and conventional accelerators, there is no problem in co-vulcanising the components of a blend into a homogeneous whole.

Because of the high degree of stereoregularity in the synthetic isoprene rubbers, they are able to crystallise on stretching. As a consequence, gum vulcanisates exhibit high tensile strength. As the results summarised in Table 6.4 show, the tensile strengths of gum vulcanisates from synthetic isoprene rubbers are nearly as high as those of gum vulcanisates from natural rubber. The results summarised in this table suggest that gum

TABLE 6.4

MECHANICAL PROPERTIES OF TYPICAL GUM VULCANISATES FROM SYNTHETIC POLY-ISOPRENES AND NATURAL RUBBER[2]

	Synthetic polyisoprene		Natural rubber
	96% cis	*92% cis*	
Compounds		*Parts by weight*	
Isoprene rubber	100	100	100
Zinc oxide	6	3	5
Stearic acid	4	3	2
Sulphur	3	1·5	2·75
Zinc 2-mercaptobenzthiazolate	1	0	0
N-cyclohexylbenzthiazyl-2-sulphenamide	0	0·3	0
Heptaldehyde–aniline condensation product	0	0·1	0
Amine reaction product	0	0·3	0
Di-2-benzthiazyl disulphide	0	0	1
Tetramethylthiuram disulphide	0	0	0·1
Antioxidant	1	1	1
Cure: minutes/°F	60/250	15/293	15/287
Mechanical properties			
Tensile strength (lb in^{-2})	3 900	3 710	4 300
% Elongation at break	735	910	730
Modulus at 300% extension (lb in^{-2})	—	160	300
Modulus at 500% extension (lb in^{-2})	750	—	—
Hardness (Shore A)	42	35	41

vulcanisates from a 92 % cis-1,4-polyisoprene are rather softer and of lower modulus than are those from natural rubber, although, of course, it must be remembered that the stiffness of a gum rubber vulcanisate is greatly affected by the curing system which has been used to effect vulcanisation, and also by the extent to which the rubber has been vulcanised.

Considerable interest has been shown in the matter of whether synthetic isoprene rubbers can be used as a straight replacement for natural rubber in filled rubber compounds, expecially in those reinforced with carbon black. Table 6.5 gives results for the mechanical properties of vulcanisates reinforced with a high-abrasion-furnace black and based upon two synthetic isoprene rubbers. It also gives results for the mechanical properties of a similar vulcanisate based upon natural rubber. The main conclusion which emerges from these results is that, for a given loading of a given carbon black, the synthetic polyisoprenes give rather softer vulcanisates of lower modulus than does natural rubber. If modulus and

TABLE 6.5

MECHANICAL PROPERTIES OF HAF-BLACK-REINFORCED VULCANISATES FROM SYNTHETIC POLYISOPRENES AND NATURAL RUBBER[2]

	Synthetic cis-1,4 polyisoprene	Oil-extended synthetic cis-1,4 polyisoprene	Natural rubber
Compounds		Parts by weight	
Isoprene rubber	100	100	100
HAF carbon black	50	50	50
Processing oil	5	0	5
Sulphur	2·75	2·75	2·75
plus conventional amounts of zinc oxide, stearic acid, accelerators and antioxidant			
Mechanical properties			
Tensile strength (lb in^{-2})	3 750	3 080	3 960
Modulus at 300% extension (lb in^{-2})	1 630	1 330	1 990
% Elongation at break	520	530	550
Hardness (Shore A)	56		63
Angle tear (lb in^{-1})	380	430	600
% Yarzley resilience	71	73	72
Heat build-up (Goodrich flexometer ΔT (°C))	19	22	18·5

hardness similar to that of a natural rubber vulcanisate are required, then one or more of the following adjustments must be made to the compound: (a) less processing oil should be incorporated, (b) more of the given type of carbon black should be incorporated, or (c) a more reinforcing carbon black should be used.

6.3 CHLOROPRENE RUBBER (CR)

6.3.1 Introduction

Chloroprene rubber, whose nominal repeat unit is as shown at (XLV), has the distinction of being the earliest commercially-successful synthetic rubber. First introduced in 1932 by Du Pont under the trade name 'Duprene', later sold under the trade name 'Neoprene', and designated as GR–M (Government Rubber–Monovinylacetylene) by the U.S. Government during World War II, it is still produced in quantity today by several large chemical companies, including Du Pont.

$$\cdots-CH_2-\underset{\underset{Cl}{|}}{C}=CH-CH_2-\cdots$$

(XLV)

It has been justly claimed that chloroprene rubber provides a very useful all-round balance of properties; this is undoubtedly the reason for it having been produced commercially for so many years. It is a stereoregular rubber, being, in fact, largely *trans*-1,4-polychloroprene. It is therefore able to crystallise on stretching. In consequence, its gum vulcanisates have high tensile strength. Its vulcanisates show moderate resistance to swelling in hydrocarbon oils and greases, combined with good resistance to low-temperature stiffening. They also show good resistance to heat ageing, to oxidation and, in particular, to attack by ozone. The chemical resistance of the vulcanisates is also generally good.

In view of the foregoing remarks, it may be wondered why it is that chloroprene rubber has not achieved wider application than it has, for it is still regarded as being in the category of special-purpose synthetic rubbers. One reason is, of course, cost; chloroprene rubber is considerably more expensive than are the general-purpose hydrocarbon rubbers. Another reason is that, useful though the balance of properties offered by chloroprene rubber may be for some applications, there are many applications for which one particular property of chloroprene rubber is insufficiently developed. Thus, for instance, for many sealing applications, the resistance to swelling in hydrocarbon oils is inadequate. Or again, for other applications, the resistance to high temperature may be insufficient.

6.3.2 Production of Chloroprene Rubber

Chloroprene rubber is produced exclusively by the free-radical emulsion polymerisation of chloroprene. It is possible to polymerise chloroprene by cationic initiation, by anionic initiation, and also by Ziegler–Natta catalysis, but the reactions take place more slowly than does free-radical polymeris-ation. Furthermore, the products of such reactions lack useful properties.

Emulsion-polymerisation processes for the production of chloroprene rubber can be classified broadly into two types:

1. those in which control of the polymer molecular weight is achieved by copolymerisation with sulphur followed by reaction in which the polymer chains are cleaved at the sites where the sulphur is copolymerised; and

2. those in which control of the polymer molecular weight is achieved by the inclusion within the polymerisation recipe of a chain-transfer agent.

Chloroprene rubbers produced by the first type of process are usually described as being *sulphur-modified*, although they are also sometimes described as being *thiuram-modified* because, as will appear subsequently, the reagent which is commonly used to cleave the polychloroprene at the sulphur sites is a thiuram sulphide. Chloroprene rubbers produced by the second of these processes are usually described rather negatively as being *non-sulphur-modified*. Little published information seems to be available concerning these latter types. It is known that at least some of them are produced using mercaptans as regulators of molecular weight. Grades produced in this way are described as being *mercaptan-modified*.

When chloroprene is emulsion polymerised in the absence of either sulphur or a chain-transfer modifier, the product which forms even at low conversion is a tough, insoluble, non-plastic material. If sulphur to the extent of $0.5–1.5\%$ is dissolved in the chloroprene monomer prior to polymerisation, the product of the polymerisation is still a tough, insoluble material which is unsuitable for use as an elastomer. If, however, the polymer still in the form of its latex is heated in an alkaline medium with a 'thiophilic' reagent, such as tetraethyl thiuram disulphide, the polymer becomes 'plasticised' or 'peptised' to a soft, plastic, soluble polychloroprene which is suitable for use as an elastomer. The chemistry of the copolymerisation of sulphur with chloroprene is not fully understood, nor are the reactions which occur during the subsequent peptisation. The product of the copolymerisation reaction is believed to have the following type of structure:

$$\cdots \text{---}(CH_2\text{---}CCl\text{=}CH\text{---}CH_2)_{\overline{n}}S_x\text{---}(CH_2\text{---}CCl\text{=}CH\text{---}CH_2)_{\overline{n}}S_x\text{---}\cdots$$

The value of x is thought to be in the range 2–6; n is thought to be of the order of 100-times x. It has been suggested that the peptisation reaction involves interaction between the polysulphide links in the copolymer and dithiocarbamate anions $(R_2NCS.S^{\ominus})$ derived from the thiuram sulphide $(R_2NCS.S_y.S.CSNR_2)$.

The emulsion polymerisation of chloroprene can be carried out industrially either by batch process or by continuous process. For the production of a sulphur-modified chloroprene rubber, the chloroprene is emulsified in water using a rosin-acid soap, and then polymerisation is initiated by potassium persulphate in the presence of sulphur. A typical recipe for the production of a sulphur-modified chloroprene rubber is given in Table 6.6. The sulphur and the rosin acid are first dissolved in the chloroprene. This solution is then emulsified in water in which the sodium hydroxide and the sodium salt of the naphthalenesulphonic acid–formaldehyde condensation product have been dissolved. The rosin acid and sodium hydroxide react to form a soap, sodium resinate, *in situ* at the interface between the chloroprene droplets and the aqueous phase. This soap then stabilises the droplets of chloroprene against coalescence. In this way, a stable emulsion of chloroprene in water is formed. The function of

TABLE 6.6

RECIPE FOR PRODUCTION OF SULPHUR-MODIFIED CHLOROPRENE RUBBER BY EMULSION POLYMERISATION AT $40\,^{\circ}C^3$

	Parts by weight
Chloroprene	100
N wood resin	4 ⎱ dissolved in
Sulphur	0·6 ⎰ the monomer
Water	150
Sodium alkylnaphthalenesulphonates	0·7
Sodium hydroxide	0·8
Potassium persulphate	0·2–1·0

the sodium salt of a naphthalenesulphonic acid–formaldehyde condensate is partly to assist with the breaking up of chloroprene droplets initially, but primarily subsequently to stabilise the polymer latex when it is acidified. Emulsification is achieved by recirculating the chloroprene–water mixture through a centrifugal pump. The monomer emulsion contains some 38 % of chloroprene. After emulsification, the emulsion is pumped into a jacketed, glass-lined kettle equipped with a glass-coated agitator. An aqueous solution of potassium persulphate is then added in order to initiate the polymerisation. The temperature of the reaction mixture is maintained at the required temperature (e.g., 40 °C, but see further below) by circulating cooled brine through the jacket of the kettle and by varying the agitator speed. The progress of the polymerisation is monitored by way of the changes which occur in the specific gravity of the emulsion. At approximately 90 % conversion, the polymerisation is stopped by adding tetraethyl thiuram disulphide as an aqueous emulsion of a solution in xylene. The latex is then cooled to 20 °C and allowed to stand for about 8 h. During this time, the thiuram disulphide cleaves some of the chloroprene–sulphur linkages and so peptises the polymer. The thiuram disulphide also acts as a stabiliser for the finished dry rubber. After having stood for about 8 h, the alkaline latex is then acidified to pH 5·5–5·8 by addition of 10 % acetic acid. The purpose of the acidification is (a) to arrest the peptising action of the thiuram disulphide (which takes place at a reasonable rate in alkaline medium only), (b) to convert the sodium resinate back to rosin acid, which then dissolves in the polymer, and (c) to prepare the latex for the subsequent separation of the polymer.

Chloroprene rubber is separated from its latex by the continuous coagulation of a polymer film on a frozen drum, followed by washing and drying. The dry rubber is then rolled into a kind of rope which is cut into sections and bagged. The success of the separation process depends upon the latex being completely coagulated within a few seconds of being cooled to a temperature within the range -10 to -15 °C, and furthermore upon the latex when so coagulated giving a film which is strong enough to withstand the stresses imposed upon it during washing and drying. In a typical isolation process, the acidified latex is fed through porcelain pipes to a pan in which a stainless-steel freeze roll, 9 ft in diameter, rotates partly immersed in the latex at a peripheral speed of about 36 ft min^{-1}. The roll is cooled to about -15 °C by circulating brine. The polymer in the frozen film of latex deposited on the roll is coagulated as the drum makes part of a revolution. The film is stripped from the roll by a stationary knife, and falls on to a continuous woven stainless-steel belt, where it is thawed and washed. Water is sprayed on to the film and is then forced through the film by suction applied underneath it. The washed film is then fed to squeeze rolls where a pressure of about 15 lb in^{-2} is applied to reduce the water content to 25–30 % of the dry rubber weight. The film is dried in a current of air at about 120 °C as it is carried through a dryer on an endless conveyor. It is then cooled to 50 °C and the dried film discharged over a driven stripper roll. The film of rubber is then fed over a water-cooled roll to a 'roper', from which the rope is conveyed to a cutter and bagger.

Data for gel content as a function of conversion for chloroprene rubbers prepared by emulsion polymerisation at 40 °C (a) without added sulphur and (b) in the presence of 0·6 % added sulphur are summarised in Table 6.7. In the case of (b), the data refer to the polychloroprenes before treatment with a peptiser such as tetraethyl thiuram disulphide. In both cases, polymer gel begins to form quite early in the reaction. However, these results indicate that the sulphur does show a slight tendency to act as a modifier during the polymerisation, in that the onset of gel formation is delayed when sulphur is present. Also delayed is the point in the reaction at which the polymer is virtually entirely gel. The data given in Table 6.7

TABLE 6.7

VARIATION OF GEL CONTENT WITH CONVERSION FOR CHLOROPRENE RUBBERS PREPARED BY EMULSION POLYMERISATION AT 40 °C[4]

(a) *Chloroprene rubbers prepared without added sulphur*

% Conversion	% Gel
5	9
15	37
26	86
36	98
45	96
56	97

(b) *Chloroprene rubbers prepared in presence of 0·6% added sulphur, before peptisation*

% Conversion	% Gel
5	0
11	45
19	48
26	62
50	86
85	98

illustrate that the emulsion polymerisation of chloroprene in the absence of a conventional modifier produces a polymer which is extensively gelled, especially if the reaction is taken to complete conversion. They also illustrate the susceptibility of sulphur-modified polychloroprene to post-polymerisation plasticisation, since the sulphur-modified polymers could be rendered completely soluble (and thus all the gel fraction destroyed) by treatment with tetraethyl thiuram disulphide, as long as the initial conversion of monomer to polymer did not exceed about 90%.

The molecular-weight distribution for a thiuram-peptised sulphur-modified chloroprene rubber polymerised at 40 °C is shown in Fig. 6.2. Also included in Fig. 6.2 are distributions for the sol fraction of a typical natural rubber, and also for natural rubber peptised by treatment with phenylhyd-razine. The natural rubber sol has a higher modal molecular weight than does the chloroprene rubber, but the distribution of molecular weights in the case of natural rubber is somewhat narrower and more nearly symmetrical than in the case of the chloroprene rubber.

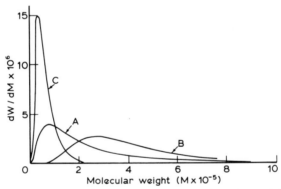

FIG. 6.2.   Molecular weight distributions for a thiuram-peptised sulphur-modified chloroprene rubber polymerised at 40 °C (A), a typical natural rubber sol fraction (B), and natural rubber peptised by treatment with phenylhydrazine (C).[5]

6.3.3 Effect of Polymerisation Temperature upon Polychloroprenes Produced by Emulsion Polymerisation

The temperature at which the emulsion polymerisation of chloroprene is carried out is an important variable in determining the character of the polymer produced. The effects of polymerisation temperature can be conveniently discussed under three headings:

1.   effect upon gel content and molecular-weight distribution;
2.   effect upon stereoregularity; and
3.   effect upon tendency of polymer to crystallise.

Effect of polymerisation temperature upon gel content and molecular-weight distribution

Reducing the polymerisation temperature reduces the tendency for polymer gel to form during the polymerisation. This is illustrated in Fig. 6.3 for sulphur-modified chloroprene rubbers produced with 0·6 % of sulphur, before peptisation of the polymer with a thiuram disulphide. Results are given for polymers produced at two polymerisation temperatures, namely, 40 °C and 10 °C. It is evident that the onset of polymer gel formation occurs much later in the reaction when the polymerisation temperature is lowered. Whereas a substantial proportion of the polymer is gel at only 10 % conversion when the polymerisation is carried out at 40 °C, the polymer is essentially gel-free up to 40 % conversion in the case of polymerisation at 10 °C.

It might be thought that, as a consequence of the reduced tendency to polymer-gel formation at lower temperatures, the average molecular weights of the low-temperature polymers *after* peptisation with a thiuram

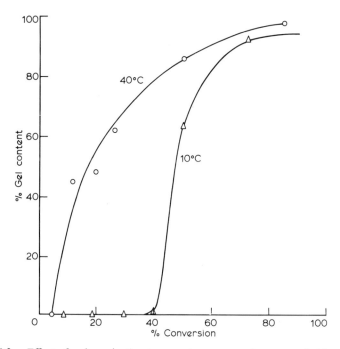

FIG. 6.3.   Effect of polymerisation temperature upon gel content of chloroprene rubber produced by emulsion polymerisation in presence of 0·6 % sulphur and before peptisation with a thiuram disulphide.[4]

sulphide would be lower than those of high-temperature polymers similarly treated. This does not appear to be the case. This is illustrated by the two molecular-weight distributions given in Fig. 6.4. These distributions are for two chloroprene rubbers made using essentially similar recipes, the only significant difference being that in one case the polymerisation was carried out at 40 °C and in the other at 10 °C; both were sulphur-modified, and both were peptised after polymerisation by treatment with tetraethyl thiuram disulphide. It appears that the high-temperature polymer has the lower modal molecular weight and also the narrower distribution of molecular weights.

Effect of polymerisation temperature upon stereoregularity
It has already been noted that polychloroprene produced by emulsion polymerisation is essentially the *trans*-1,4 polymer. This is broadly true whatever is the polymerisation temperature, although the details of the microstructure do vary somewhat with polymerisation temperature. Typical results for microstructure analysis are given in Table 6.8. From these results, it is seen that the ratio *cis*-1,4/*trans*-1,4 decreases as the polymerisation temperature is reduced, but the overall content of 1,4 microstructures increases. The balance is made up of 1,2 and 3,4 microstructures in approximately equal amounts (except for polymers produced at very low temperatures where the 1,2 microstructures pre-

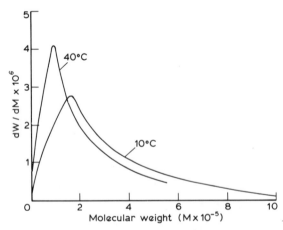

FIG. 6.4. Effect of polymerisation temperature upon molecular-weight-distribution of sulphur-modified chloroprene rubbers produced by emulsion polymerisation, after peptisation with a thiuram disulphide.[6]

TABLE 6.8

EFFECT OF POLYMERISATION TEMPERATURE UPON MICROSTRUCTURE OF CHLOROPRENE
RUBBERS PREPARED BY EMULSION POLYMERISATION[7]

Polymerisation temperature (°C)	Polychloroprene microstructure			
	% 1,2	% 3,4	% cis-1,4	% trans-1,4
−40	0·9	0·3	5	94
−10	—	—	7	—
10	1·1	1·0	9	84
40	1·6	1·0	10	86, 81
50	—	—	11	—
80	2·0	2·1	—	—
100	2·4	2·4	13	71

dominate over 3,4). A consequence of the overall content of 1,4 microstructures increasing with decreasing polymerisation temperature is that the content of 1,2 plus 3,4 microstructures must decrease. It will be seen subsequently that, although those microstructures are present in relatively low concentration, their presence is very important for the technology of chloroprene rubbers.

Effect of polymerisation temperature upon tendency of chloroprene rubbers to crystallise

As the polymerisation temperature is reduced, so there is an increase in the tendency of the resultant chloroprene rubber to crystallise. This may be illustrated for the case of two chloroprene rubbers made by essentially identical polymerisation recipes, the one being made by polymerisation at 40 °C and the other by polymerisation at 10 °C. Both were sulphur-modified and peptised by post-polymerisation treatment with tetraethyl thiuram disulphide. The rubbers produced at 10 °C hardened by crystallisation after only a few hours standing at 25 °C. Several days were required for similar changes to occur in the rubber produced at 40 °C. Furthermore, the gum vulcanisates from the low-temperature rubber had higher tensile strengths than did those from the high temperature rubber. A further illustration of the effect of polymerisation temperature upon the crystallisation tendency of chloroprene rubbers is to be found in the observation that the melting point of the crystallites in crystallised chloroprene rubber falls linearly as the polymerisation temperature is increased (Fig. 6.5).

The effect of polymerisation temperature upon crystallisation tendency

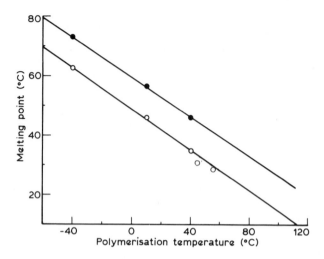

FIG. 6.5.   Effect of polymerisation temperature upon crystalline melting point of chloroprene rubbers produced by emulsion polymerisation.[8] ● Highest observed value, ○ lowest observed value.

is interpreted principally in terms of the effect of polymerisation temperature upon the stereoregularity of the polychloroprene produced. It appears that, for chloroprene homopolymers, other variables such as molecular weight and crosslink density have little effect upon the tendency to crystallise. Reference to Table 6.8 shows that it is probably the enhanced concentrations of cis-1,4 microstructures in chloroprenes produced at higher temperatures which are responsible for the retardation of crystallisation, rather than the presence of 1,2 and 3,4 microstructures.

We may note in passing that the crystallisation tendency of a rubber is important for several reasons. As implied above, it affects the rate at which the raw polymer hardens during storage at room temperature and below. It also affects the tensile strength of gum rubber vulcanisates: the greater the tendency to crystallise on stretching, the higher will be the tensile strength. Crystallisation is also important for the rapid development of strength in films which are cast from rubber solution; this is important in certain applications of rubbers as adhesives and cements.

6.3.4 Copolymerisation of Chloroprene with Other Monomers

The structure and properties of chloroprene rubbers can be modified to some extent by copolymerisation with minor amounts of a second monomer. In particular, the rate of crystallisation of the raw polymer at ambient temperatures can be reduced to very low levels; grades of chloroprene rubber having this property are sometimes described as being 'crystallisation resistant'. The effect of the comonomer units is, of course, to disturb the regularity of the polymer sequence, and so discourage the polymer molecules from participating in the formation of a crystal lattice. However, it is not easy to find suitable comonomers for use with chloroprene, because of the high reactivity of polychloroprenyl radicals with chloroprene relative to their reactivities with other monomers. Thus there is a strong tendency for chloroprene homopolymers to form rather than copolymers. The comonomers which have been used with most success are those which have chemical structures similar to that of chloroprene itself. The most widely-used comonomer is probably 2,3-dichloro-1,3-butadiene (XLVI). Other suitable comonomers include 2-cyano-1,3-butadiene (XLVII) and 1,1,3-trifluoro-1,3-butadiene (XLVIII).

$$CH_2\!\!=\!\!C\!\!-\!\!C\!\!=\!\!CH_2 \qquad CH_2\!\!=\!\!C\!\!-\!\!CH\!\!=\!\!CH_2 \qquad CF_2\!\!=\!\!CH\!\!-\!\!C\!\!=\!\!CH_2$$
$$\quad\;\; |\;\;\; | \qquad\qquad\quad\; | \qquad\qquad\qquad\qquad |$$
$$\quad\;\; Cl\;\; Cl \qquad\qquad\quad CN \qquad\qquad\qquad\qquad F$$

| XLVI | XLVII | XLVIII |

6.3.5 Compounding Principles

Vulcanisation

The vulcanisation chemistry of chloroprene rubbers is fundamentally different from that of the other diene rubbers. The carbon–carbon double bonds in the polychloroprene chain are deactivated by the presence of the electronegative chlorine atoms, and one consequence of this is that vulcanisation by heating with sulphur occurs to a very limited extent only. The vulcanising agents which are usually used for chloroprene rubbers are the oxides of divalent metals such as zinc and magnesium. It is usually assumed that crosslinking of the polychloroprene molecules occurs by way of ether formation and accompanying loss of chlorine atoms. The chlorine atoms which participate in the crosslinking reaction are believed to be those which are present in the small proportion of chloroprene units which became enchained in the polymer sequence by 1,2 and 3,4 addition. The chlorine atoms present in the 1,4 units are thought to be insufficiently reactive to participate in vulcanisation reactions, their reactivity being reduced because of their positions relative to the carbon–carbon double bonds. The chlorine atom in a 1,4 unit is commonly described as being *vinylic*, because it is present as part of a C=C—Cl unit. The chlorine atom in a 1,2 or 3,4 unit is described as being *allylic*, because it is present as part of a C=C—C—Cl unit. It is known that vinylic chloride atoms are considerably *less* reactive than are those in an alkyl chloride, and that allylic chlorine atoms are considerably *more* reactive than those in an alkyl chloride. There is thus a large difference in reactivity between the majority of the chlorine atoms in a chloroprene rubber and the small minority (*c*, 1·5–2 %) which are present in the 1,2 and 3,4 units. It thus appears that the possibility of vulcanising a chloroprene rubber by heating it with a metal oxide depends upon the somewhat fortuitous presence in the polymer molecule of a few allylic chlorine atoms arising from 1,2 and 3,4 additions during polymerisation.

The chemistry of the vulcanisation of chloroprene rubber by metal oxides is conventionally represented as follows:

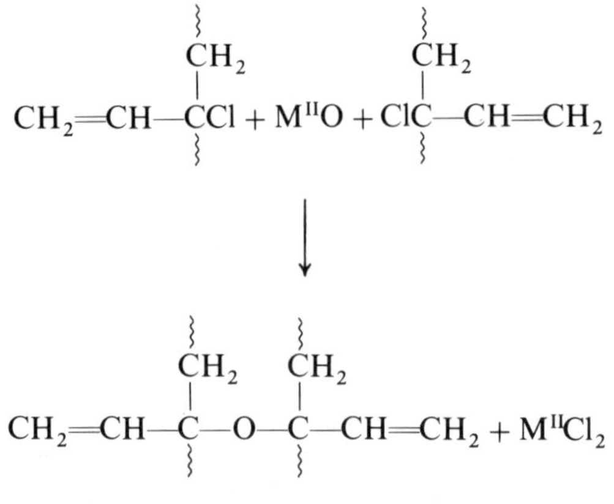

$$CH_2=CH-\underset{\underset{\}{CH_2}}{\overset{\overset{\}{CH_2}}{C}}Cl + M^{II}O + Cl\underset{\underset{\}{}}{\overset{\overset{\}{CH_2}}{C}}-CH=CH_2$$

$$\downarrow$$

$$CH_2=CH-\underset{\underset{\}{}}{\overset{\overset{\}{CH_2}}{C}}-O-\underset{\underset{\}{}}{\overset{\overset{\}{CH_2}}{C}}-CH=CH_2 + M^{II}Cl_2$$

In the above reaction scheme, the participating chloroprene units are shown as having arisen from 1,2 addition. An exactly analogous reaction scheme can be formulated involving the chlorine atoms in 3,4 units. However, it must be pointed out that, although the vulcanisation reaction for chloroprene rubber is conventionally represented as above, the reality of the postulated ether crosslinks does not ever seem to have been clearly established. What does seem to be quite clear is that, regardless of the detailed chemistry of the reactions which occur, the presence of metal oxides is essential if controlled vulcanisation is to take place and a vulcanisate having good ageing behaviour is to be obtained.

A typical vulcanising combination for the sulphur-modified chloroprene rubbers is 5 pphr of zinc oxide plus 4 pphr of a specially-calcined magnesium oxide. No organic accelerator need be added; it appears that residual thiuram sulphide and its decomposition products in the polymer from the peptisation step function as vulcanisation accelerators. However, if need be, the rate of vulcanisation can be enhanced by the addition of the same types of organic accelerator as are used for the non-sulphur-modified grades of chloroprene rubber (see below). It is interesting to observe that a combination of two metal oxides is normally used for the vulcanisation of chloroprene rubbers. The use of zinc oxide alone produces a scorchy flat-curing rubber compound which gives a vulcanisate of rather low ultimate state of cure. On the other hand, magnesium oxide used alone gives a slow rate of cure, but, if sufficient time is allowed, eventually a high state of cure is attained. It is found that suitable combinations of the two oxides (such as that suggested above) give a satisfactory balance between processing safety, rate of vulcanisation, and ultimate state of cure. Curves illustrating the effect of altering the balance of zinc and magnesium oxides upon the vulcanisation behaviour of a sulphur-modified chloroprene rubber are shown in Fig. 6.6. The grade of magnesium oxide used for the vulcanisation of chloroprene rubber has an important effect upon processing. The so-called 'light' magnesium oxide of high specific surface area made by lightly

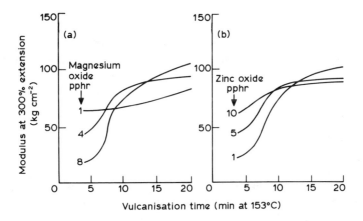

Fig. 6.6. Effect of zinc oxide/magnesium oxide balance upon vulcanisation of a sulphur-modified chloroprene rubber. (a) Effect of varying magnesium oxide level at constant zinc oxide level of 5 pphr; (b) effect of varying zinc oxide level at constant magnesium oxide level of 4 pphr.[9] Compound: chloroprene rubber 100, stearic acid 0·5, SRF carbon black 30, antioxidant 2, zinc oxide and magnesium oxide as indicated.

calcining precipitated magnesium hydroxide is preferred to coarser grades of lower purity made by heavily calcining magnesium minerals. Contamination with moisture and carbon dioxide should also be avoided. The grade of zinc oxide is not critical. It has been suggested that, when used in conjunction with zinc oxide, magnesium oxide improves processing safety by removing the zinc chloride which is formed, by means of the following reaction:

$$2ZnCl_2 + MgO + H_2O \longrightarrow 2Zn(OH)Cl + MgCl_2$$

According to this view, zinc chloride is able to catalyse the crosslinking reaction, and this is why the vulcanisation reaction is scorchy in the absence of the moderating influence of the magnesium oxide.

The only other metal oxides which are used to any appreciable extent for the vulcanisation of chloroprene rubbers are those of lead. Oxides of lead (litharge and red lead, of which the latter is preferred because it imparts great processing safety) are used where optimum water resistance is required in the vulcanisate. The reason why the lead oxides give a more water-resistant vulcanisate than do the zinc and magnesium oxides is probably to be found in the fact that lead chloride is much less soluble in water than are the chlorides of zinc and magnesium. None of the other metal oxides used as curatives for chloroprene rubber gives the same balance of processing safety, rate and state of vulcanisation, and desirable vulcanisate properties as do the oxides of zinc, magnesium and lead.

The non-sulphur-modified chloroprene rubbers vulcanise only very slowly if metal oxides alone are used. In order to obtain rates of vulcanisation which are practically useful, it is necessary to use an organic accelerator in addition to the combination of zinc and magnesium oxides. A well-known example of an organic accelerator for the vulcanisation of chloroprene rubber is 2-mercaptoimidazoline, also known as ethylene thiourea (with which it is tautomeric). Levels are approximately 0·5 pphr. This accelerator can either be used alone or in conjunction with a thiazole or a thiuram disulphide accelerator. Sulphur can also be used to assist the vulcanisation of the non-sulphur-modified chloroprene rubbers. It is commonly used in conjunction with a thiazole and a thiuram disulphide; 2-mercaptoimidazoline may also be added.

The following scheme of reactions has been proposed to explain the ability of substituted thioureas (and therefore 2-mercaptoimidazoline) to accelerate the vulcanisation of chloroprene rubber by metal oxides. Again, vulcanisation is believed to occur by way of the allylic chlorine atoms:

$$\underset{\substack{|\\CH_2\\|}}{ClC}-CH{=}CH_2 + \underset{\substack{|\\NR_2}}{\overset{NR_2}{S{=}C}} \longrightarrow \underset{\substack{|\\CH_2\\|}}{C}{=}CH-CH_2-S-\underset{\substack{|\\NR_2}}{\overset{NR_2}{C}}-Cl$$

$$\Big\downarrow M^{II}O$$

$$\underset{\substack{|\\CH_2\\|}}{C}{=}CH-CH_2-S-\underset{\substack{|\\NR_2}}{\overset{NR_2}{C}}-O-M^{\oplus}Cl^{\ominus}$$

$$\Big\downarrow$$

$$\underset{\substack{|\\CH_2\\|}}{C}{=}CH-CH_2-S-M^{\oplus}Cl^{\ominus} + O{=}\underset{\substack{|\\NR_2}}{\overset{NR_2}{C}}$$

$$\Big\downarrow \; \underset{\substack{|\\ClC-CH{=}CH_2\\|}}{CH_2}$$

$$\underset{\substack{|\\CH_2\\|}}{C}{=}CH-CH_2-S-CH_2-CH{=}\underset{\substack{|\\CH_2\\|}}{C} + M^{II}Cl_2$$

Because of serious reservations which have arisen concerning the toxicity of 2-mercaptoimidazoline, various alternative thioureas have been introduced as accelerators for the vulcanisation of non-sulphur-modified chloroprene rubbers. These include trimethyl thiourea, tributyl thiourea, diethyl thiourea, and diphenyl thiourea. Other compounds which have been used as accelerators for the vulcanisation of non-sulphur modified chloroprene rubbers include the di-o-tolyl guanidine salt of dicatechol borate, which is used in conjunction with tetramethyl thiuram disulphide, and dimethyl thiocarbamoyl-2-imidazolidenethione.

Chloroprene rubbers can also be crosslinked by heating with poly-amines. The reaction probably involves the allylic chlorine atoms again. The following reaction mechanism involving 1,3 allylic shifts has been suggested:

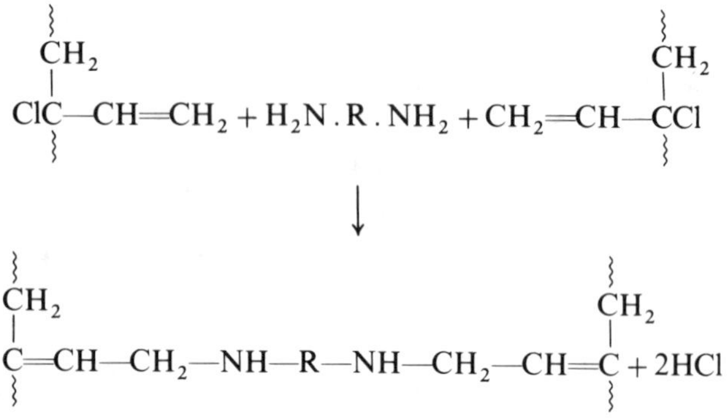

$$
\begin{array}{ccc}
\overset{\displaystyle \}}{\underset{\displaystyle |}{CH_2}} & & \overset{\displaystyle \}}{\underset{\displaystyle |}{CH_2}} \\
ClC\!-\!CH\!=\!CH_2 + H_2N.R.NH_2 + CH_2\!=\!CH\!-\!CCl
\end{array}
$$

$$\downarrow$$

$$
\begin{array}{ccc}
\overset{\displaystyle \}}{\underset{\displaystyle |}{CH_2}} & & \overset{\displaystyle \}}{\underset{\displaystyle |}{CH_2}} \\
C\!=\!CH\!-\!CH_2\!-\!NH\!-\!R\!-\!NH\!-\!CH_2\!-\!CH\!=\!C + 2HCl
\end{array}
$$

The hydrochloric acid formed in this reaction then reacts with zinc or magnesium oxides if these are present in the rubber compound.

Plasticisation
Unplasticised chloroprene rubber vulcanisates can be cooled to about $-20\,°C$ before appreciable stiffening occurs. Below that temperature, the stiffness increases very rapidly, and below $-40\,°C$ the vulcanisate is brittle and shatters when struck. The flexibility at low temperatures can be improved by compounding the rubber with a suitable plasticiser. These are usually esters. A wide range is available. Dioctyl sebacate is particularly useful. Butyl oleate is a rather cheaper but quite efficient low-temperature plasticiser. Figure 6.7 shows the effect of increasing levels of butyl oleate upon the brittleness temperature and the temperature at which the torsional modulus is $700\,kg\,cm^{-2}$, for a typical chloroprene rubber gum vulcanisate.

Petroleum oils are used as processing aids, but they do not in general improve the low-temperature flexibility of chloroprene rubber vulcanisates. Amounts ranging from 10 to 20% by weight of the filler loading are generally required for processing purposes. The non-sulphur-modified types require more than the sulphur-modified types. Provided that no more than 25 pphr of oil is to be added, then a naphthenic oil may be used. Much use is made of naphthenic 'light process oils' of relatively low molecular weight. However, these have a strong tendency to evaporate when the vulcanisate is exposed to hot air, and they can exude from the vulcanisate if

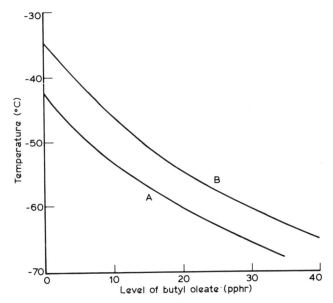

Fig. 6.7. Effect of butyl oleate upon brittleness temperature (A) and low-temperature stiffening temperature (B) of a non-sulphur-modified chloroprene–rubber vulcanisate.[9] Curve B gives the temperature at which the torsional modulus reached $700\,kg\,cm^{-2}$. Compound: chloroprene rubber 100, zinc oxide 5, magnesium oxide 4, SRF carbon black 60, 2-mercaptoimidazoline 0·5, antioxidant 2, butyl oleate variable. Vulcanisation: 30 min at 153 °C.

used in large quantities. Furthermore, they are rather expensive. The less expensive aromatic petroleum oils can be used at higher loadings, these being more compatible with chloroprene rubber than are the naphthenic type. However, they have the disadvantage that they darken light-coloured vulcanisate and can also stain surfaces with which they come into contact.

For improving the flame resistance of chloroprene rubber vulcanisates, an ester of phosphoric acid (e.g., trixylyl phosphate) or a chlorinated hydrocarbon oil can be used as a replacement for the petroleum oil. Other softeners which are sometimes added to chloroprene rubbers to improve their processing behaviour include stearic acid, microcrystalline waxes, and low-molecular-weight polyethylenes. Hydrogenated rosin esters and coumarone–indene resins are used as 'tackifiers'.

Reinforcement
It has already been noted that gum vulcanisates from chloroprene rubber

generally have high tensile strength. In this respect they resemble vulcanisates from natural rubber. It is not therefore necessary to incorporate
reinforcing fillers in order to obtain high tensile strength. The principles
which govern the use of reinforcing fillers in chloroprene rubber (indeed the
use of fillers in general in chloroprene rubber) are similar to those which
apply for natural rubber. The effects of increasing loadings of various types
of carbon black upon certain mechanical properties of chloroprene rubber
vulcanisates are summarised in Fig. 6.8. As usual, the 'harder' finer-particle
carbon blacks provide more effective reinforcement of the vulcanisate than
do the 'softer' larger-particle blacks.

Protection against deteriorative influences
An antioxidant must be incorporated in a chloroprene rubber compound if
the vulcanisate is to have good ageing resistance. Phenyl-1-naphthylamine
and phenyl-2-naphthylamine have been recommended as good general-
purpose antioxidants for vulcanisates where discolouration and possible
staining can be tolerated. For light-coloured vulcanisates, hindered
phenols can be used.

   Since the inherent ozone resistance of chloroprene rubber is good, it is
sometimes used for applications where this property is important (e.g. for
weather-sealing strip). In such cases, it may be desirable to optimise the
resistance to deterioration by ozone by incorporating an antiozonant. The
compounds commonly used for this purpose are the alkylaryl-*p*-phenylene
diamines and the diaryl-*p*-phenylene diamines. Likewise, although the heat
resistance of chloroprene rubber is good, it may be desirable to enhance
it for some applications. This objective can be attained by compounding
with a suitable additive, e.g., nickel dibutyl dithiocarbamate.

Choice of polymer type
Many types of solid chloroprene rubber are now available commercially.
Some of these are intended for special applications such as for the
preparation of fast-setting adhesives and cements. The general-purpose
grades of chloroprene rubber fall broadly into the two categories of sulphur
modified and non-sulphur-modified. The sulphur-modified types are apt to
undergo changes of molecular weight during storage and during mastication. They therefore show limited storage stability and can be softened by
mastication. They can be compounded to give fast-curing stocks which are
nevertheless safe during processing. As has already been pointed out, it is
not necessary to add organic accelerators in order to achieve practicable

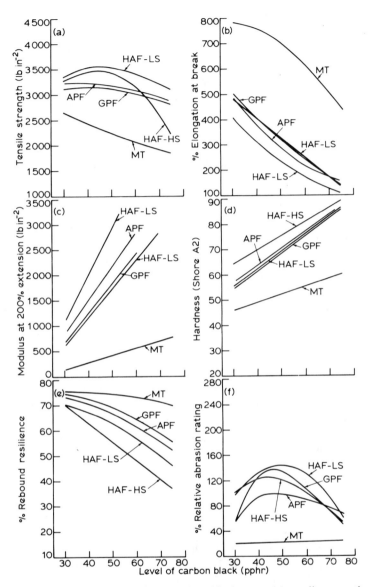

FIG. 6.8. Effect of various types of carbon black upon (a) tensile strength, (b) elongation at break, (c) modulus, (d) hardness, (e) rebound resilience and (f) abrasion resistance of vulcanisates from a sulphur-modified chloroprene rubber.[10] Compound: chloroprene rubber 100, stearic acid 0·5, zinc oxide 5, magnesium oxide 4, light process oil 7·5, 2-mercapto-imidazoline 0·5, antioxidant 2, carbon black variable. Significance of abbreviations for types of carbon black: MT = medium thermal; GPF = general-purpose furnace; APF = all-purpose furnace; HAF–LS = high-abrasion furnace—low structure; HAF–HS = high-abrasion furnace—high structure.

rates of vulcanisation. The sulphur-modified type of chloroprene rubber is ·
used when it is desirable to optimise such properties as tear strength, flex-
cracking resistance and resilience. The non-sulphur-modified chloroprene
rubbers, on the other hand, do not undergo such pronounced changes of
molecular weight during storage and mastication. Their storage stability is
therefore much better than that of the sulphur-modified grades, and they
do not soften during mastication. Accelerators are required in order to
achieve practicable rates of vulcanisation. The non-sulphur-modified type
of chloroprene rubber is recommended for applications where optimum
heat ageing and resistance to compression set are required.

6.3.6 Processing Behaviour

The most important general problem which can arise during the processing
of chloroprene rubbers is that of scorch. Compounds should be mixed at as
low a temperature as possible and for as short a time as possible. The
magnesium oxide (which retards the scorching tendency, as has been
discussed above) should be added early in the mixing programme, whereas
the zinc oxide and accelerators (if any) should be withheld until as late as
possible. The mixing temperature should not be allowed to exceed 130 °C.
If compounded stocks are to be held in storage bins before further
processing, it is good practice to omit the addition of the zinc oxide and
accelerator until the stock is to be processed further. It is important to note
that the effect of heating the uncured stock is cumulative. Thus, if all the
curatives are present, the processing safety of the stock may be reduced
progressively, even although no appreciable changes in the plasticity of the
stock appear to have occurred.

The processing behaviour of the non-sulphur-modified chloroprene
rubbers differs in certain important respects from that of the sulphur-
modified chloroprene rubbers. It has already been noted that the non-
sulphur-modified types show less tendency to break down during milling.
As compared with the sulphur-modified types, they mix faster and develop
less heat during mixing. Their uncured extrusions are more resistant to
distortion, and they offer greater latitude in respect of cure rate and
processing safety because of the need to add an organic accelerator. Stocks
based upon sulphur-modified chloroprene rubbers have less nerve than
have those based upon the non-sulphur-modified types. Consequently,
they shrink less and sheet out more smoothly on calendering. Extruded
surfaces tend to be smoother and better defined. The sulphur-modified
types have better self-adhesive tack.

GENERAL BIBLIOGRAPHY

Bean, A. R., Himes, G. R., Holden, G., Houston, R. R., Langton, J. A. and Mann, R. H. (1967). Isoprene polymers. In: *Encyclopedia of Polymer Science and Technology*, Vol. 7, John Wiley and Sons, New York, pp. 782f.

Schoenberg, E., Marsh, H. A., Walters, S. J. and Saltman, W. M. (1979). Polyisoprene, *Rubb. Chem. Technol.*, **52**, 526.

Saltman, W. M. (Ed.) (1977). *The Stereo Rubbers*, John Wiley and Sons, New York, Chapters 2 and 4.

Whitby, G. S., Davis, C. C. and Dunbrook, R. F. (Eds.) (1954). *Synthetic Rubber*, John Wiley and Sons, New York, Chapter 22.

Johnson, P. R. (1976). Polychloroprene rubber, *Rubb. Chem. Technol.*, **49**, 650.

Hargreaves, C. A. and Thompson, D. C. (1965). 2-Chlorobutadiene polymers. In: *Encyclopedia of Polymer Science and Technology*, Vol. 3, John Wiley and Sons, New York, pp. 705f.

Murray, R. M. and Thompson, D. C. (1964). *The Neoprenes*, Elastomer Chemicals Department, E. I. Du Pont de Nemours & Co. (Inc.), Wilmington, Delaware.

REFERENCES

1. Duck, E. W. and Locke, J. M. (1977). *The Stereo Rubbers*, Saltman, W. M. (Ed.) John Wiley and Sons, New York, p. 139.
2. Bean, A. R. *et al.* (1967). *Encyclopedia of Polymer Science and Technology*, Vol. 7, John Wiley and Sons, New York, 782.
3. Neal, A. M. and Mayo, L. R. (1954). *Synthetic Rubber*. Whitby, G. S., Davis, C. C. and Dunbrook, R. F. (Eds.), John Wiley and Sons, New York, p. 767.
4. Mochel, W. E. (1952). *J. Polym. Sci.*, **8**, 583.
5. Mochel, W. E., Nichols, J. B. and Mighton, C. J. (1948). *J. Amer. Chem. Soc.*, **70**, 2185.
6. Mochel, W. E. and Nichols, J. B. (1949). *J. Amer. Chem. Soc.*, **71**, 3435.
7. Maynard, J. T. and Mochel, W. E. (1954). *J. Polym. Sci.*, **13**, 251.
8. Maynard, J. T. and Mochel, W. E. (1954). *J. Polym. Sci.*, **13**, 235.
9. Murray, R. M. and Thompson, D. C. (1964). *The Neoprenes*, Elastomers Chemicals Department, E. I. Du Pont de Nemours & Co. (Inc.), Wilmington, Delaware.
10. Hansen, E. B. and Church, F. C. *A Practical Approach to Carbon Black Selection for Rubber Compounding*, Cabot Corporation Technical Report RG-127.

Chapter 7

Rubbers Derived from Olefins and from Olefin Oxides

7.1 INTRODUCTION

The theme of this chapter is rubbers which are obtained either directly or indirectly from olefins. We begin with a description of the oldest type of olefin rubber which has been produced commercially, namely, that which is obtained by copolymerising isobutene with minor amounts of a diolefin. It is convenient to note here also the existence of certain other similar olefin rubbers of very minor importance, and also to deal at the same time with the rubbers which are obtained from isobutene–diolefin rubbers by halogenation. We next consider the important family of elastomers which is obtained by copolymerising ethylene and propylene as the main comonomers, and also the rubbers which are obtained by copolymerising ethylene with vinyl acetate and with acrylic monomers. Next to be considered is the rubber which is obtained from ethylene homopolymers by the process known as 'chlorosulphonation', and also a rubber which is obtained from ethylene homopolymers by chlorination. The chapter concludes with a description of those rubbers which are obtained either directly or indirectly from olefin oxides; these materials are sometimes known as polyether rubbers because they contain ether linkages within the backbones of their polymer chains.

The main feature which all these rubbers have in common—apart from being derived directly or indirectly from olefins—is that they are almost entirely chemically saturated. Such unsaturation as they do contain is induced deliberately by copolymerising with a minor amount of, for example, a multiolefin. A consequence of their being essentially chemically saturated is that they are inherently more resistant to various deteriorative influences, such as oxygen, heat and ozone, than are most of the rubbers obtained by the polymerisation or copolymerisation of dienes.

Susceptibility to attack by oxygen, and especially by ozone, is associated with the presence of carbon–carbon double bonds in the main chain of the polymer. Such double bonds are either susceptible to direct attack themselves (as in the case of reaction with ozone), or they increase the susceptibility of adjacent sites to attack (as is probably the case with oxidation). Whatever may be the detailed mechanism by which carbon–carbon double bonds sensitise a polymer chain to attack by oxygen or ozone, it is the case that polymer chains which do not contain such double bonds are inherently less prone to degradation by oxygen and ozone. A further consequence of the rubbers being chemically saturated is that, as a group, their general chemical resistance is good.

Another general characteristic of these rubbers as a group is that they retain their flexibility at low temperatures. This is because the linkages which make up their main chains show high rotational flexibility. As might be expected, the resistance of the purely hydrocarbon rubbers in this group to swelling in hydrocarbon oils is poor. The ether rubbers have somewhat better resistance to swelling in oils.

7.2   ISOBUTENE(ISOBUTYLENE)–ISOPRENE RUBBERS (IIR)

7.2.1   Introduction

Isobutene–isoprene rubber first became commercially available during the early part of World War II. It was designated by the U.S. Government as GR–I (Government Rubber–Isobutylene), but was commonly known as 'butyl' rubber. Since the end of World War II, this latter name has been retained to denote the whole family of rubbers of this type which are produced by various manufacturers.

The principal repeat unit in these rubbers, shown at (XLIX), is derived from isobutene. These rubbers also contain minor amounts (0.5–2.5 mole %) of units derived from a diene. The diene is normally isoprene, and so the units derived from the diene can be conventionally represented as shown at (L). The purpose of introducing the minor amount of unsaturation arising from copolymerisation with the diene is to make the

$$\cdots -CH_2-\underset{\underset{CH_3}{|}}{\overset{\overset{CH_3}{|}}{C}}- \cdots \qquad\qquad \cdots -CH_2-\underset{\underset{CH_3}{|}}{C}=CH-CH_2- \cdots$$

(XLIX)   (L)

rubber susceptible to vulcanisation with sulphur and conventional accelerators. Although the minor amount of unsaturation is usually introduced by copolymerisation with a diene, and the diene is usually isoprene, it should be noted that other multiply-unsaturated comonomers have been used. Thus, for example, interest is currently being shown in a rubber of the isobutylene type in which the reactive functionality is a conjugated diene.

Apart from possessing the general properties noted in the introduction to this chapter, in particular, chemical inertness arising from the low level of unsaturation contained in the polymer, isobutene–isoprene rubbers have one unusual property, and it was this property which first made them of great interest as synthetic rubbers. This property is low permeability to gases, in particular, to air. Figure 7.1 shows the air permeability of an isobutene–isoprene rubber vulcanisate as a function of temperature. Data for similar vulcanisates from other rubbers are shown for purposes of comparison. It is seen that the permeability to air increases with increasing temperature in such a way that the logarithm of the permeability decreases linearly with increasing reciprocal absolute temperature. At $c.$ 25 °C, the air permeability of a natural rubber vulcanisate is some 25 times that of a similar isobutene–isoprene rubber vulcanisate. At the same temperature, the corresponding factor for a styrene–butadiene rubber vulcanisate is approximately 10. This property of low permeability to air has made isobutene–isoprene rubber particularly useful for the manufacture of tyre inner tubes. This possibility was of considerable interest during World War II and in the years immediately following the war, because in those days all pneumatic tyres were used with inner tubes. The advent of the tubeless tyre had serious implications for the manufacture of isobutene–isoprene rubbers, but, notwithstanding the disappearance of part of the inner-tube market, this type of rubber still finds widespread application in various areas because of its inherently good ozone resistance and chemical resistance.

The low permeability of isobutene–isoprene rubbers to gases is attributed to mutual interference between the methyl groups of the isobutene units. Although not greatly affecting the glass-transition temperature of the rubber (which is $c.$ −72 °C), such interference does slow down the segmental mobility of the polymer chain at normal temperatures. This is thought to be the cause of the low gas permeability, since the mechanism of the permeation of gases through rubbers is though to be diffusion aided by the segmental motion of the polymer chains. Another consequence of retarded segmental motion is that isobutene–isoprene rubber vulcanisates are rather sluggish in responding to mechanical deformation; they

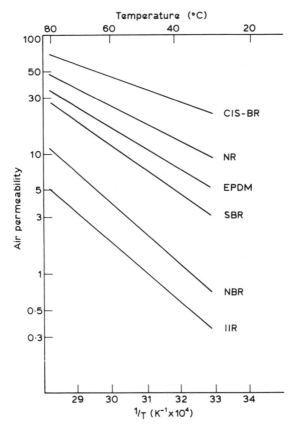

FIG. 7.1.   Permeability to air of various vulcanisates containing 50 pphr of carbon black as function of temperature.[1]

therefore have low resilience at ambient temperatures (although the resilience does rise markedly with increasing temperature), and much of the energy used in deforming the rubber is dissipated in viscous processes rather than being stored for subsequent recovery.

7.2.2 Production of Isobutene–Isoprene Rubber

Isobutene is not susceptible to free-radical polymerisation. It is not therefore possible to produce isobutene–isoprene rubber by emulsion polymerisation. Nor is isobutene susceptible to polymerisation by anionic initiators. It can, however, be readily polymerised by a cationic mechanism using initiators such as boron trifluoride and aluminium chloride.

Isobutene–isoprene rubber is produced by copolymerising the two monomers in solution in methyl chloride using aluminium chloride as the initiator. The polymerisation is carried out at low temperature (below $-90\,°C$). The heat of polymerisation is removed by causing it to evaporate a liquefied hydrocarbon gas, such as ethylene, which is being circulated through a jacket which surrounds the reactor. Although the two monomers are readily soluble in methyl chloride, the rubber derived from them is not. The rubber therefore precipitates as a slurry of fine particles as it forms. This slurry is passed to a flash tank which contains water heated by steam. The methyl chloride and residual monomers vaporise, and are then compressed, dried and purified for recycling. The rubber is left as a coarse slurry in hot water. A small amount of an antiagglomerant (usually a mixture of stearic acid and zinc stearate in amounts ranging from 0·4 to 1·0 % of the weight of rubber hydrocarbon) is added at this stage, and also sufficient sodium hydroxide to decompose the aluminium chloride initiator. A small amount of a polymer stabiliser (such as alkylated diphenylamines, alkylated phenyl phosphites or alkylated phenols to the extent of 0·02–0·15 % of the rubber hydrocarbon) is also usually added at this stage to protect the rubber from degradation during the drying process, during storage, and during the early stages of compounding. The rubber is separated from most of the water by screening, and is then fed successively into dewatering and drying extruders where most of the remaining water is squeezed out. On leaving the drying extruder, the residual water, which has been heated under pressure, forms steam which explodes the rubber into a dry, fluffy, porous crumb. The dry crumb is then cooled on a conveyor system and fed into balers. Initially the bales are opaque because of the air entrapped in the porous crumb; however, they tend to become transparent on standing.

7.2.3 Effect of Polymerisation Temperature and Monomer Concentration upon Polymer Produced

The reason why isobutene rubber is produced by polymerisation at such a low temperature is that only if the reaction is carried out at low temperature is a product of high molecular weight obtained. The logarithm of the degree of polymerisation of the polymer increases linearly with the reciprocal absolute polymerisation temperature, and the molecular weight of the product is influenced by the monomer concentration. At low temperatures (that is, below about $-45\,°C$), low monomer/diluent ratio favours high molecular weight, whereas at higher temperatures the converse is the case. Thus it is possible to produce isobutene–isoprene rubbers of widely varying molecular weights and plasticities by appropriate choice of polymerisation temperature and monomer/diluent ratio.

7.2.4 Types of Isobutene–Isoprene Rubber Available Commercially

The various grades of isobutene–isoprene rubber which are available commercially differ amongst themselves as regards the level of unsaturation (determined by the ratio of isoprene to isobutene in the monomer feed) as well as by their plasticity. It is therefore possible in principle to have a very large number of different grades. In fact, the grades which are available commercially tend to fall into fairly well-defined areas as regards unsaturation and plasticity. These areas can be conveniently delineated as in Fig. 7.2, in which plasticity and unsaturation are indicated by means of vertical and horizontal scales respectively. Two scales have been used for the representation of plasticity: for the grades of higher plasticity, the Mooney viscosity at 100 °C (ML1 + 8 min) is used as the index, whereas for the grades of lower plasticity, the index is Mooney viscosity at 125 °C (ML1 + 12 min).

The reason why grades of differing plasticities are available commercially is that the offer of a choice enables the rubber processor to select a raw rubber which has a plasticity appropriate to the product he wishes to make and also to the process he intends to use for its manufacture. It will be seen

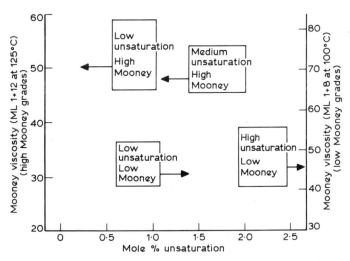

FIG. 7.2.   Illustrating broad classification of commercially-available grades of isobutene–isoprene rubber, according to unsaturation and plasticity.[1]

subsequently that, in common with many other synthetic rubbers, the isobutene–isoprene rubbers do not normally break down appreciably during mastication. It is not therefore normally possible to reduce the plasticity of an isobutene–isoprene rubber by mastication; instead, it is necessary to select a grade of rubber which has an appropriate initial plasticity. The reason why rubbers having a range of levels of unsaturation are available is that the rubber processor is thereby offered a choice in respect of the inherent rate of vulcanisation of the rubber with sulphur/ accelerator systems. As the level of unsaturation in the rubber increases, so (a) the inherent rate of vulcanisation with sulphur increases (this matter will be dealt with in more detail subsequently), and (b) the inherent chemical reactivity of the rubber (particularly towards ozone) also increases. The rubber processor usually has to compromise in the choice of the grade of isobutene–isoprene rubber with respect to level of unsaturation, the compromise being between vulcanisation rate on the one hand and chemical inertness on the other. In respect of ozone resistance, it may be noted that the ozonolysis of isoprene units which were enchained in the 1,4 mode leads to main-chain scission, and therefore to degradation. Ozonolysis of units enchained in the 1,2 and 3,4 modes does not lead to main-chain scission.

In addition to the grades summarised in Fig. 7.2, crosslinked grades of isobutene–isoprene rubbers are also available. These are produced by including in the monomer feed mixture a minor proportion of a tetrafunctional crosslinking monomer. Such rubbers can be used to produce stocks which resist flow under low shear; stocks of this type are required for certain applications such as caulking. Normal isobutene–isoprene rubbers are free of gel, and in consequence are entirely soluble in hydrocarbon solvents such as hexane, benzene and toluene.

7.2.5 Compounding Principles

Vulcanisation

Because of the presence of the small amount of unsaturation arising from the copolymerised isoprene, isobutene–isoprene rubbers can be vulcanised with sulphur and conventional accelerators. This is the curing system which is used for most applications of these rubbers. Compared to diene rubbers such as natural rubber and styrene–butadiene rubber, isobutene–isoprene rubbers are rather sluggish in vulcanisation behaviour. They therefore require more active accelerator systems and lower levels of sulphur than do the diene rubbers in order to achieve a satisfactory balance of vulcanisate mechanical properties. Combinations of thiuram sulphides and thiazole accelerators are commonly used, especially where cost is important and the vulcanisate is not required to be used at high temperature. Dithiocarbamate accelerators give fast cures and a vulcanisate with good ageing behaviour, but tend to give rather scorchy stocks. A combination of a dithiocarbamate and a delayed-action thiazole can be used to provide a balance of fast cure and processing safety. A typical general-purpose vulcanisation system would consist of 1·25 pphr of sulphur in combination with 1·5 pphr 2-mercaptobenzthiazole and 1·0 pphr of tetramethyl thiuram disulphide. Isobutene–isoprene rubbers can also be vulcanised by means of sulphur donors, such as the thiuram polysulphides.

As has already been noted, in accordance with expectation the reactivity of isobutene–isoprene rubbers towards sulphur and sulphur donors increases as the level of unsaturation in the rubber increases. This is illustrated in Fig. 7.3, which shows the progress of crosslink insertion in

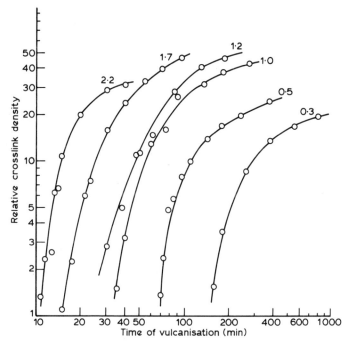

FIG. 7.3.   Crosslink density as function of time of vulcanisation for vulcanisation of various isobutene–isoprene rubbers with tetramethylthiuram disulphide at 261 °F.[2] The numbers appended to the curves indicate the mole % of unsaturation of the rubber.

rubbers containing various levels of unsaturation, as revealed by measurement of equilibrium swelling in cyclohexane at 25 °C. The vulcanising agent
was tetramethyl thiuram disulphide, and the level of unsaturation ranged
from 0·3 to 2·2 mole %. It has been found that, for a given sulphur
vulcanisation system, the logarithm of the time required to reach a given
state of vulcanisation, as judged by equilibrium swelling, decreases linearly
with the logarithm of the concentration of carbon–carbon double bonds in
the polymer chain. This is illustrated in Fig. 7.4 for two vulcanisation
systems.

There are many other compounds besides sulphur and sulphur donors
which can be used for effecting the vulcanisation of isobutene–isoprene
rubbers. Of these, many are aromatic polynitroso compounds and quinone
oximes. Two such vulcanising agents have attained some practical
importance. These are *p*-benzoquinone dioxime (LI) and dibenzoyl-*p*-

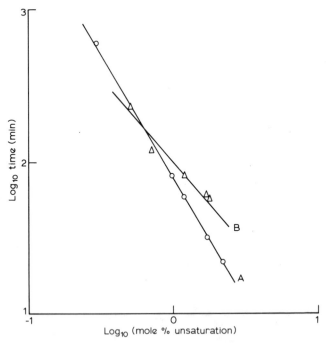

FIG. 7.4.   Time of vulcanisation required to attain a fixed level of crosslink density
as a function of unsaturation of isobutene–isoprene rubber for vulcanisation by
tetramethylthiuram disulphide (A) and by sulphur plus *N*-cyclohexyl-2-benzthiazyl
sulphenamide (B).[2]

benzoquinone dioxime (LII). These substances are used in conjunction with an inorganic oxidising agent such as red lead. A typical combination would be 6 pphr of (LII) plus 10 pphr of red lead.

NOH	NO.CO. ϕ	NO
NOH	NO.CO. ϕ	NO
(LI)	(LII)	(LIII)

Vulcanisation systems of this type give very fast cure rates and are well-suited to continuous vulcanisation processes for, say, the production of insulated cables. The crosslinks produced by this type of vulcanisation system are thermally more stable than are those produced by sulphur systems; the vulcanisates therefore age less rapidly at high temperatures. The quinone dioximes are not able to effect the vulcanisation of isobutene–isoprene rubbers in the absence of an oxidising agent, but the aromatic polynitroso compounds are. It is therefore thought that the first step of vulcanisation by the quinone dioxime system is oxidation of the quinone dioxime to a dinitroso compound, e.g., (LIII).

Isobutene–isoprene rubbers can also be vulcanised by heating with suitable phenol–formaldehyde resins. The process is slow and requires a high temperature. However, the vulcanisates exhibit very good resistance to ageing at high temperatures in air and in steam. This vulcanisation system is therefore mainly used for the production of curing bags and bladders (for use in tyre production).

Normal grades of isobutene–isoprene rubber cannot be vulcanised satisfactorily by heating with organic peroxides. The crosslinked grades are, however, vulcanisable with peroxides, although the vulcanisates tend to have a rather rough surface finish.

Reinforcement
Gum vulcanisates from isobutene–isoprene rubbers have high tensile strength. However, in order to achieve high resistance to tear and abrasion, these rubbers are normally compounded with a reinforcing filler such as carbon black. The effects observed are similar to those observed with other hydrocarbon rubbers. But in the case of isobutene–isoprene rubbers, optimal response to reinforcing blacks of the furnace type is achieved by

heating the mixture of rubber and carbon black to a temperature in the range 160–200 °C. It is found that the tendency of the vulcanisate to dissipate energy during mechanical deformation diminishes with successive cycles of heating and milling of the rubber/carbon black mixture from which the vulcanisate was made. As well as imparting high resilience, heat treatment gives vulcanisates which have higher modulus and higher tensile strength.

The changes which accompany heat treatment and milling of the rubber/carbon black mixtures are attributed to improved dispersion of the carbon black and to interaction between the black particles and the rubber matrix. It has been found that various substances can be added to the rubber/carbon black mixture as 'promoters' of these changes. These substances include p-dinitrosobenzene (LIII), p-nitrosophenol (LIV), N-methyl-N, 4-dinitrosoaniline (LV) and N-(2-methyl-2-nitropropyl)-4-nitrosoaniline (LVI). These compounds vary widely in their effectiveness as promoters, (LV) being particularly efficient. However, they all have certain features in common. Thus they are all p-substituted aromatic nitroso compounds, and they are all crosslinking agents for isobutene–isoprene rubbers.

NO   NO   NO

OH CH_3—N—NO $NH.CH_2.C.CH_3$ with CH_3 above and NO_2 below the C

(LIV)   (LV)   (LVI)

7.2.6 Processing

Stabilised grades of isobutene–isoprene rubber are resistant to breakdown during mastication. The unstabilised grades do, however, undergo breakdown when masticated at high temperatures. Figure 7.5 illustrates the effect of milling at 150 °C in air upon the two types of grade. Raw isobutene–isoprene rubber is nervy and may be difficult to band on the mill. This problem can be overcome by banding a rubber/filler masterbatch set aside from a previous batch of the same mix, and then adding alternate increments of polymer and filler.

Contamination with conventional diene rubbers, such as natural rubber and styrene–butadiene rubber, during mixing must be avoided. This is because such rubbers interfere with the vulcanisation of isobutene–

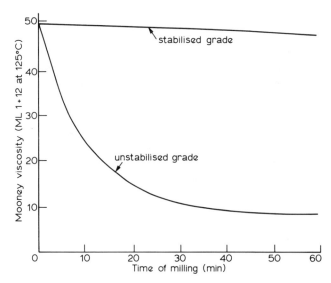

FIG. 7.5.   Effect of milling in air at 150 °C upon stabilised and unstabilised grades of isobutene–isoprene rubber.[1]

isoprene rubbers by sulphur and accelerators. This is a consequence of the low unsaturation of isobutene–isoprene rubbers as compared with conventional diene rubbers; the large numbers of unsaturated sites in small amounts of a diene rubber are able to compete successfully for the curatives with the small number of unsaturated sites in the isobutene–isoprene rubber. For the same reason, it is not possible to co-vulcanise blends of these rubbers and diene rubbers; indeed, such blends are not usually contemplated.

A further problem in connection with the processing of isobutene–isoprene rubbers arises from their low permeability to gases and vapours. Any moisture or air which is entrapped in the rubber stock has difficulty in escaping during vulcanisation. This is particularly the case during moulding. Precautions which can be taken to avoid porosity and surface imperfections arising from this cause include (a) adequate 'bumping' of the press before final closure of the mould, and (b) ensuring that all the compounding ingredients are thoroughly dry before mixing.

7.2.7. Other Related Rubbers

It is convenient to note here that interest has been shown in certain other olefin rubbers which are closely related to the isobutene–isoprene rubbers.

For example, 3-methyl-1-butene (LVII) and 4-methyl-1-pentene (LVIII) have been polymerised to rubbers by means of cationic catalysts at low temperatures. Rather surprisingly, the repeat units which are present in these rubbers are not predominantly those which would be expected to arise from 1,2 addition. Instead, they are the repeat units which are shown at (LIX) and (LX) respectively. Thus the polymer obtained by this means from 4-methyl-1-pentene can be regarded as being an alternating copolymer of ethylene and isobutene, whereas that obtained from 3-methyl-1-butene can be regarded as being an alternating sequence of methylene units and units derived from isobutene. It has been suggested that repeat units of the types (LIX) and (LX) occur because most of the carbonium ions which form in

$$
\begin{array}{c}
\overset{\displaystyle CH_3}{\underset{\displaystyle CH_3}{\overset{\displaystyle |}{\underset{\displaystyle |}{CH_2\!=\!CH\!-\!CH}}}}
\end{array}
\qquad\qquad
\begin{array}{c}
\overset{\displaystyle CH_3}{\underset{\displaystyle CH_3}{\overset{\displaystyle |}{\underset{\displaystyle |}{CH_2\!=\!CH\!-\!CH_2\!-\!CH}}}}
\end{array}
$$

$$
\text{(LVII)} \qquad\qquad\qquad\qquad \text{(LVIII)}
$$

$$
\cdots\!-\!CH_2\!-\!CH_2\!-\!\overset{\displaystyle CH_3}{\underset{\displaystyle CH_3}{\overset{\displaystyle |}{\underset{\displaystyle |}{C}}}}\!-\!\cdots
\qquad\qquad
\cdots\!-\!CH_2\!-\!CH_2\!-\!CH_2\!-\!\overset{\displaystyle CH_3}{\underset{\displaystyle CH_3}{\overset{\displaystyle |}{\underset{\displaystyle |}{C}}}}\!-\!\cdots
$$

$$
\text{(LIX)} \qquad\qquad\qquad\qquad\qquad \text{(LX)}
$$

the course of the polymerisation are subject to a rapid 'hydride ion' shift, which takes place before the carbonium ion which first forms can react further. According to this view, repeat units of the type (LIX) are formed from 3-methyl-1-butene by the following sequence of reactions (in which the associated counterions are omitted for simplicity):

$$
\overset{\displaystyle CH_3}{\underset{\displaystyle CH_3}{\overset{\displaystyle |}{\underset{\displaystyle |}{CH_2\!=\!CH\!-\!CH}}}} + H^{\oplus} \longrightarrow CH_3\!-\!\overset{\displaystyle CH_3}{\underset{\displaystyle CH_3}{\overset{\displaystyle |}{\underset{\displaystyle |}{\overset{\oplus}{CH}\!-\!CH}}}} \longrightarrow CH_3\!-\!CH_2\!-\!\overset{\displaystyle CH_3}{\underset{\displaystyle CH_3}{\overset{\displaystyle |}{\underset{\displaystyle |}{C^{\oplus}}}}}
$$

$$
CH_3\!-\!CH_2\!-\!\overset{\displaystyle CH_3}{\underset{\displaystyle CH_3}{\overset{\displaystyle |}{\underset{\displaystyle |}{C^{\oplus}}}}} + CH_2\!=\!\overset{\displaystyle CH_3}{\underset{\displaystyle CH_3}{\overset{\displaystyle |}{\underset{\displaystyle |}{CH\!-\!CH}}}}
$$

$$
\downarrow
$$

$$
\underset{\underset{CH_3}{|}}{CH_3-CH_2-\overset{\overset{CH_3}{|}}{C}-CH_2-}\overset{\oplus}{CH}-\underset{\underset{CH_3}{|}}{\overset{\overset{CH_3}{|}}{CH}}
$$

$$\downarrow$$

$$
\underset{\underset{CH_3}{|}}{CH_3-CH_2-\overset{\overset{CH_3}{|}}{C}-CH_2-CH_2-}\underset{\underset{CH_3}{|}}{\overset{\overset{CH_3}{|}}{C}}{}^{\oplus}
$$

A similar scheme of reactions will explain the formation of repeat units of the type LX from 4-methyl-1-pentene.

7.3 HALOGENATED ISOBUTENE–ISOPRENE RUBBERS (CIIR AND BIIR)

Two types of halogenated isobutene–isoprene rubber are currently available commercially:

1. those produced by chlorination of isobutene–isoprene rubber; and
2. those produced by bromination of isobutene–isoprene rubber.

Chlorination and bromination can be carried out by adding a solution of the halogen to a solution of the rubber in an inert solvent. The corresponding halogen halide is evolved almost immediately. Some 3 % by weight of halogen can be reacted with the rubber without serious degradation of the rubber taking place. More care in avoiding degradation is necessary in the case of chlorination than in the case of bromination, because the chlorination reaction is more inclined to be accompanied by main-chain scission than is the bromination reaction.

It might be thought that halogenation involves the simple addition of the halogen molecule to the carbon–carbon double bonds of the isoprene units of the isobutene–isoprene rubber. In fact, the mechanism of the bromination is believed to be as follows:

$$
\sim\!\!CH_2-\underset{\underset{CH_3}{|}}{C}\!\!=\!\!CH-CH_2\!\sim\; +\; Br_2 \longrightarrow \sim\!\!CH_2-\underset{\underset{CH_3}{|}}{\overset{\oplus}{C}}-CHBr-CH_2\!\sim\; +\; Br^{\ominus}
$$

$$\updownarrow$$

$$
\sim\!\!CH_2-\underset{\underset{CH_2}{\|}}{C}-CHBr-CH_2\!\sim\; \overset{-H^{\oplus}}{\longleftarrow}\; \sim\!\!CH_2-\underset{\underset{CH_3}{|}}{\overset{\overset{Br}{\diagup\overset{\oplus}{\diagdown}}}{C}}-CH-CH_2\!\sim
$$

Thus virtually all of the bromination does take place at the isoprene units of the copolymer, but steric hindrance of the carbon–carbon double bonds favours substitution rather than addition. It appears that approximately 90% of the bromine in the final product is 'allylic' to the carbon–carbon double bonds. It is seen from the above mechanism that much of the original unsaturation is retained, although it has become largely isomerised. Presumably the mechanism of chlorination is similar to that of bromination. The extents of halogenation in the commercially-available rubbers of this type are in the range 1·0–1·5 mole %.

The advantages which halogenated isobutene–isoprene rubbers offer over the conventional types are as follows:

1. They can be vulcanised by several reagents which do not vulcanise the conventional types.
2. In the case of vulcanisation systems which vulcanise both types, the halogenated rubbers vulcanise more rapidly.
3. They can be co-vulcanised with conventional diene rubbers.
4. They adhere well to themselves and to other rubbers when they are vulcanised in contact with each other.
5. For a given vulcanising system, they show better heat resistance.

Halogenated isobutene–isoprene rubbers can be vulcanised by all the reagents which vulcanise the conventional rubbers. In addition, several other vulcanising systems can be used. Thus the halogenated rubbers can be vulcanised by heating with organic peroxides. Although peroxides alone can be used, high degrees of vulcanisation can be achieved only if a co-vulcanising agent is used, an example of which is *N,N'-m*-phenylene-dimaleimide (LXI). A typical combination would be 1·5 pphr of dicumyl peroxide plus 1·5 pphr of (LXI). Halogenated isobutene–isoprene rubbers can also be vulcanised by heating with metal oxides, in particular, with zinc oxide. This is probably the method of vulcanising these rubbers which is

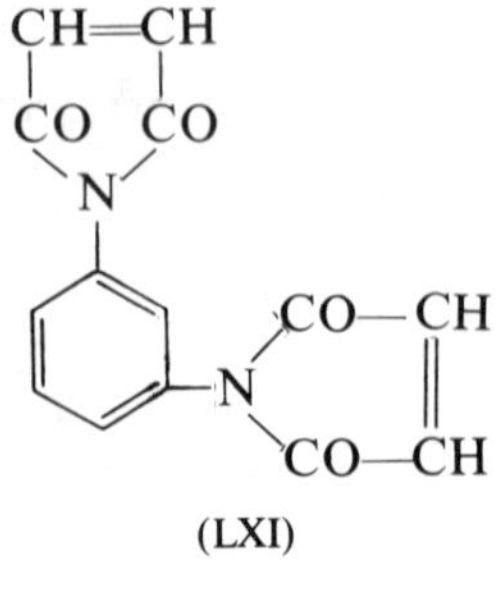

(LXI)

most commonly used. The reaction which leads to the formation of crosslinks is presumably similar to that which leads to the crosslinking of chloroprene rubber by heating with metal oxides (see Chapter 6). It will be recalled that it is the small number of reactive allylic chlorine atoms in chloroprene rubber which is believed to be responsible for its ability to be crosslinked with metal oxides. It is therefore probably significant that most of the halogen atoms in halogenated isobutene–isoprene rubbers are believed to be of the allylic type. The vulcanisation of chlorinated isobutene–isoprene rubber by heating with zinc oxide is illustrated in Fig. 7.6. There are several other combinations of additives which can be used to vulcanise the halogenated isobutene–isoprene rubbers which are not available for the non-halogenated rubbers. These include zinc oxide plus accelerator, zinc oxide plus resin, and sulphur plus a base such as calcium hydroxide.

The ability to co-vulcanise with, or perhaps to vulcanise in the presence of, conventional diene rubbers is important for two reasons. In the first place, it eliminates the problems which can arise when conventional isobutene–isoprene rubbers become accidentally contaminated with small amounts of diene rubbers during mixing. But, secondly, it permits the

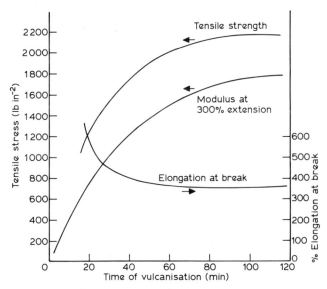

FIG. 7.6.   Vulcanisation of chlorinated isobutene–isoprene rubber by heating with zinc oxide (5 pphr) at 307°F.[2]

exploitation of blends of isobutene–isoprene rubbers with conventional diene rubbers. The reason why the halogenated isobutene–isoprene rubbers are cure-compatible with diene rubbers, whereas the non-halogenated rubbers are not, is probably two-fold: (a) the halogenated rubbers vulcanise much faster with sulphur and accelerators than do the non-halogenated rubbers, and therefore the vulcanisation reaction is not upset to the same extent by the presence of the diene rubber; and (b) other vulcanisation mechanisms are available to halogenated rubbers, e.g., reaction with the zinc oxide which is inevitably present as part of the sulphur-curing system. Whatever may be the true chemical explanation, the vulcanisation behaviour of blends of halogenated isobutene–isoprene rubbers and diene rubbers is markedly different from that of similar blends based upon non-halogenated rubbers. This is illustrated in Fig. 7.7, which shows the tensile strength of

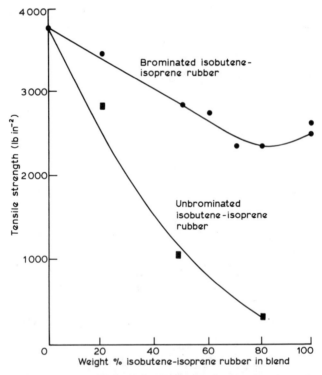

FIG. 7.7. Tensile strengths of vulcanisates from blends of (a) brominated isobutene–isoprene rubber plus natural rubber, and (b) conventional isobutene–isoprene rubber plus natural rubber.[2]

vulcanisates obtained from the two types of blend as a function of the blend composition. The two types of blend to which these results refer are (a) brominated isobutene–isoprene rubber plus natural rubber, and (b) conventional isobutene–isoprene rubber plus natural rubber.

The chlorinated and brominated rubbers are generally very similar in behaviour. The brominated rubbers are, however, rather more reactive than are the chlorinated rubbers. Thus, for example, vulcanisation by heating with zinc oxide occurs more rapidly in the case of the brominated rubber.

7.4 ETHYLENE–PROPYLENE RUBBERS (EPM AND EPDM)

7.4.1 Introduction

We come now to consider the important class of general-purpose hydrocarbon synthetic rubbers which are made by the copolymerisation of ethylene and propylene as the main monomers. The principal repeat units in these rubbers are shown at (LXII) and (LXIII) below. Interest in these comparatively late additions to the range of general-purpose synthetic rubbers has its origin principally in two considerations. In the first place, these rubbers are comparatively cheap, their price reflecting the widespread availability of the two main monomers. Secondly, in common with other rubbers derived from olefins, the majority of the repeat units in the polymer sequence are saturated. These rubbers are in consequence rather inert chemically. They show very good resistance to ageing, and outstanding resistance to deterioration by ozone. They also show good resistance to attack by chemical reagents generally.

$$\cdots -CH_2-CH_2- \cdots \qquad \cdots -CH_2-\underset{\underset{\displaystyle CH_3}{|}}{CH}- \cdots$$

(LXII) (LXIII)

A copolymer which consists entirely of units derived from ethylene and propylene is, of course, entirely saturated, apart from the possible presence of carbon–carbon double bonds in end groups. Such rubbers are available commercially, and are known generically as EPM (in which the E and the P denote, of course, ethylene and propylene respectively, and the M denotes a rubber which has a saturated chain of the polymethylene type). They are also sometimes referred to as ethylene–propylene 'bipolymers'. These rubbers can be vulcanised by heating with organic peroxides, but they cannot be vulcanised by heating with sulphur and conventional accelerators. For this reason, a second family of ethylene–propylene rubbers has been developed. In these rubbers, the majority of the repeat units are still derived from ethylene and propylene, but there is also a minority of units derived from a third monomer. This third monomer is chosen in such a way that the units derived from it are unsaturated, and of appropriate reactivity to enable the rubber subsequently to be vulcanisable by heating with sulphur and accelerators. Thus the third monomer in these rubbers fulfils the same role as does the isoprene in the isobutene–isoprene rubbers. The nature of the third monomer will be considered subsequently. We merely note here that these latter rubbers are sometimes known generally as EPT (ethylene–propylene terpolymers), but more specifically as EPDM (in which again the E and the P signify ethylene and propylene respectively, the D signifies that the rubber contains units derived from a diene monomer (more strictly that the diene units are enchained in such a way that the residual unsaturation is in a side chain), and the M has the same significance as in EPM).

There is one further general matter which requires to be dealt with in this introduction to the ethylene–propylene rubbers. This is the question of why these rubbers were comparatively late additions to the range of general-purpose synthetic rubbers. It was not until the early 1960s that these rubbers became commercially available, notwithstanding that ethylene and propylene had been readily available for many years previously. The reason for the late development of these rubbers is that it was not until the late 1950s that suitable catalyst systems were discovered for the copolymerisation of ethylene and propylene to give rubbery products. Interest in these rubbers has increased steadily since they first became available commercially. There are currently approximately ten large companies producing them. The interest is now mainly in the terpolymer grades, rather than in ethylene–propylene bipolymers; according to a recent estimate, some 95% of the ethylene–propylene rubbers used are now the terpolymer grades.

7.4.2 Effect of Varying the Ethylene/Propylene Ratio

The carbon–carbon bonds in a sequence of ethylene units have a high degree of rotational flexibility. The same is true to a lesser extent of sequences of propylene units. Notwithstanding this, polypropylene and (to a lesser extent) polyethylene are rigid plastics at normal temperatures. This illustrates the fact that rotational flexibility of some or all the bonds in the main chain of a polymer sequence does not of itself guarantee that the polymer in bulk will exhibit the long-range reversible elasticity which is the characteristic of rubbery materials. Other factors are also important. In the case of polyethylene and polypropylene, it is the tendency of the successive repeat units to arrange themselves on a crystal lattice which is the cause of the absence of the expected elastomeric properties.

The crystallisability of polyethylene arises from the essential simplicity of the repeat unit; in the case of an entirely linear polyethylene, the repeat unit is simply $-CH_2-$. The regularity of a polyethylene sequence is disturbed by the presence of both short and long chains, and the extent to which such chains occur depends upon the catalyst system which was used for the production of the polymer. In general, the more nearly linear is a polyethylene sequence, the greater is its tendency to crystallise and the higher is the proportion of the polymer which is present as crystalline material at room temperature. The consequences of this for appearance and mechanical properties are that, as the proportion of crystalline polymer increases, so the polymer becomes harder, more rigid, less extensible, and less rubbery. A range of polyethylenes is produced commercially, the various types differing both as regards linearity and molecular weight, and therefore as regards hardness and rigidity. They are all, however, sufficiently rigid to be regarded and used as plastics rather than as rubbers. (However, it is interesting to note in passing that, if a few crosslinks are introduced into polyethylene by, say, high-energy irradiation, then the ability to crystallise is severely reduced and the product is rubbery in character.)

The crystallisability of polypropylene also arises from the simplicity of the repeat unit. However, in this case there are factors other than branching which can inhibit the tendency to crystallise. Chief amongst these is irregularity of the polymer sequence arising from the random enchainment of successive units having D- and L-configurations. The extent to which the sequence is stereoregular in this respect is determined largely by the catalyst system which was used to effect the polymerisation of the propylene. In general, the more stereoregular the polymer sequence, the greater is the tendency to crystallise and the more rigid is the polypropylene.

The possibility of making elastomeric copolymers of ethylene and propylene depends upon the fact that, although linear sequences of either type of unit separately have a strong tendency to crystallise, the two types of unit are not able to co-crystallise satisfactorily. Thus each type of unit is able to inhibit the crystallisation tendency of the other, thereby allowing the inherent rotational flexibility of the two types of main-chain bond to be manifest as elastomeric character in the bulk material.

From what has been said so far, it will be apparent that the rubberiness of an ethylene–propylene copolymer will depend upon the extent to which the copolymer is random, as well as upon the ethylene/propylene ratio. More will be said about the former aspect in the next section; we merely note here that the longer are the successive blocks of ethylene and propylene units in the copolymer, the less will be the inhibition of crystallisation, and the harder and less elastomeric will be the copolymer in bulk at normal temperatures.

The concern here is primarily with the effect of ethylene/propylene ratio upon the properties of the copolymer. The copolymers are rubbery over a wide range of ethylene/propylene ratio. The rubbers which are available commercially range from about 30 % to 70 % in respect of ethylene content by weight. The effect of ethylene content upon the tensile stress–strain behaviour of unvulcanised ethylene–propylene rubbers is shown schematically in Fig. 7.8. Rubbers which are rich in either ethylene or propylene have higher tensile strength and (particularly) elongation at break in the unvulcanised state than do rubbers which contain approximately equal amounts of the two comonomers. This is a reflection of the ability of ethylene- and propylene-rich copolymers to crystallise on stretching. Over the range 20–50 % polypropylene content by weight, the 'green strength' of unvulcanised compounds falls rapidly with increasing polypropylene content.

The main effect upon vulcanisate properties of increasing the ethylene/propylene ratio over the range for which the copolymers are rubbery is to increase the degree of rubberiness and, specifically, the rebound resilience.

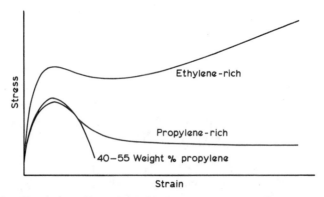

FIG. 7.8.   Illustrating effect of ethylene content upon tensile stress–strain behaviour of unvulcanised ethylene–propylene rubbers.[3]

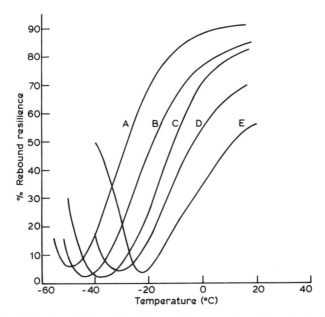

FIG. 7.9.   Variation of rebound resilience with temperature for vulcanisates from ethylene–propylene rubbers of various ethylene contents as follows: A 72·5, B 61·5, C 52, D 44 and E 30 mole % of ethylene.[4]

This effect is illustrated in Fig. 7.9, which shows rebound resilience as a function of temperature for ethylene–propylene copolymer vulcanisates of various ethylene contents. As the ethylene content decreases, so too does the resilience of the vulcanisate at normal temperatures. In common with all elastomers, the rebound resilience falls as the temperature is reduced, and passes through a minimum. The temperature at which this minimum occurs is very close to the glass-transition temperature of the rubber. It is evident from Fig. 7.9 that the temperature at which the rebound resilience is a minimum falls as the ethylene content increases. In fact, this temperature falls linearly with increasing ethylene content; this is illustrated in Fig. 7.10. These changes which are observed as the ethylene/propylene ratio is varied are consequences of the fact that the rubberiness of ethylene–propylene rubbers is due primarily to the inherent flexibility of sequences of methylene groups whose tendency to crystallise has been suitably discouraged. As long as crystallisation does not occur to any appreciable extent, the flexibility of the polymer sequence increases with the proportion of ethylene units.

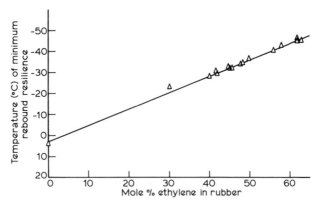

FIG. 7.10.   Effect of ethylene content upon temperature at which rebound resilience of vulcanisates from ethylene–propylene rubbers is a minimum.[4]

If the ethylene/propylene weight ratio is increased much above 70/30 (which corresponds to a mole ratio of about 78/22), then products are obtained which are not very rubbery in character. The temperature at which the rebound resilience is a minimum is considerably higher than that which would be predicted by extrapolation of the straight line shown in Fig. 7.10. In fact, as the ethylene/propylene mole ratio is increased above 60/40, so the temperature at which the rebound resilience is a minimum itself passes through a minimum. The ethylene/propylene mole ratio for which the minimum-resilience temperature is itself a minimum is in the region of 70/30, and this corresponds to a weight ratio of approximately 60/40. The existence of a maximum in the mobility of ethylene–propylene copolymer sequences as the ethylene content is increased is clearly demonstrated by the existence of a shallow minimum in glass-transition temperature as the copolymer composition is varied, as is illustrated in Fig. 7.11.

As well as having higher rebound resilience, vulcanisates from rubbers of high ethylene content tend to be rather stiffer at room temperature, and thus have higher modulus and hardness at a given crosslink density. They also tend to have higher tensile strengths. These effects are attributed to an enhanced tendency to crystallise during deformation. In general, the static and dynamic mechanical properties of vulcanisates improve as the ethylene content rises, provided that the ethylene content is not so high that appreciable crystallisation occurs in the undeformed state.

If two or more ethylene–propylene copolymers of different overall compositions are blended, then the mechanical properties of the blend, and

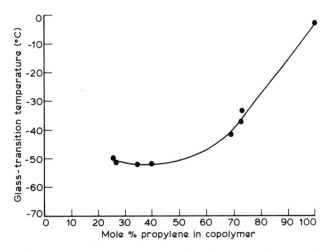

FIG. 7.11.   Glass-transition temperatures of ethylene–propylene copolymers as function of propylene content.[5]

of vulcanisates obtained from it, are usually those which would be predicted from the overall ethylene/propylene ratio in the blend. In particular, the temperature at which the rebound resilience is a minimum is usually near to that which is predicted from the overall ethylene content using the line of Fig. 7.10. However, such predictions may be invalidated if one of the components of the blend has a distribution of monomer sequences which enables it to crystallise within the blend. In such cases, the temperature corresponding to minimum rebound resilience is markedly higher than that which is predicted from the overall composition. A corollary of these considerations is that the mechanical properties of an ethylene–propylene rubber do not depend greatly upon the way in which the overall ethylene/propylene ratio is distributed amongst the various molecules which make up the rubber, so long as the distribution is not so uneven as to give rise to effects which arise from crystallisation and immiscibility.

7.4.3 Effect of Sequence Distribution

The nature of the distribution of ethylene–propylene sequence lengths in an ethylene–propylene rubber has little effect upon the mechanical properties unless the sequence lengths are sufficient to permit crystallisation to occur either as a result of stretching or, in more extreme cases, in the undeformed condition. The effect of the sequence-length distribution upon the tensile stress–strain curve of an unvulcanised ethylene–propylene rubber containing approximately 25 mole % of propylene is illustrated in Fig. 7.12. It is difficult to quantify the sequence-length distribution precisely, and no attempt will be made to discuss this problem here. It is sufficient to note that the curves shown relate to three types of copolymer, namely, one in which

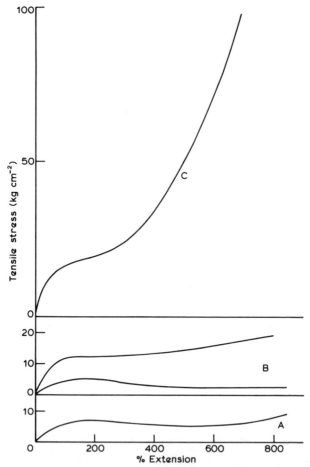

FIG. 7.12.   Illustrating effect of sequence-length distribution upon tensile stress–strain curve of an unvulcanised ethylene–propylene rubber containing approximately 25 mole % of propylene.[3] Curve A refers to a copolymer in which the ethylene and propylene units essentially alternate with one another, curve B to copolymers in which the distribution of the two types of unit is essentially random, and curve C to a copolymer in which the two types of unit occur in long blocks.

the ethylene and propylene units essentially alternate with one another, one in which the distribution is essentially random, and one is which the two types of unit occur in long blocks. The effect of the long blocks in promoting crystallisation during stretching is evident.

7.4.4 Third Monomers

The principal requirements for a third monomer which is to be incorporated to make an ethylene–propylene rubber sulphur-vulcanisable are as follows:

1. It should be multiply unsaturated, but only one of the carbon–carbon double bonds should be polymerisable by the catalyst system to be used for the manufacture of the rubber.
2. It should not seriously reduce the activity of the catalyst system, e.g., by complexing strongly with it.
3. It should copolymerise readily with ethylene and propylene under the polymerisation conditions being used for the production of the rubber. To the extent that it does not copolymerise readily, it should be easily recoverable and recyclable.
4. The residual unsaturated site which results from its incorporation in the polymer chain should be capable of reacting rapidly and efficiently with sulphur and accelerators during vulcanisation.

When attempts were first being made to develop sulphur-vulcanisable ethylene–propylene terpolymers, the use of simple, acyclic dienes as third monomers was investigated, because of their cheapness and availability. However, they were soon found to be unsuitable because they gave rise to problems of catalyst deactivation and homopolymer formation. Furthermore, quite apart from these problems, it was realised that 1,4 enchainment of such dienes would reduce the ozone resistance of the resultant rubber, although 1,2 and 3,4 enchainment would not. (This point has been discussed in more detail in connection with the isobutene–isoprene rubbers—see Section 7.2 of this chapter.) It soon became clear that, for an acyclic third monomer to be satisfactory, its carbon–carbon double bonds should be unconjugated, in order to avoid the possibility of 1,4 addition. Furthermore, one of the carbon–carbon double bonds should be readily polymerisable by the catalyst systems used for the production of ethylene–propylene rubbers, but the other should not. These requirements can be met by having one of the double bonds as a terminal vinyl group, and the other as an internal 1,2-disubstituted or more highly substituted double bond, as in 1,4 hexadiene (LXIV). Alternatively, the second double bond could be terminal but substituted, that is of the vinylidene type, as in 2-methyl-1,4-pentadiene (LXV).

$$CH_2\!\!=\!\!CH\!-\!CH_2\!-\!CH\!\!=\!\!CH\!-\!CH_3 \qquad\qquad CH_2\!\!=\!\!\underset{\underset{CH_3}{|}}{C}\!-\!CH_2\!-\!CH\!\!=\!\!CH_2$$

(LXIV)   (LXV)

A large number of cyclic and polycyclic dienes have also been examined as possible third monomers for the production of sulphur-vulcanisable ethylene–propylene rubbers. These include 1,4-cycloheptadiene (LXVI), dicyclopentadiene (LXVII), and 2-ethylidene-bicyclo(2,2,1)-5-heptene (LXVIII), the latter being commonly known as ethylidene norbornene (more precisely, 5-ethylidene-2-norbornene). Of these, (LXVIII) is said to be the most widely used. It is thought to incorporate mainly through the carbon–carbon double bond in the ring. It incorporates readily by copolymerisation, and the residual unsaturated site promotes rapid vulcanisation. (LXVII) is relatively cheap, but causes some problems because it can lead to branching. Furthermore, the residual unsaturated site is not so effective in promoting vulcanisation as is that derived from (LXVIII). (LXVI) is also used commercially, but is not so reactive as either (LXVII) or (LXVIII).

(LXVI)   (LXVII)   (LXVIII)

The proportion of third monomer used in commercial ethylene–propylene terpolymer rubbers varies widely, the range being about 3–25% by weight. However, because of the high molecular weights of the third monomers relative to those of ethylene and propylene, the molar proportion of third monomer is much lower than the weight content seems to indicate at first sight. Thus 25% by weight of ethylidene norbornene corresponds to approximately 10% of the enchained units, the exact proportion depending upon the ethylene/propylene ratio.

The presence of units derived from the third monomer has little effect upon the processing behaviour of ethylene–propylene rubbers. Nor do these units have much effect upon the mechanical properties of the vulcanisate, except insofar as these are affected indirectly by the susceptibility of the rubber to vulcanisation by sulphur and accelerators.

7.4.5 Production of Ethylene–Propylene Rubbers

Ethylene–propylene rubbers are produced by solution copolymerisation using catalysts of the Ziegler–Natta type. There are two basic types of process which are described broadly as the *solution process* and the *slurry* or *precipitation process* respectively.

In the *solution process*, the monomers and a solvent for both the monomers and the eventual polymer are cooled and fed into a reactor together with the catalyst components and a modifier (if used). The reaction mixture remains homogenous. The extent of the ethylene conversion is generally very high. The problem of monomer recovery and recycling is confined mainly to the propylene and any third monomer which may be used. The solution which effluxes from the reactor is treated with a catalyst deactivator, such as water or ethyl alcohol. The solution is then mixed with water, an antioxidant added, and the mixture heated to remove solvent and any unreacted monomers. At this stage, the polymer is in the form of a slurry of particles suspended in water, and a substantial part of the catalyst residue has been extracted into the water phase. The process of mixing the polymer solution with water is referred to as 'de-ashing', because the major component of the ash content of the final rubber is derived from the catalyst. (It should be noted that, as in the case of diene rubbers produced using Ziegler–Natta catalysis, the presence of catalyst residues in the rubber is disadvantageous principally because they are able subsequently to catalyse the oxidation of the rubber.) Most of the water in the polymer–water slurry is removed by screening; the remainder is removed by squeezing the rubber through an extruder and by drying. Because of the high viscosity of solutions of rubbery ethylene–propylene copolymers, a fairly high ratio of solvent to rubber produced is necessary. Thus the solvent is the major component of the mixture present in the reactor, and large quantities of solvent have to be recovered and purified relative to the quantity of rubber produced. Furthermore, the purification of the solvent before re-use is of critical importance, because the catalyst systems are so sensitive to poisoning.

The *slurry process* is analogous to that which is used for the production of isobutene–isoprene rubber. The polymerisation is carried out in an organic diluent which, at the polymerisation temperature, is a solvent for the monomers but not for the polymer. Typical diluents are methyl chloride and methylene dichloride. The low reactivity of propylene relative to ethylene also permits one of the main monomers to be used as a diluent. The advantage of the slurry process in comparison with the solution process lies in the comparatively low viscosity of the reaction medium, permitting the use of lower diluent/polymer ratios. There are certain disadvantages. Thus branching is more likely to occur during polymerisation because the polymer has precipitated from solution, and de-ashing is more difficult than in the solution process.

The catalyst system which seems to be most widely used for the production of ethylene–propylene rubber is derived from an alkylaluminium and a hydrocarbon-soluble vanadium compound.|Examples|of the former include ethylaluminium dichloride, diethylaluminium chloride and triisobutylaluminium. Examples of the latter include vanadyl chloride, vanadium tetrachloride, vanadium triacetylacetonate, and vanadyl diacetylacetonate. These catalysts are essentially homogeneous, and yield polymers with good elastomeric properties. It is also possible to use vanadium compounds which are not soluble in hydrocarbons, such as vanadium trichloride, but rather inhomogeneous polymers tend to be contained.

Unlike free-radical copolymerisation, the composition of the copolymers obtained from ethylene and propylene by Ziegler–Natta catalysis depends upon the nature of the catalyst as well as upon the concentrations of the two monomers in the liquid phase of the reaction system. Thus, for example, a catalyst derived from trihexylaluminium plus vanadyl chloride has been found to produce a copolymer which is richer in ethylene than that produced from a monomer mixture of the same composition by a catalyst derived from trihexylaluminium plus vanadium tetrachloride. If all other conditions remain unchanged, variables such as copolymerisation time, catalyst concentration and catalyst ageing have little effect upon the composition of the copolymer formed. The molar ratio of alkylaluminium compound to transition-metal compound does not usually affect the copolymer composition, although some effects have been reported for the system triisobutylaluminium plus vanadyl chloride in aliphatic hydrocarbon solvents.

The molecular weight of the rubber produced can be controlled to some extent by varying the reaction conditions in respect of temperature, catalyst concentration, catalyst type and catalyst age. Molecular-weight regulators are also used. Of the various compounds which have been proposed, hydrogen and the alkylzincs appear to be preferred. They seem to act as chain-transfer agents, rather like the mercaptan in a styrene–butadiene emulsion copolymerisation, and not as chain terminators like some other additives, e.g., the alkyl halides. They do not in consequence seriously reduce the rate of polymerisation.

7.4.6 Compounding Principles

Vulcanisation

Ethylene–propylene bipolymers are usually vulcanised by heating with organic peroxides, usually in the presence of other substances which are variously known as 'co-agents', 'promoters' or 'adjuvants'. In the absence of a co-agent, the mechanism by which crosslinking occurs is extraction of hydrogen from the polymer chain by reaction with the radicals which are derived from the peroxide by thermal decomposition, followed by mutual combination in pairs between the polymeric radicals so formed. Denoting by $R\cdot$ the radical formed by decomposition of the peroxide, and by PH the polymer chain, the crosslinking process can be represented as follows:

$$R\cdot + PH \longrightarrow RH + P\cdot$$

$$P\cdot + \cdot P \longrightarrow P-P$$

However, other reactions involving the polymeric radical $P\cdot$ can also occur. To the extent that these do not lead to the formation of crosslinks, they represent a wastage of the initial peroxide as a vulcanising agent. To the extent that they lead to main-chain scission, the peroxide has, in effect, acted as an agent of network degradation rather than of network formation.

The other reactions in which $P\cdot$ can participate include the following:

1. unimolecular chain scission without loss of radical activity;
2. bimolecular disproportionation without chain scission, but with loss of radical activity;
3. intrachain cyclisation with loss of radical activity; and
4. bimolecular combination with radicals derived from the peroxide.

Unimolecular chain scission is particularly prone to occur in polymeric radicals which have resulted from the abstraction of a tertiary hydrogen atom:

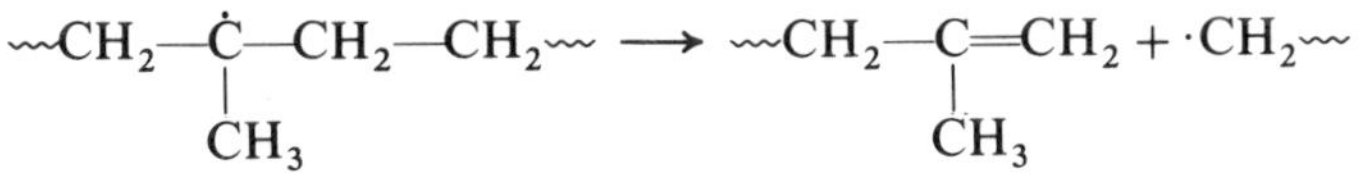

$$\sim CH_2-\overset{\displaystyle |}{\underset{\displaystyle CH_3}{\dot{C}}}-CH_2-CH_2\sim \longrightarrow \sim CH_2-\overset{\displaystyle |}{\underset{\displaystyle CH_3}{C}}=CH_2 + \cdot CH_2\sim$$

An example of bimolecular disproportionation is:

$$\sim CH_2-\overset{\displaystyle .}{C}-CH_2-CH_2\sim + \sim CH_2-\overset{\displaystyle .}{C}-CH_2-CH_2\sim$$
$$\qquad\quad \underset{\displaystyle CH_3}{|} \qquad\qquad\qquad\quad \underset{\displaystyle CH_3}{|}$$

$$\downarrow$$

$$\sim CH_2-CH-CH_2-CH_2\sim + \sim CH_2-C{=}CH-CH_2\sim$$
$$\qquad\quad \underset{\displaystyle CH_3}{|} \qquad\qquad\qquad\qquad \underset{\displaystyle CH_3}{|}$$

An example of cyclisation is:

$$\sim\overset{\displaystyle .}{C}H-(CH_2)_4-\overset{\displaystyle .}{C}H\sim \longrightarrow$$

$$
\begin{array}{c}
CH_2 \\
CH_2 \quad CH\sim \\
CH_2 \quad CH\sim \\
CH_2
\end{array}
$$

Bimolecular combination with radicals derived from the peroxides can be represented simply as:

$$P\cdot + \cdot R \longrightarrow P{-}R$$

In addition to these reactions, transfer reactions can occur in which the radical activity is transferred by hydrogen abstraction from one site on a polymer chain to another which may be on the same chain or on a neighbouring chain.

What are the factors which influence the extents to which these many possibilities occur? The first point to note is that the ease of abstraction of hydrogen atoms from alkanes decreases in the order tertiary > secondary > primary, and the reactivities of the radicals produced are in the reverse order. The proportion of the various types of radical which form will clearly depend upon the ethylene/propylene ratio in the polymer, since this ratio will affect the relative populations of the various types of carbon–hydrogen bond. In fact, reference to the structures (LXII) and (LXIII) will show that the ratio of primary:secondary:tertiary carbon–hydrogen bonds is $3:2 + 4(E/P):1$, where E/P is the mole ratio of propylene to ethylene units. However, because of the relative ease of abstraction of the various kinds of hydrogen atom, and also the tendency of primary radicals to transfer to secondary, and secondary radicals to transfer to tertiary, tertiary radicals are relatively more numerous than the ratios of bond populations would suggest.

The second point to note is that, broadly speaking, the ease with which the bimolecular combination of two radicals occurs decreases as the extent of substitution at the carbon atoms in the vicinity of the radical site increases. This is because the greater the extent of substitution, the greater is the degree of steric hindrance to the close approach of the radical sites which is necessary as a precursor to combination. Thus, quite apart from any inherent differences in radical activity, it will be expected that the tendency of radicals to combine and form crosslinks will decrease in the order primary > secondary > tertiary. As has been stated above, this is also the order of decreasing radical reactivity, and so considerations of both inherent radical reactivity and of steric hindrance lead to the conclusion that crosslink formation will occur most readily between pairs of primary radical sites and least readily between pairs of tertiary sites. As a further broad generalisation, it can be said that the less readily radicals tend to combine with one another in pairs, the more opportunity they will have to participate in unimolecular chain-scission reactions and bimolecular disproportionation reactions. The former reactions lead to reduction of molecular weight and network degradation; the latter lead to wastage of radical activity as regards crosslinking.

From the foregoing discussion, it is possible to given reasonable explanations for the observations which have been made concerning the vulcanisation of ethylene/propylene bipolymers by organic peroxides. There is first the matter of crosslinking efficiency, that is, the number of crosslinks formed as a fraction of the number of peroxide molecules decomposed. The mechanism given above predicts a maximum value of unity for the crosslinking efficiency defined in this way. It is known that the crosslinking efficiency falls as the propylene content of the bipolymer increases; figures of $0 \cdot 70$ and $0 \cdot 10$ have been reported for bipolymers which contain respectively 25 and 90 mole % of propylene units. The low crosslinking efficiency of peroxides in propylene-rich bipolymers is attributed to the preponderance of chain scission relative to crosslinking, this preponderance arising from the conjunction of two circumstances: (a) tertiary radicals are more likely to form than primary or secondary radicals; and (b) the combination of such radicals is sterically hindered. A further factor which probably contributes to low crosslinking efficiency in propylene-rich bipolymers is that the steric hindrance of a tertiary radical will be greater when the propylene unit on which it occurs is part of a sequence of propylene units than when the propylene unit is isolated. Such sequences are more likely to occur in propylene-rich bipolymers than in ethylene-rich bipolymers. Thus it is to be expected that most of the crosslinks present in a peroxide-vulcanised ethylene–propylene bipolymer will be of the types (LXIX), (LXX) and (LXXI) shown below. Few will be expected to be of the type (LXXII), and even fewer of the type (LXXIII), especially if either or both of the propylene units involved is part of a sequence of propylene units.

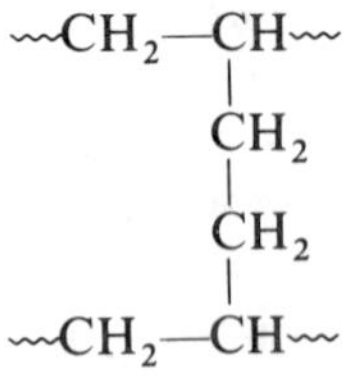

(LXIX) 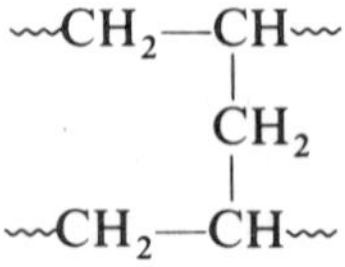   (LXX) 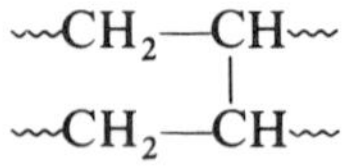   (LXXI)

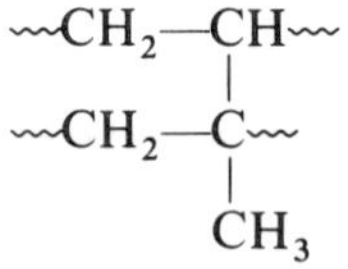

(LXXII) 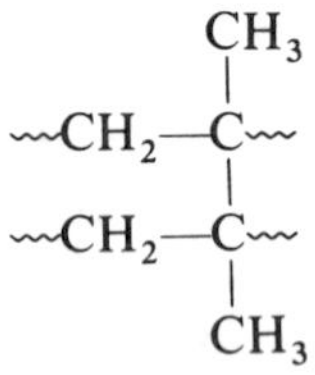   (LXXIII)

Also to be considered is the matter of the role of the co-agent, if one is used. Many substances have been proposed for use in conjunction with organic peroxides for the vulcanisation of ethylene–propylene rubbers. These include sulphur, olefinic substances such as divinyl benzene (structure (XXX) of Chapter 5), p-benzoquinone dioxime (structure (LI) of this chapter), triallyl cyanurate, ethylene glycol dimethacrylate, etc. Of these, sulphur is the most widely used. The effect of added sulphur is to increase the efficiency of the organic peroxide. Its effect depends upon the ratio of sulphur to peroxide, the greatest effect being observed when approximately one atom of sulphur is present per molecule of peroxide. This is illustrated in Fig. 7.13, from which it is seen that, although the vulcanisate tensile strength is little affected by sulphur/peroxide ratio after an initial sharp rise, the modulus goes through a well-defined maximum in the region of one atom of sulphur per molecule of peroxide. It is the vulcanisate modulus, rather than its tensile strength, which is a measure of the concentration of crosslinks in the rubber. The sulphur is believed to exert its effect by combining with tertiary radicals, thereby in effect converting them to primary radicals:

$$\text{\textasciitilde CH}_2\text{---}\overset{\cdot}{\text{C}}\text{\textasciitilde} + S_x \longrightarrow \text{\textasciitilde CH}_2\text{---}\overset{\overset{\displaystyle \dot{S}_x}{|}}{\underset{\underset{\displaystyle CH_3}{|}}{C}}\text{\textasciitilde}$$

The steric barrier to bimolecular combination at these sites is thus eliminated, and the opportunities for main-chain scission and disproportionation are consequently diminished. This explanation is consistent with the observation that the effect of sulphur is most marked in propylene-rich rubbers where, in the absence of sulphur, the tendency to scission is greatest. That the effectiveness of sulphur in enhancing crosslinking efficiency depends upon the sulphur/peroxide ratio is probably to be explained by the fact that di- and higher-sulphide crosslinks are themselves prone to participate in crosslink scission reactions.

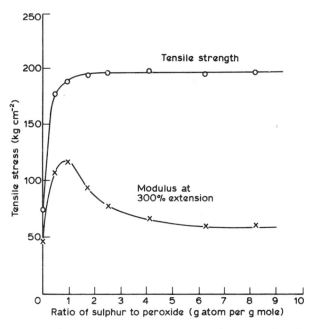

FIG. 7.13.   Illustrating effect of sulphur on the tensile properties of a peroxide-cured ethylene–propylene bipolymer vulcanisate.[3]

Ethylene–propylene terpolymers can also be vulcanised by heating with organic peroxides. In this case, the crosslinking efficiency is much higher than with the corresponding bipolymers. This has been attributed to the initiation of free-radical chain-addition through the unsaturated sites of the terpolymer. This view is supported by the observation that sulphur can reduce the crosslinking efficiency of peroxides in ethylene–propylene terpolymers, presumably because the propagating radical can be stabilised by reaction with sulphur.

The usual method of vulcanising ethylene–propylene diene terpolymers is by reaction with sulphur and conventional accelerators; indeed, as has been pointed out above, it is to impart the facility of sulphur-vulcanisability that the third monomer is introduced. The rate of crosslink formation with a given vulcanising system depends upon the nature of the third monomer, as well as upon its concentration. This is illustrated in Fig. 7.14, which shows quite clearly the advantage of ethylidene norbornene in imparting a fast cure rate. Ethylidene norbornene has the further advantages of encouraging high crosslinking efficiency and of giving networks which have a reduced tendency to undergo reversion on prolonged vulcanisation.

As in the case of isobutene–isoprene rubbers, difficulties are experienced when attempts are made to co-vulcanise blends of ethylene–propylene

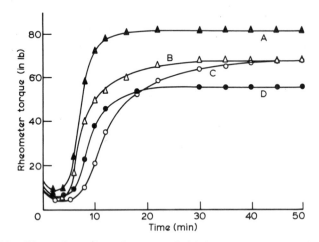

FIG. 7.14.   Illustrating effect of nature of third monomer upon vulcanisation behaviour of ethylene–propylene terpolymer rubbers.[3] The ordinates are the torque developed in an oscillating-disc rheometer. The curves refer to rubbers containing different third monomers as follows: A ethylidene norbornene, B methyl tetra-hydroindene, C dicyclopentadiene and D 1,4-hexadiene.

terpolymers and conventional diene rubbers. These difficulties are illustrated in Fig. 7.15 for blends of ethylene–propylene terpolymer and styrene–butadiene rubber. The addition of a small amount of the diene rubber brings about a marked reduction in vulcanisate crosslink density, and this is accompanied by reduction in modulus, tensile strength and (rather surprisingly) elongation at break. These difficulties are believed to arise from several causes, such as the low unsaturation of the ethylene–propylene terpolymer, the lack of miscibility of the two rubbers on the molecular scale, and the tendency of the curatives to partition in the diene-rubber phase rather than in the ethylene–propylene-terpolymer phase.

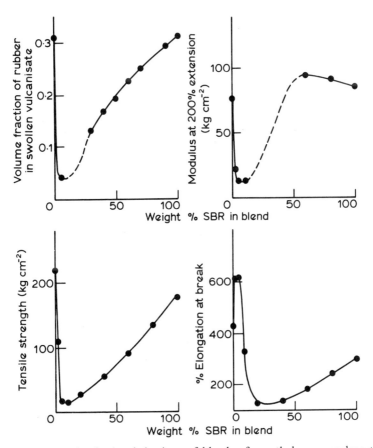

FIG. 7.15.   Covulcanisation behaviour of blends of an ethylene–propylene terpolymer rubber and a styrene–butadiene rubber.[3]

Some attention has been directed towards the overcoming of these problems, because of the desire to blend diene rubbers with ethylene–propylene rubbers in order to increase the ozone resistance of the former. The most fruitful approach appears to be to increase the level of third monomer in the ethylene–propylene terpolymer.

Some interest has been shown in recent years in ethylene–propylene rubbers which can be vulcanised by reaction with moisture. Processes of this type have been successfully exploited on a large scale with polyethylene to produce crosslinked polyethylene for cable insulation and other purposes. They have also been developed for use with ethylene-rich copolymers. In one such process, the ethylene–propylene copolymer is graft-copolymerised with a trialkoxysilane which also contains an unsaturated hydrocarbon moiety. Hydrolysis of the alkoxysilane side chains occurs when the graft-copolymer is subsequently exposed to moisture, the consequence being that crosslinks are formed between the polymer molecules. The process of graft-copolymerisation and the subsequent crosslinking can be represented as follows:

$$\xrightarrow{\text{radicals}} \cdot \xrightarrow{\text{CH}_2\text{:CHSi(OR)}_3} -\text{CH}_2\dot{\text{C}}\text{HSi(OR)}_3 \xrightarrow[\substack{\text{from another} \\ \text{molecule}}]{+\text{H abstracted}} -\text{CH}_2\text{CH}_2\text{Si(OR)}_3$$

$$-\text{CH}_2\text{CH}_2\text{Si(OR)}_3 + \text{H}_2\text{O} + (\text{RO})_3\text{SiCH}_2\text{CH}_2-$$

$$\downarrow$$

$$-\text{CH}_2\text{CH}_2\overset{\overset{\displaystyle \text{OR}}{|}}{\underset{\underset{\displaystyle \text{OR}}{|}}{\text{Si}}}-\text{O}-\overset{\overset{\displaystyle \text{OR}}{|}}{\underset{\underset{\displaystyle \text{OR}}{|}}{\text{Si}}}\text{CH}_2\text{CH}_2- + 2\text{ROH}$$

The graft-copolymerisation is accomplished by heating the ethylene–propylene copolymer with the silane in the presence of a free-radical initiator such as an organic peroxide. The use of an excess of the silane relative to the initiator ensures that graft-copolymerisation occurs preferentially to crosslinking of the polymer molecules by the usual radical-coupling reaction. The crosslinking of the grafted rubber is accomplished either by immersing the object to be vulcanised in hot water, or by exposing it to a hot humid atmosphere. In the absence of a suitable catalyst, the crosslinking reaction is slow. In order to attain practicable rates of crosslinking, it is necessary to incorporate c. 0·1 % by weight of a compound such as dibutyltin dilaurate (structure (XCI) of Chapter 8) in order to accelerate the reaction. It is, of course, necessary to store the grafted rubber in a dry atmosphere, but storage problems can be minimised by omitting the catalyst until it is desired to process and vulcanise the rubber.

Reinforcement

The response of the ethylene–propylene rubbers to compounding with reinforcing fillers is similar to that of other rubbers. However, it has been claimed that ethylene–propylene bipolymers have an ability to disperse low-structure carbon blacks which is superior to that shown by any other type of rubber. An important feature of ethylene–propylene rubbers is that they can be very highly loaded with fillers and extender oils without suffering too serious a deterioration of desirable vulcanisate properties. The reason for this behaviour seems to be obscure. One important consequence of compounding with high levels of oils is that, for a given grade of ethylene–propylene rubber and a given curing system, crosslinking by sulphur and conventional accelerators is retarded. This is partly because the concentrations of vulcanising agents are reduced, but principally because dilution of the rubber with oil favours the formation of *intra*-molecular linkages rather than the desired *inter*-molecular crosslinks. This effect can be at least partially offset by using a grade of ethylene–propylene rubber which has a higher level of unsaturation than is usual for rubbers to be used in conventional stocks.

7.4.7 Processing Behaviour

In common with other rubbers, the processing behaviour of ethylene–propylene rubbers depends very much upon their average molecular weight and molecular-weight distribution, and also upon the degree of branching and the gel content of the rubber. Wide variations in these molecular characteristics are possible through the choice of appropriate catalyst systems for their production, and, in consequence, rubbers can be produced which have widely-varying rheological characteristics and processing behaviour.

As regards behaviour on a two-roll mill, ethylene–propylene rubbers of low Mooney viscosity and broad molecular-weight distribution are the easiest to handle. Rubbers of higher Mooney viscosity and narrow molecular weight distribution tend to become lacey during the initial stages of milling, although they become smoother and more coherent as the mix warms up and as the proportion of rubber in the mix decreases because fillers and oils are being incorporated. As a general rule, the higher the

Mooney viscosity of the rubber, the higher are the loadings of fillers and oils which it can tolerate. It has been recommended that the fillers and oils should be mixed separately before they are added to the rubber.

The recommended procedure for the mixing of stocks containing high levels of fillers and oils in an internal mixer is that of 'upside-down' mixing. The fillers and oils are first charged to the mixer, whereupon the oils are absorbed by the filler powder. The rubber is then added and mixing continued until the filler/oil mixture has been thoroughly assimilated by the rubber. Mixing in internal mixers can be accomplished in short times. A 10–20 % volume overload of the mixer is recommended in order to ensure that the mix is kept under a positive ram pressure and the mixing process thereby facilitated. In the case of stocks which contain relatively low levels of filler and oil, stepwise mixing in an internal mixer is recommended in order to ensure good dispersion. Part of the filler and all the polymer are charged initially. The remainder of the filler is added when the mix has integrated.

Ethylene–propylene rubber stocks calender and extrude well, and this is partly a reflection of the high loadings of fillers and oils which these stocks frequently contain. Vulcanisation by means of sulphur and conventional accelerators can be effected in open steam, hot air, or by platen heating, and also by the newer continuous methods such as microwave heating and the use of a fluidised bed. However, vulcanisation by means of organic peroxides places some limitations on the methods which can be used. Hot-air vulcanisation is not satisfactory, because oxygen interferes with peroxide-vulcanisation; the product of peroxide-vulcanisation in hot air has a surface which is poorly cured and in consequence tacky. Vulcanisation by peroxides also places limitations on the formulation of the mix. Thus fillers of acidic reactions should be avoided, as should high levels of oils, because the oils are frequently as reactive towards radicals derived from peroxides as is the rubber itself.

7.5 ETHYLENE–VINYL ACETATE RUBBERS (EAM) AND ETHYLENE–ACRYLATE RUBBERS

Both polyethylene and polyvinyl acetate are rigid plastics at normal ambient temperatures. It has already been noted that, in the case of polyethylene, the rigidity arises from the fact that the material is partially crystalline, notwithstanding that the glass-transition temperature of the amorphous polymer is well below room temperature. In the case of polyvinyl acetate, the material is rigid because it is slightly below its glass-transition temperature (*c.* 30 °C) at room temperature. Random copolymers of ethylene and vinyl acetate, however, can be rubbery over a certain range of composition. The vinyl acetate units are able to inhibit the crystallisation of the ethylene units, and the ethylene units can be regarded as acting as an internal plasticiser for the vinyl acetate sequences.

The commercially-available ethylene vinyl acetate rubbers have vinyl acetate contents ranging from 40 to 60% by weight. They are chemically saturated, and do not contain functional groups other than those derived from the two monomers. In consequence, they are vulcanised by heating with organic peroxides. The vulcanisates have the resistance to heat, oxygen and ozone which is expected in a rubber whose main chain is chemically saturated. The oil and solvent resistance of ethylene–vinyl acetate rubber vulcanisates is greater than that of the ethylene–propylene rubbers. This is to be expected; the improved resistance to swelling arises from the presence in the rubber of polar units derived from vinyl acetate (LXXIV). As expected, the resistance to swelling increases as the vinyl acetate content of the rubber increases. So, too, does the resistance to heat ageing.

$$\cdots-CH_2-CH-\cdots$$
$$|$$
$$O.CO.CH_3$$

(LXXIV)

Useful rubbers can also be produced by copolymerising ethylene with acrylic ester monomers. One group of such rubbers which is currently available commercially is produced by copolymerising ethylene with methyl acrylate and a small proportion of a third monomer which contains a carboxylic-acid group. Thus in addition to ethylene units, this type of rubber also contains units derived from methyl acrylate (LXXV) and a small proportion of units which contain free carboxylic acid. In these

$$\cdots-CH_2-CH-\cdots$$
$$|$$
$$CO.OCH_3$$

(LXXV)

rubbers, the tendency of the ethylene units to crystallise is inhibited by the presence of the methyl acrylate units. Furthermore, like polyvinyl acetate, polymethyl acrylate has its glass-transition temperature near to normal room temperature, and thus the ethylene units can be regarded as functioning as an internal plasticiser for the methyl acrylate sequences. The purpose of incorporating a minor proportion of a monomer which contains a carboxylic-acid group is to provide the facility for the rubber to be crosslinked by means other than heating with organic peroxides or exposure to high-energy radiation. This matter will be dealt with in more detail in Chapter 9 (Section 9.1), where the so-called 'carboxylated rubbers' will be considered as a class. It is sufficient to note here that the presence of carboxylic-acid groups in the polymer sequence permits the possibility of crosslinking by reaction with certain difunctional reagents such as diamines and dihydroxy compounds. In the case of the carboxylated ethylene–methyl acrylate rubbers, a typical vulcanising combination comprises 1·25 pphr of methylene dianiline plus 4 pphr of diphenyl guanidine.

Ethylene–acrylate rubbers offer an interesting combination of properties, namely, high heat resistance (intermediate between that of chlorosulphonated polyethylene and that of the silicone rubbers), weather resistance, ozone resistance, resistance to swelling in oils and solvents, and good resistance to low-temperature stiffening. Heat resistance is adversely affected by the presence of metal oxides in the vulcanisate.

7.6 CHLOROSULPHONATED (CHLOROSULPHONYL) POLYETHYLENE RUBBERS (CSM) AND CHLORINATED POLYETHYLENE RUBBERS (CM)

The chlorosulphonated polyethylene rubbers are unusual amongst synthetic rubbers in that they are produced by chemical modification of an existing polymer. Although this is also true of the urethane rubbers, the unusual feature of the rubbers under consideration here is that the chemical modification involves certain of the repeat units of the starting polymer, and is not (unlike the urethane rubbers) concerned with increasing the molecular weight of the original polymer.

As their name implies, the chlorosulphonated polyethylene rubbers are made from polyethylene. Polyethylene is reacted in solution with a mixture of chlorine and sulphur dioxide in appropriate ratio. The product of the reaction is a polymer having the general type of structure shown at (LXXVI). Chlorosulphonation brings about two important changes in polyethylene. In the first place, it destroys the regularity of the polyethylene chains, thereby inhibiting the tendency of the polymer to crystallise. The inherent mobility of the polymer chain, arising from the rotational flexibility of the majority of the carbon–carbon bonds, is not therefore restricted by the formation of crystallites. In practical terms, this means that the polymer exhibits rubbery character instead of being a rather rigid plastic like polyethylene. But secondly, chlorosulphonation introduces

$$\left\{\left(CH_2-CH_2-\underset{\underset{Cl}{|}}{CH}-CH_2\right)_x CH\underset{\underset{SO_2Cl}{|}}{}\right\}_y$$

(LXXVI)

into the polymer chain reactive sites by means of which the rubber can be vulcanised. The process of chlorosulphonation therefore changes a non-vulcanisable, rather rigid thermoplastic polymer into a vulcanisable rubber which can be processed by techniques which are familiar to the rubber industry.

The chlorine content of commercially-available chlorosulphonated polyethylenes ranges from about 25 % to about 45 % by weight. The range of sulphur contents is about $1 \cdot 0$–$1 \cdot 5 \%$ by weight, the majority having a sulphur content of approximately $1 \cdot 0 \%$. For a chlorosulphonated polyethylene which contains 35 % by weight of chlorine and 1 % by weight of sulphur, it can be calculated that approximately 1 in 5 of the carbon atoms of the initial polyethylene chain has been either chlorinated or chlorosulphonated, and that, of these, approximately 1 in 30 has been chlorosulphonated. Chlorosulphonated polyethylene rubbers are available in a range of plasticities, this property being determined mainly by the molecular weight of the polyethylene from which the rubber was made.

The vulcanisation of chlorosulphonated polyethylene is normally effected by heating with metal oxides. Magnesium oxide is used unless a vulcanisate of high water resistance is required, in which case lead oxide (lithage) is recommended. Typical proportions are 4 pphr and 25 pphr for the two oxides respectively. The crosslinking reaction probably involves interaction between the metal oxide and the sulphonyl chloride groups of the rubber molecule, with the formation of metal sulphonate bridges. The presence of traces of an active-hydrogen compound, such as water, is essential for satisfactory crosslinking. For this reason, a hydroxy compound such as pentaerythritol is sometimes included in the compound at a level of about 3 pphr. The state of cure can be increased by incorporating conventional sulphur donors and accelerators, such as the thiuram polysulphides and di-2-benzthiazyl disulphide. The compound N,N'-m-phenylenedimaleimide, whose chemical structure has been given at (LXI), has also been recommended as an aid for the vulcanisation of chloro-sulphonated polyethylene rubber.

The tensile strength of gum vulcanisates from chlorosulphonated polyethylene rubbers depends upon the grade of rubber which was used for their preparation. Some have high tensile strength ($c.$ 4500 lb in^{-2}), some have medium tensile strength ($c.$ 2500 lb in^{-2}), whilst others have rather low tensile strength ($c.$ 1200 lb in^{-2}). These differences are probably reflections of differences in the ability of the vulcanisate to crystallise on stretching. The moduli of the gum vulcanisates obtained using a given curing system also vary widely with the grade of rubber used. The tensile strengths and low-extension moduli of gum vulcanisates do not appear to correlate with one another very closely. Chlorosulphonated polyethylene rubbers can be compounded with conventional rubber fillers; in general, the expectations concerning the enhancement of properties such as modulus, tensile strength, tear strength and abrasion resistance are fulfilled.

Chlorosulphonated polyethylene is a special-purpose rubber which gives vulcanisates having outstanding resistance to deterioration by ozone. Its oxidation rate is also very low, and it is very little affected by deteriorative influences such as sunlight and ultraviolet light. It has been claimed that chlorosulphonated polyethylene, when appropriately compounded, is one of the most weather-resistant rubbers currently available. Furthermore, it resists discoloration by light. Its resistance to swelling in hydrocarbon oils is good; as expected, resistance to swelling in these oils increases with the chlorine content of the rubber. Also, its general resistance to attack by chemicals is good. Other advantages include resistance to combustion (because of its high chlorine content), and resistance to heat ageing. General-purpose vulcanisates can be used continuously at temperatures up to about 150 °C, and specially-designed vulcanisates can be used intermittently at temperatures up to 163 °C. Vulcanisates have generally satisfactory mechanical properties, but the compression set tends to be rather high, the actual value depending upon the grade of rubber used, the design of the compound, the vulcanising system, and the degree of vulcanisation. The brittle point of these rubbers again depends upon the grade, as well as upon the design of the compound. The range is $c.$ −60 °C to $c.$ −40 °C. Depending upon the grade of rubber and the way in which it has been compounded, the lowest temperature at which conventional vulcanisates can be used is in the range -18 to $-23\,^\circ$C. If specially compounded, vulcanisates can retain their flexibility down to $-40\,^\circ$C.

The chlorinated polyethylene rubbers are also produced by chemical modification of an existing polymer. In this case, the polyethylene is merely chlorinated. Earlier grades were produced by the chlorination of low-density polyethylene dissolved in a chlorinated hydrocarbon solvent. The more recent grades have been produced by the chlorination of high-density polyethylenes obtained, for example, by Ziegler–Natta catalysis. Because the polyethylene is obtained from such processes as a fine powder, it is possible to carry out the chlorination in aqueous suspension as well as in solution.

Replacement of some of the hydrogen atoms of polyethylene by bulky chlorine atoms provides yet another way of reducing the tendency of polyethylene sequences to crystallise, thereby permitting the inherent flexibility of the polyethylene chain to be manifest as rubbery character in the bulk material. Chlorinated polyethylenes having chlorine contents in the range 25–40% by weight are rubbery. Chlorination to higher levels produces more rigid materials which resemble unplasticised polyvinyl chloride.

Vulcanisation of chlorinated polyethylene rubbers is best achieved by heating with organic peroxides. The use of suitable co-agents with peroxides is recommended in order to increase both the rate of curing and the ultimate state of cure. An example of a suitable co-agent for this purpose is triallyl phosphate (LXXVII); it is used at a level of 2–4 pphr. The

$$OP(OCH_2CH:CH_2)_3$$

(LXXVII)

presence of zinc compounds in chlorinated polyethylene rubbers must be avoided because, in an acidic environment, zinc compounds act as catalysts of polymer decomposition through dehydrochlorination. Chlorinated polyethylene rubbers should be compounded with an efficient heat stabiliser such as magnesium oxide, lithage or calcium oxide. Vulcanisates from chlorinated polyethylene rubber display good resistance to weathering. Because of their high chlorine content, they also have the advantage that they burn more slowly than many other rubbers. They invite immediate comparison with plasticised polyvinyl chloride, over which they have two important advantages: their retention of flexibility at low temperatures is better, and problems associated with plasticiser loss are eliminated because no plasticisers are required in their compounding.

7.7   OLEFIN OXIDE RUBBERS (CO, ECO AND GPO)

A few special-purpose synthetic rubbers produced from olefin oxides are currently available commercially. They exhibit excellent resistance to swelling in oils. Resistance to chemicals is generally good; so also is resistance to ozone and weathering. Heat resistance is also good. One of the available types (CO) is a homopolymer of epichlorohydrin (LXXVIII), its repeat unit having the structure shown at (LXXIX). This rubber has a

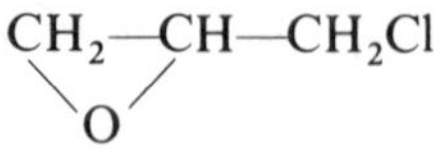

$$CH_2\underset{\displaystyle O}{\diagup\diagdown}CH-CH_2Cl \qquad\qquad \cdots-CH_2-\underset{\displaystyle CH_2Cl}{\overset{\displaystyle |}{CH}}-O-\cdots$$

(LXXVIII) 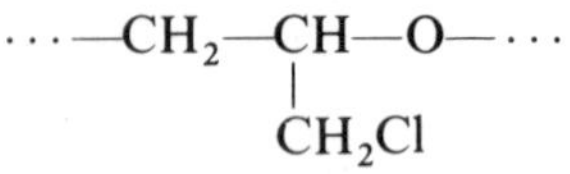   (LXXIX)

rather high brittle point of approximately $-15\,°C$. The olefin oxide rubbers as a group have low permeability to gases. The permeability of epichlorohydrin rubber is especially low, being approximately one-third that of isobutene–isoprene rubber. As in the case of the latter rubber, this property is attributed to the sluggish movements of the segments which make up the polymer chain. Like isobutene–isoprene rubber, epichlorohydrin rubber has low rebound resilience at normal temperatures.

The high brittle temperature of epichlorohydrin rubber is associated with the inhibition of segmental motions by electrostatic attractions between the dipoles associated with the carbon–chlorine bonds. The low-temperature flexibility of these rubbers can be improved if the epichlorohydrin is copolymerised with a less polar monomer. A modified epichlorohydrin rubber (ECO) containing about 40% by weight of ethylene oxide is also currently available commercially. In addition to the repeat unit shown at (LXXIX), it also contains the repeat unit shown at (LXXX). Copolymerisation with this amount of ethylene oxide reduces the

$$\cdots-CH_2-CH_2-O-\cdots$$

(LXXX)

brittle point to $c.\ -40\,°C$ with little loss of oil resistance. The copolymer is therefore recommended for applications which call for reasonable low-temperature flexibility. The low-temperature behaviours of the homopolymer and copolymer are compared in Fig. 7.16. This figure shows the variation with temperature of the torsional rigidities of similar vulcanisates from the two rubbers. The resilience of the copolymer is higher than that of epichlorohydrin homopolymer.

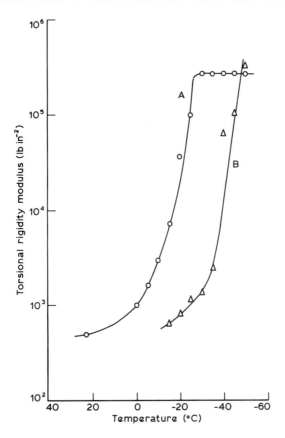

FIG. 7.16.   Variation of torsional rigidity with temperature for vulcanisates from an epichlorohydrin homopolymer rubber (A) and an epichlorohydrin–ethylene oxide copolymer rubber (B).[6]

Epichlorohydrin rubbers are vulcanised with difunctional reagents which react with the pendant chloromethyl groups of the rubber. Such reagents include diamines, ureas and thioureas. 2-Mercaptoimidazoline (ethylene thiourea) has been found to be a particularly effective vulcanising agent. It is most effective when used in conjunction with an acid acceptor such as red lead. The vulcanisation behaviour of an epichlorohydrin rubber containing 5 pphr of red lead and various levels of 2-mercaptoimidazoline is illustrated in Fig. 7.17. There have also recently become available two new vulcanising agents for epichlorohydrin rubbers. Their compositions are as yet undisclosed. They are supplied under the trade names 'Echo S' and 'Echo P'

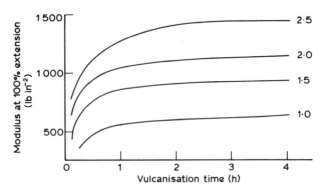

FIG. 7.17.   Vulcanisation behaviour of an epichlorohydrin rubber.[7] Compound: epichlorohydrin–ethylene oxide rubber 100, FEF carbon black 50, red lead 5, zinc stearate 0·75, nickel dibutyl dithiocarbamate 1, 2-mercaptoimidazoline variable (levels indicated by numbers appended to curves). Vulcanisation carried out at 160 °C.

by Hercules Powder Company. Amongst the advantages claimed for these curatives over 2-mercaptoimidazoline are improved processing safety, faster vulcanisation, and reduced compression set. It is recommended that they be used in conjunction with an inorganic base, such as barium carbonate, as an acid acceptor. The low-temperature flexibility of epichlorohydrin rubbers can be improved by compounding them with plasticisers. Dioctyl phthalate is very suitable for this purpose.

A sulphur-vulcanisable olefin oxide rubber is also currently available commercially. It is a copolymer of propylene oxide and allyl glycidyl ether (LXXXI). As a family, rubbers of this type are designated by the initials GPO. The repeat units which they contain are shown at (LXXXII) and (LXXXIII). In the type of rubber which is of current interest, the proportion of units derived from allyl glycidyl ether is understood to be

$$CH_2-CH-CH_2-O-CH_2-CH=CH_2$$
$$\diagdown O \diagup$$

(LXXXI)

$$\cdots-CH-CH_2-O-\cdots$$
$$\qquad |$$
$$\qquad CH_3$$

(LXXXII)

$$\cdots-CH-CH_2-O-\cdots$$
$$\qquad |$$
$$\qquad CH_2$$
$$\qquad |$$
$$\qquad O-CH_2-CH=CH_2$$

(LXXXIII)

approximately 10% by weight. The facility of being vulcanisable with sulphur and conventional accelerators is associated, of course, with the unsaturation arising from the presence of the allyl side chains. A typical vulcanising system comprises 1·25 pphr sulphur plus 1·5 pphr 2-mercaptobenzthiazole plus 1·5 pphr tetramethyl thiuram monosulphide. The low-temperature flexibility of these rubbers is good (stiffening temperature $c. -60\,°C$). They combine excellent ozone resistance with moderate resistance to swelling in hydrocarbon oils.

GENERAL BIBLIOGRAPHY

Whitby, G. S., Davis, C. C. and Dunbrook, R. F. (Eds). (1954). *Synthetic Rubber*, John Wiley and Sons, New York, Chapter 24.

Buckley, D. J. (1959). Elastomeric properties of butyl rubber, *Rubb. Chem. Technol.*, **32**, 1475.

Buckley, D. J. (1965). Butylene polymers. In: *Encyclopedia of Polymer Science and Technology*, Vol. 2, John Wiley and Sons, New York, pp. 754f.

Baldwin, F. P. and ver Strate, G. (1972). Polyolefin elastomers based on ethylene and propylene, *Rubb. Chem. Technol.*, **45**, 709.

Natta, G., Crespi, G., Valvassori, A. and Sartori, G. (1963). Polyolefin elastomers, *Rubb. Chem. Technol.*, **36**, 1583.

Saltman, W. M. (Ed.) (1977). *The Stereo Rubbers*, John Wiley and Sons, New York, Chapter 7.

du Puis, I. C. *A Guide to the types of 'Hypalon'*, Bulletin HP-200.1, Elastomers Chemicals Department, E. I. Du Pont de Nemours & Co. (Inc.), Wilmington, Delaware.

Johnson, J. B. (1975). Chlorinated polyethylene rubbers, *Rubber Age*, **107**(3), 29.

Hsieh, H. L. and Wright, R. F. (1972). Properties of poly(alkylene oxide) elastomers, *Rubb. Chem. Technol.*, **45**, 900.

REFERENCES

1. *Polysar Butyl Handbook*, Polysar Ltd, Sarnia, Canada.
2. Buckley, D. J. (1959). *Rubb. Chem. Technol.*, **32**, 1475.
3. Crespi, G., Valvassori, A. and Flisi, U. (1977). *The Stereo Rubbers*, Saltman, W. M. (Ed.), John Wiley and Sons, New York, p. 365.
4. Natta, G., Crespi, G., Valvassori, A. and Sartori, G. (1963). *Rubb. Chem. Technol.*, **36**, 1583.
5. Baldwin, F. P. and ver Strate, G. (1972). *Rubb. Chem. Technol.*, **45**, 709.
6. *Herclor Elastomers*, Bulletin ORH-8C, Hercules Powder Company Ltd.
7. *Herclor Elastomers*, Bulletin ORH-3B, Hercules Powder Company Ltd.

Chapter 8

Plasticised Polyvinyl Chloride (PVC)

8.1 INTRODUCTION

In this chapter, brief consideration will be given to a rubber-like product which is obtained by plasticising the rigid plastic known as polyvinyl chloride. The principal repeat unit present in polyvinyl chloride is shown at (LXXXIV). Polyvinyl chloride can be obtained in several forms, depending upon the method by which it is prepared. The common form is produced by

$$\cdots -CH_2-CH- \cdots$$
$$|$$
$$Cl$$

(LXXXIV)

the free-radical polymerisation of vinyl chloride, and in the unplasticised state has a glass-transition temperature in the region of 80–90 °C. It is therefore apparent that, at normal temperatures, unplasticised polyvinyl chloride will behave as a rigid, glassy material. Unplasticised PVC is resistant to most acids and alkalis, although it is decomposed by sulphuric, nitric and chromic acids. It is soluble in a few organic solvents, such as tetrahydrofuran, and is swollen to varying extents by polar solvents such as ketones and esters, and by aromatic and chlorinated hydrocarbons. It is essentially unaffected by aliphatic hydrocarbons and oils. The solubility behaviour of polyvinyl chloride is determined primarily by the presence of the dipoles, associated with C–Cl bonds, at regular intervals throughout the polymer sequence. In addition to its general chemical resistance, an important attribute of polyvinyl chloride is its low flammability. This also is a consequence of its high chlorine content.

If polyvinyl chloride is compounded with sufficient plasticiser, the glass-transition temperature is reduced to below room temperature and the

244

product is rubber-like at normal temperatures. It is customary to regard this material as a kind of synthetic rubber, partly because it does exhibit something of the reversible long-range elasticity which is conventionally associated with a rubber, and partly because it competes with other synthetic rubbers, such as chloroprene rubber, in certain applications such as belting and cable sheathing. Considered as a synthetic rubber, polyvinyl chloride presents at least three unusual features:

1. plasticisation is essential;
2. it is not normally crosslinked; and
3. stabilisation against dehydrochlorination is necessary.

Concerning the first of these distinguishing features, it is merely necessary to remind the reader that, for almost all other rubbers, plasticisation is optional. It is normally carried out in order either to improve the low-temperature flexibility and processability of the rubber, or to cheapen the rubber. In the latter case, the plasticiser is more usually referred to as an 'extender'. But in the case of polyvinyl chloride to be used as a rubber, plasticisation is not optional; the presence of plasticiser is essential if the product is to be rubbery.

Concerning the second of the distinguishing features listed above, it is the case that polyvinyl chloride does not contain sites of suitable reactivity to enable it conveniently to be crosslinked by reaction with common reagents. Consequently, plasticised polyvinyl chloride is more prone to flow under sustained stress than are conventional crosslinked rubbers. Furthermore, the rate of flow increases more rapidly with increasing temperature than in the case of crosslinked rubbers. In more practical terms, this means that plasticised polyvinyl chloride suffers from excessive creep and stress relaxation if subjected to sustained stress or strain, especially at elevated temperatures. It is therefore unsuited for application where sustained stress or strain has to be borne, unless fabric or other reinforcement is provided in order to restrict reversible flow.

Concerning the third of these distinguishing features, it is merely necessary to note here that polyvinyl chloride suffers from a tendency to degrade at elevated temperatures by dehydrochlorination. Furthermore, the hydrogen chloride which is evolved may be a catalyst for the further degradation of the polymer. Consequently, it is necessary to compound the polymer with a stabiliser which affords protection against this form of degradation. Dehydrochlorination is a problem which occurs mainly during the processing of the polymer.

There remains one further general aspect of the technology of plasticised polyvinyl chloride which must be mentioned in this introduction. In order to achieve miscibility between polyvinyl chloride and the substances which are commonly used to plasticise it, it is necessary to raise the temperature considerably above ambient. It is in consequence possible to prepare dispersions of polyvinyl chloride powder in plasticiser which are moderately stable at room temperature, but which gel to form a homogeneous mass of plasticised polyvinyl chloride when the temperature is raised. Such dispersions of polyvinyl chloride powder in plasticiser are known as polyvinyl chloride *pastes* or *plastisols*. They form the basis of a rather unusual variant of rubber technology which has been widely exploited in applications such as dipping and rotational casting, and in the coating of wire meshes. This technology will be considered in Section 8.7 of this chapter.

8.2   PRODUCTION OF POLYVINYL CHLORIDE

Most, if not all, commercial polyvinyl chloride is made by the free-radical addition polymerisation of vinyl chloride. Bulk, solution, suspension and emulsion polymerisation procedures are all used. In the past, the most important processes have been emulsion and suspension polymerisations. In recent years, processes for the bulk and solution polymerisation of vinyl chloride have assumed increasing importance.

In a typical early process for the production of polyvinyl chloride by emulsion polymerisation, vinyl chloride was added to an equal weight of a 4% aqueous solution of a sodium C_{15}-sulphonate and 0·4 parts by weight per 100 parts of monomer of 40% aqueous hydrogen peroxide. Polymerisation was carried out in either nickel or glass-lined steel autoclaves at 40–50 °C. After 24-h reaction, when the vapour pressure of vinyl chloride had started to fall, the contents of the autoclave were discharged, and the polyvinyl chloride obtained by spray-drying the latex. The extent of conversion of monomer to polymer was c. 90%, and no attempt was made to recover the unreacted monomer. In more recent processes, more efficient initiators, such as potassium persulphate, have been used. Greater attention is paid to the control of the reaction conditions, and the unpolymerised monomer is recovered. However, hydrogen peroxide is preferred as the initiator when it is desired to produce emulsion polymers which have superior electrical properties.

Emulsion polymerisation has been especially used for the production of polyvinyl chlorides for use as pastes. The principal requirement of a good paste polymer is that it should be able to form a stable low-viscosity plastisol when blended with the minimum of plasticiser. To this end, it is generally necessary that the particles of the paste polymer should have diameters within the range 5000–20 000 Å (500–2000 nm). Emulsion polymerisation provides a very convenient method for producing polymer particles which meet this requirement. For the production of paste polymers, vinyl chloride can be emulsion polymerised in a reaction system which contains two emulsifiers, one of which is soluble in the aqueous phase and the other of which is soluble in the monomer. Typical examples of such emulsifiers are sodium dodecyl sulphate and cetyl alcohol respectively. In at least one such published polymerisation recipe for the production of vinyl chloride paste polymers, the initiator (azobisisobutyronitrile) is insoluble in the aqueous phase but soluble in the monomer. It is therefore to be questioned whether such reaction systems really are emulsion polymerisations in the conventional sense of the term, or whether they are better described as micro-suspension polymerisations. That the latter may well be the case is supported by the observation that emulsifier combinations such as sodium dodecyl sulphate plus cetyl alcohol are known to be capable of causing water-insoluble liquids to form micro-emulsions of the oil-in-water type. Whatever may be the exact mechanism of the reaction by which polymerisation occurs, the product is a fine dispersion of polyvinyl chloride particles in an aqueous medium. The polymerisation is carried out at c. 50 °C and taken substantially to completion. The particle size of the final dispersion can be controlled by the addition of inorganic electrolytes to the polymerisation system, so that the reaction is carried out in a state of incipient coagulation. The polymer is isolated from the dispersion either by coagulation and filtration, or by spray-drying.

In a typical early process for the production of polyvinyl chloride by suspension polymerisation, vinyl chloride was polymerised in a stainless-steel reactor using a reaction system which comprised monomer, water (200 pphm), dibenzoyl peroxide (0·13 pphm), and a 5 % aqueous solution of polyvinyl alcohol (3·3 pphm). The function of the latter component (which is obtained from polyvinyl acetate by partial hydrolysis) was to form a protective layer around the droplets of polymerising monomer, thereby preventing them from coalescing. The reaction mixture was agitated at 40 rpm, and polymerisation allowed to occur for about 50 h at 40 °C, by which time the conversion was about 65 %. After recovery of the unreacted monomer, the slurry was filtered, and the polymer washed and centrifuged to give a cake containing 35 % moisture, which was then dried.

Because of the great importance of the suspension polymerisation process for the production of polyvinyl chloride, many variants of the process have been developed since the early processes. The general principles underlying the process have, however, remained unaltered. A protective colloid is always required, and amongst the many alternatives to polyvinyl alcohol are methyl celluloses, gelatin, and vinyl acetate–maleic anhydride copolymers. Secondary emulsifiers can also be used to modify the shape of the particles produced. Other initiators besides dibenzoyl peroxide have been used; one example is diisopropyl percarbonate.

Bulk polymerisation processes for vinyl chloride have assumed increasing importance in recent years. They have been developed almost exclusively by the Péchiney–St. Gobain company in France. In considering these processes, it is important to realise that, unlike many other polymer–monomer pairs, polyvinyl chloride is inherently almost insoluble in vinyl chloride. As the polymerisation proceeds, the polymer therefore precipitates from the monomer from an early stage of the reaction. Both one-stage and two-stage processes have been described for the bulk polymerisation of vinyl chloride. In one example of a one-stage process, vinyl chloride is polymerised with 0·8 pphm of dibenzoyl peroxide for 17 h at 58 °C in a rotating cylinder containing steel balls. One function of the steel balls is to facilitate the removal of heat from the reaction system. Difficulties of process control (in particular, in respect of the particle size of the product, which is affected by the grinding action of the tumbling balls) have led to the development of two-stage processes. In one example of such a process, vinyl chloride containing 0·02 pphm of azobisisobutyronitrile and 5×10^{-5} pphm of diisopropyl percarbonate is charged to a small reactor which is fitted with a high-speed agitator. Polymerisation is allowed to proceed at 50–60 °C. Polyvinyl chloride begins to precipitate from the monomer almost immediately (at less than 1 % conversion). Polymerisation is continued until about 10 % conversion (c. 3 h reaction), when the entire contents of the reactor are charged into a large horizontal autoclave and further monomer added. At 20 % conversion, the polymerising mass is a wet powder. At 40 % conversion, the residual monomer is completely absorbed by the polymer, so that the reaction mixture is now apparently a dry powder. The conversion is taken to c. 90 %, the reaction conditions being such that the reaction mixture remains as a powder. The residual monomer is removed under vacuum, and the polymer then discharged from the reactor.

Processes have also been developed for the polymerisation of vinyl chloride in solution. The solvent is usually chosen so that the polymer precipitates from the reaction mixture as the polymerisation proceeds.

Typical solvents are *n*-butane and cyclohexane. The nature of the solvent affects both the point at which the polymer precipitates and the physical nature of the precipitate.

One of the most important respects in which the various commercially available grades of polyvinyl chloride differ from one another is molecular weight. The number-average molecular weight of most commercial grades falls within the range 3×10^4–1×10^5. The average molecular weight of vinyl chloride polymers is commonly characterised by means of the so-called *Fikentscher K-value* of the polymer. This is defined by means of the equation

$$\log_{10} \eta_r = \left\{ \left(\frac{7 \cdot 5 \times 10^{-5} K^2}{1 + 1 \cdot 5 \times 10^{-3} Kc} \right) + 10^{-3} K \right\} c$$

where η_r is the *relative viscosity* of a solution containing the polyvinyl chloride at a concentration of c grams per decilitre. By the term *relative viscosity* is meant the ratio η/η_0, where η is the viscosity of the solution and η_0 is the viscosity of the solvent at the same temperature. The K-value is somewhat dependent upon the conditions of measurement, in particular, the nature of the solvent, the concentration, the temperature of measurement, and the method of measuring the viscosity. Typically, K-values are determined using 0·5 % solutions in ethylene dichloride at 25 °C.

8.3   PLASTICISATION OF POLYVINYL CHLORIDE

The number of plasticisers which can be used for making polyvinyl chloride flexible is legion. A convenient chemical classification is as follows:

1. dialkyl esters of phthalic acid;
2. esters of orthophosphoric acid;
3. dialkyl esters of aliphatic dicarboxylic acids;
4. low-molecular-weight polyesters; and
5. miscellaneous compounds such as epoxidised oils and esters.

The general chemical structures of plasticisers in classes (1), (2) and (3) are shown at (LXXXV), (LXXXVI) and (LXXXVII) respectively.

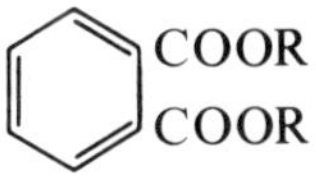

COOR⟨⟩COOR	OP(OR)₃	ROOC(CH₂)ₙCOOR
(LXXXV)	(LXXXVI)	(LXXXVII)

Before considering each of these groups of plasticiser in more detail, it is desirable to enumerate some of the more important factors which have to be taken into account in selecting a polyvinyl chloride plasticiser for a given application. The obvious economic factors such as cost and availability can be dismissed without further comment. Rather less obvious is the requirement that the plasticiser and the grade of polyvinyl chloride to be used should be completely miscible with each other at the level at which the plasticiser has to be present in order to meet the other requirements, and at the temperature at which the plasticised material is to be used. Any lack of compatibility will be manifest by the plasticiser tending to exude from the polymer. A primary consideration which influences the choice of a particular plasticiser is its efficiency as a flexibilising agent for polyvinyl chloride. This characteristic may be assessed by way of the effect of the plasticiser upon properties such as modulus, brittle temperature and glass-transition temperature. The effect of the plasticiser upon other mechanical properties such as tensile strength, elongation at break, tension and compression set, resilience, creep and stress relaxation, etc., is also important. If the plasticised material is to be used in electrical applications, the effect of the plasticiser upon electrical properties such as electrical resistivity will clearly be very important. Apart from the initial effect of a plasticiser upon electrical properties, its ability to withstand hydrolysis can be an important factor in determining long-term suitability. The tendency of the plasticiser to be extracted by solvents, oils and detergent solutions will be important if the plasticised material is to be used in contact with these media; extraction of the plasticiser will lead to changes in the flexibility of the material. Volatility of the plasticiser is always important, but especially if the plasticised material is to be used at elevated temperatures. Not only will the material stiffen if the plasticiser is lost by evaporation; it will also shrink in size, and this can be a serious problem if appreciable quantities of plasticisers are lost in this way. A related matter is that of odour, which depends, in part at least, upon volatility; the rather strong odour imparted to polyvinyl chloride by certain plasticisers is objectionable in some applications. The flammability of the plasticiser has to be taken into account in applications where fire is a potential hazard. Whilst polyvinyl chloride itself is inherently self-extinguishing when ignited, its flame-retardant properties are generally reduced by compounding with plasticisers. Quite apart from their effects upon flammability, different plasticisers have different effects upon the heat stability of polyvinyl chloride, and this is a factor which must be taken into account in determining suitability for certain applications. Toxicity is always an important consideration, especially if the plasticised material is to come into prolonged contact with human beings, foodstuffs, drinking water, beverages, etc. A related matter is the ease with which the plasticiser can diffuse from plasticised polyvinyl chloride into its surroundings.

The relevant efficiencies of representative members of classes (1), (2) and (3) above as plasticisers for polyvinyl chloride are illustrated in Fig. 8.1(a), (b) and (c). Figure 8.1(a) shows the effects of increasing levels of the various plasticisers upon the cold-flex temperature of the product. As a class, the phosphoric-acid esters are relatively inefficient (on a weight basis) as improvers of low-temperature flexibility. They do, however, have the very important attribute of reducing the flammability of plasticised polyvinyl chloride. By contrast, the dialkyl esters of aliphatic dicarboxylic acids are very effective in imparting resistance to stiffening at low temperatures. The dialkyl phthalates are intermediate in behaviour. Figure 8.1(a) also illustrates that, within the family of dialkyl phthalates, the ability to impart low-temperature flexibility increases as the size of the alkyl group is reduced. Thus dioctyl phthalate is a rather more efficient low-temperature

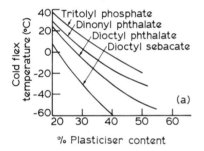

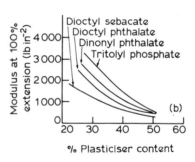

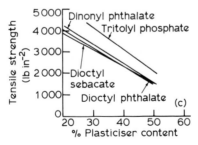

FIG. 8.1. Effect of increasing levels of various plasticisers upon mechanical properties of polyvinyl chloride.[1]

plasticiser than is dinonyl phthalate. Furthermore, Fig. 8.1(a) gives an indication of the actual levels of plasticiser which have to be added in order to achieve a product which will retain its flexibility down to a given temperature. The minimum level of plasticiser which is required depends very much upon the particular plasticiser. As a guide, it may be noted that some 40 parts of dioctyl phthalate per 100 parts of polyvinyl chloride are required in order to obtain a material which remains flexible down to about $-20\,^{\circ}\mathrm{C}$.

Figure 8.1(b) shows the effects of increasing levels of the various plasticisers upon the modulus at 100% extension of plasticised polyvinyl chloride. The results summarised in this diagram enable comparisons to be made of the relative efficiencies of the various classes of plasticiser as flexibilising agents for polyvinyl chloride at room temperature. Comparing Fig. 8.1(b) with Fig. 8.1(a) shows that the relative ability of a plasticiser to impart flexibility at room temperature broadly parallels its ability to impart flexibility at low temperatures. Thus the dialkyl esters of aliphatic dicarboxylic acids are the most efficient flexibilisers, the phosphate esters are the least efficient, and the dialkyl phthalates are of intermediate efficiency. This broad correlation between ability to impart flexibility at room temperature and ability to impart resistance to low-temperature stiffening is exactly what would be expected on general grounds.

Figure 8.1(c) shows the effects of increasing levels of the various plasticisers upon the tensile strength of plasticised polyvinyl chloride. The tensile strength falls approximately linearly with increasing level of plasticiser. Those plasticisers which are most effective in reducing modulus and imparting resistance to low-temperature stiffening tend to have the largest effect in reducing tensile strength. However, the differences in this latter respect between the dialkyl phthalates on the one hand and the dialkyl esters of aliphatic dicarboxylic acids on the other are not so marked as they are in respect of effects upon modulus and low-temperature flexibility.

Whilst considering the general effects of plasticisers upon the mechanical properties of polyvinyl chloride, it is appropriate to note that the phenomenon of *antiplasticisation* is well-established for this polymer (as indeed it is for several others). Whereas the addition of a plasticiser to polyvinyl chloride normally brings about a reduction in modulus and tensile strength and an increase in elongation at break, the converse is sometimes observed. This phenomenon of antiplasticisation is usually observed when the levels of plasticiser addition are rather low (typically, 0–10% by weight). It is not therefore a phenomenon which is of much concern to the processor of plasticised polyvinyl chloride which is to be used as a synthetic rubber. It is, however, of some concern to the processor of rigid polyvinyl chloride, because it can give rise to a product which is unacceptably brittle. Several hypotheses have been put forward to explain the phenomenon of antiplasticisation. According to one view, it is attributed to crystallisation of the polymer segments facilitated by the increase in segmental motions which become possible when small amounts of plasticiser are present. According to a second view, the plasticiser molecules are able to attach themselves to the polyvinyl chloride chains through their polar groups, and so act as crosslinks between the chains. Both these effects are expected to be manifest at low levels of plasticiser addition only. At higher levels of addition, such effects are swamped by the normal plasticising effects which arise essentially from dilution of the polymer chains by the added plasticiser.

Apart from effects upon the mechanical properties, plasticisers also have important effects upon the electrical properties of polyvinyl chloride. This is illustrated in Fig. 8.2, which shows the volume electrical resistivity of plasticised polyvinyl chloride as a function of the level of dioctyl phthalate plasticiser. The addition of plasticiser has a pronounced effect in reducing the resistivity of the material; the addition of 40 parts of dioctyl phthalate reduces the resistivity by three orders of magnitude.

The most satisfactory general-purpose primary plasticisers for polyvinyl chloride are the dialkyl phthalates. They are also the cheapest. A wide

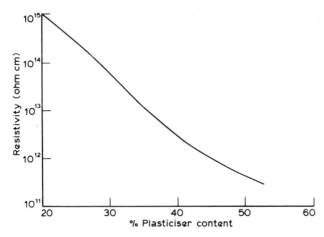

FIG. 8.2.   Effect of plasticisation of polyvinyl chloride with dioctyl phthalate upon volume electrical resistivity.[1]

range of phthalate-ester plasticisers is available commercially. As the size of the alkyl group in the dialkyl ester is increased, so the volatility, extractability and odour of the plasticiser decreases, but so too does its effectiveness as a plasticiser. The latter feature, which has been illustrated in Fig. 8.1 for dioctyl and dinonyl phthalates, arises essentially from a reducing tendency of the ester to mix with polyvinyl chloride as the size of the alkyl group is increased. The effect of size of alkyl group upon mixing tendency is illustrated in Table 8.1, which gives the fusion temperatures, minimum fluxing temperatures and clear points for polyvinyl chloride plastisols based upon a range of dialkyl phthalate plasticisers. The fusion temperature is defined as the temperature at which the torque developed in a rheometer-type test begins to increase rapidly at the onset of fusion. The minimum fluxing temperature is defined as the lowest temperature at which a plastisol develops sufficient coherence to permit it to be lifted from a heated plate. The clear point is defined as the temperature at which the plastisol first becomes clear. All three temperatures clearly increase with increasing size of the alkyl group, and this is a measure of the reduction in mixing tendency between ester and polymer. The trends which have been illustrated in Table 8.1 for the dialkyl phthalates generally apply equally to other families of ester plasticisers. Another useful generalisation is that the tendency of a plasticiser to be lost from polyvinyl chloride increases as the vapour pressure of the plasticiser increases. It has been found that, for

TABLE 8.1

FUSION TEMPERATURES, MINIMUM FLUXING TEMPERATURES AND CLEAR POINTS FOR POLYVINYL CHLORIDE PLASTISOLS BASED UPON VARIOUS DIALKYL PHTHALATE PLASTICISERS[2]

Plasticiser	Fusion temp. (°C)	Minimum fluxing temp. (°C)	Clear point (°C)
Dimethyl phthalate	59		88
Diethyl phthalate	58	54	
Dibutyl phthalate	62	68	97
Di-n-hexyl phthalate	84	75	113
Diisoöctyl phthalate	85		126
Di-2-ethylhexyl phthalate	84	105	127
Di-n-nonyl phthalate			139
Di-n-decyl phthalate	107		143
Diundecyl phthalate		136	
Didodecyl phthalate	123		
Ditridecyl phthalate	130	155	155

many plasticisers, the logarithm of the time required for a given loss to occur under standard conditions varies linearly with the logarithm of the vapour pressure of the neat plasticiser at the same temperature.

As a result of the balance of factors such as cost, availability, plasticising efficiency and volatility, the following types of phthalic-acid diester have been widely used as plasticisers for polyvinyl chloride:

1. those from individual straight-chain and branched alcohols containing between 4 and 12 carbon atoms;
2. those from mixtures of predominantly straight-chain alcohols containing between 6 and 10 carbon atoms;
3. those from mixtures of partly branched alcohols containing between 7 and 9 carbon atoms;
4. those from mixtures of predominantly straight-chain alcohols containing between 7 and 9 carbon atoms; and
5. those from mixtures of predominantly straight-chain alcohols containing between 9 and 11 carbon atoms.

In former years, dibutyl phthalate was probably the most widely-used dialkyl phthalate for this application. It suffers, however, from a marked tendency to volatilise from the plasticised polymer, especially at elevated temperature, and to be extractable by oils, solvents and detergent solutions. It has in recent years been replaced by dioctyl phthalate as the best-known plasticiser for polyvinyl chloride. The various isomers of dioctyl phthalate all behave similarly. Di-2-ethylhexyl phthalate and diisoöctyl phthalate are the two most widely-used isomers. The dioctyl phthalates are suitable as plasticisers for a wide range of applications in which the temperature does not exceed about 60 °C. For materials to be used at higher temperatures, an ester of higher molecular weight should be used, e.g. ditridecyl phthalate. The lower dialkyl esters of phthalic acid, such as dimethyl phthalate and diethyl phthalate, are excellent plasticisers for polyvinyl chloride, but are unacceptably volatile for general use.

The esters of orthophosphoric acid are not very efficient as plasticisers for polyvinyl chloride, but are widely used in applications where it is required to minimise the flammability of the plasticised material. The most widely-used plasticisers in this class are the triaryl orthophosphates. Tritolyl phosphate (formerly commonly known as tricresyl phosphate) was much used in earlier years. For economic reasons, it has been largely superseded by trixylyl phosphate. Trialkyl esters of orthophosphoric acid, such as tri-2-ethylhexyl phosphate, have also been used as plasticisers; so too have mixed alkyl aryl esters, such as 2-ethylhexyl diphenyl phosphate.

The dialkyl esters of aliphatic dicarboxylic acids are used to produce plasticised polyvinyl chloride which has good resistance to stiffening at low temperatures. The esters in this class are commonly from either adipic acid ($n = 4$ in structure LXXXVII), or azelic acid ($n = 7$), or sebacic acid ($n = 8$), or from a mixture of succinic ($n = 2$), glutaric ($n = 3$) and adipic acids. Plasticisers derived from this latter mixture are often known as 'AGS' (adipic–glutaric–succinic) plasticisers, or as 'nylonates'. It should also be noted that the esters of carboxylic acids tend to inhibit the thermal degradation of polyvinyl chloride during processing, that is, they function as stabilisers as well as plasticisers. The reasons for this are not clear. The ability to 'lubricate' the polymer molecules and reduce hot-melt viscosity seem to be important physical factors, but specific chemical effects are probably operative as well.

Polyester plasticisers of low molecular weight have been developed because of the requirement for plasticised polyvinyl chloride which will retain its flexibility and dimensions whilst in contact with organic liquids such as hydrocarbon fuel oils and dry-cleaning solvents. The first 'polymeric' plasticisers to be introduced were unmodified polyesters such as polypropylene adipate and polypropylene sebacate. These plasticisers are attended by disadvantages such as high viscosity with consequent difficulties in handling, the need for high processing temperatures if optimum physical properties are to be obtained, poor plasticising efficiency, and poor low-temperature flexibility. These disadvantages have led to the development of polyesters which have been modified by the addition of a monocarboxylic acid and/or a monohydric alcohol to the reaction system by which they were produced. The product is a polyester which is terminated by the monocarboxylic acid or monohydric alcohol, and which has a rather lower molecular weight than otherwise.

Amongst the miscellaneous plasticisers may be noted the epoxidised oils and esters. These are of particular interest because, as well as functioning as plasticisers, they improve the heat and light stability of polyvinyl chloride compounds. A typical example is butyl epoxystearate.

There are four principal ways in which the plasticisation of polyvinyl chloride is effected:

1.   by hot compounding;
2.   by dry blending;
3.   by solvent blending; and
4.   by the plastisol (or paste) technique.

In *hot compounding*, the plasticiser, polymer and other ingredients, are mixed together and fluxed on a two-roll mill, or in an internal mixer, at a temperature in the range 150–170 °C. The required temperature depends upon the polymer and upon the amount and type of plasticiser. In *dry blending*, the plasticiser is absorbed into a suitable grade of polymer at a temperature in the range 50–80 °C. The product is a dry powder which is barely distinguishable in outward appearance from the initial polymer. The dry blend, in which the polyvinyl chloride is not yet plasticised, can be stored without change at normal temperatures. In order to convert it into plasticised polyvinyl chloride, it is merely necessary to heat it for a short period at *c.* 160 °C. This can be achieved by milling, by moulding, by calendering, by extrusion or by a fluid-bed coating process. *Solvent blending* is effected by dissolving the plasticiser and the polymer in a common solvent, such as tetrahydrofuran or cyclohexanone, and then casting a film of plasticised polymer from the solution by evaporation. The *plastisol technique* will be considered separately subsequently (see Section 8.7).

It is appropriate to conclude this discussion of the plasticisation of polyvinyl chloride by referring briefly to certain substances which are known as *extenders* or *secondary plasticisers*. The best-known substances in this class are the chlorinated paraffin hydrocarbons. They are used for the partial replacement of primary ester plasticisers, mainly to reduce cost and improve flame retardancy. Chlorine contents are usually in the range 42–56 % by weight, and the average carbon chain length is in the range 15–25. As expected, the compatibility with polyvinyl chloride increases with increasing chlorine content.

8.4 STABILISATION OF POLYVINYL CHLORIDE

It has already been noted that polyvinyl chloride tends to degrade by dehydrochlorination when it is heated, and that, in consequence, it is necessary to compound the polymer with a stabiliser which protects it against this form of degradation. Thermal degradation is primarily a problem which is encountered during the processing of polyvinyl chloride, because of the high temperatures which are required for mixing, fluxing, calendering, extrusion and moulding. One important aspect of the thermal degradation concerns the hydrogen chloride which is evolved; this is able to cause serious corrosion of processing machinery. Thermal degradation is also of importance if the polymer is to be subjected to high temperatures during application. However, this is not usually the case because of the thermoplasticity of plasticised polyvinyl chloride. Of rather greater significance are two other considerations. In the first place, thermal degradation during processing may have an adverse effect upon the appearance and properties of the final product. Secondly, thermal degradation during processing may sensitise the polymer to subsequent degradation by ultraviolet light and by electromagnetic radiations of shorter wavelength. Polyvinyl chloride which has been subject to even slight thermal degradation is much more susceptible to further degradation by outdoor ageing and weathering than is polymer which has not been thermally degraded.

Much effort has been expended in seeking to understand the chemistry of the thermal degradation of polyvinyl chloride. It is not immediately obvious why the polymer should dehydrochlorinate so readily when heated. If the polymer really did consist exclusively of a sequence of repeat units of the type shown at (LXXIV) (the only structural perturbation being the presence of thermally-stable end groups), then it would be expected to possess inherently high thermal stability. Studies using low-molecular-weight 'model' compounds have indicated that the thermal instability arises from the presence of unusual units such as internal allylic chlorides (LXXXVIII), tertiary chlorides (LXXXIX), and terminal allylic chlorides (XC). The order of decreasing ability to impart thermal instability is (LXXXVIII) > (LXXXIX) > (XC). According to one view, dehydrochlorination commences at a site which either contains or is adjacent to an

$$
\cdots-CH{=}CH-CH-\cdots \qquad \cdots-CH_2-\overset{\overset{\vdots}{\underset{|}{CH_2}}}{\underset{|}{C}}-\cdots \qquad \cdots-CH-CH{=}CH_2
$$

$$
\underset{|}{} \qquad \underset{|}{} \qquad \underset{|}{}
$$

$$
Cl \qquad\qquad\qquad Cl \qquad\qquad\qquad Cl
$$

(LXXXVIII)   (LXXXIX)   (XC)

allylic or tertiary chlorine atom. If the dehydrochlorination occurs at an adjacent site, then the allylic or tertiary chlorine atom functions as an activator of dehydrochlorination. Whatever may be the exact mechanism of the reaction, the effect is that hydrogen chloride is eliminated from the polymer chain and a carbon–carbon double bond is created. The change which occurs may in general be represented by a reaction such as the following:

$$
\cdots-CH_2-\underset{\underset{Cl}{|}}{CH}-\cdots \longrightarrow \cdots-CH{=}CH-\cdots + HCl
$$

The creation of the carbon–carbon double bond has the probable consequence that the chlorine atom in one of the neighbouring units has now become an allylic chlorine atom, and so is itself reactive and also tends to activate its neighbouring unit. Thus the elimination of one hydrogen chloride molecule is thought to initiate a chain of hydrogen-chloride elimination reactions from units which would be otherwise thermally stable. In this way is explained the observation that, once dehydrochlorination commences, it proceeds quite rapidly. However, it appears that whether or not the rate of evolution of hydrogen chloride increases with time of dehydrochlorination depends upon the nature of the atmosphere which is in contact with the polyvinyl chloride. In oxygen, the rate of evolution increases with time, whereas in nitrogen it remains constant. An associated question which has been the subject of much investigation and debate is whether hydrogen chloride is itself able to catalyse the dehydrochlorination reaction. Whether or not the eliminated hydrogen chloride is itself able directly to catalyse further chemical change in the polymer, it may well attack the metal of the processing machinery and form various metal chlorides (principally iron(III) chloride). These metal chlorides are very likely to catalyse further chemical changes in the polymer.

The formation of carbon–carbon double bonds by elimination of hydrogen chloride has the consequence that the colour of polyvinyl chloride gradually deepens on heating, the progression of colour being colourless, light yellow, yellow, orange, orange–red, red, brown. These colours are attributed to the presence of increasingly long polyene sequences. The formation of visible colour is usually the first visual indication that degradation is occurring, although the evolution of hydrogen chloride can be detected by chemical methods. Changes of colour are observed long before any of the more serious manifestations of degradation become evident.

The following are the principal requirements of an ideal stabiliser for polyvinyl chloride:

1. It should absorb and neutralise the hydrogen chloride which is evolved during degradation.
2. It should deactivate labile and activating substituents in the polymer chain, such as allylic chloride.
3. It should deactivate stabiliser-degradation products (such as heavy-metal chlorides) which themselves may function as catalysts of polymer degradation.
4. It should disrupt degradative chain reactions.

5.   It should deactivate any impurities in the polymer which have a catalysing influence upon degradation.

6.   It should provide protection against degradation by ultraviolet light.

The majority of stabilisers for polyvinyl chloride contain one or more metal derivatives of a weak acid. The acid may be either inorganic or organic. Amongst the organic acids are included very weakly acidic substances such as alcohols, phenols and mercaptans. The derivative may be either basic or neutral. Such derivatives function in part at least by reacting with the hydrogen chloride which is liberated during the degradation of the polymer. Denoting the metal derivatives by MX, where M is the metal and X the moiety derived from the weak acid (X may or may not be an anion), the reaction by which hydrogen chloride is absorbed and neutralised can be represented as follows:

$$MX + HCl \longrightarrow MCl + HX$$

To be effective as a stabiliser acting in this way, the metal derivative must be able to react quickly and efficiently with hydrogen chloride, but it must not be sufficiently strongly alkaline actively to promote dehydrochlorination of the polymer itself. A further requirement is that the liberated acid HX must itself be innocuous.

Another way in which metal derivatives are able to function as stabilisers for polyvinyl chloride is by replacing reactive, labile chlorine atoms by groups which are less labile. In general, taking allylic chlorine atoms as an example, such reactions can be represented as follows:

$$MX + \cdots -CH{=}CH-CH- \cdots \longrightarrow MCl + \cdots -CH{=}CH-CH- \cdots$$
$$\qquad\qquad\qquad | \qquad\qquad\qquad\qquad\qquad\qquad\qquad |$$
$$\qquad\qquad\qquad Cl \qquad\qquad\qquad\qquad\qquad\qquad\qquad X$$

Heavy-metal carboxylates and mercaptides are thought to be especially effective in this way. However, it is to be noted that, in order to be effective in this way, it is necessary that some of the compound MX should still be present as such in the compounded materials; thus if much hydrogen chloride is present in the polymer (because partial degradation has already occurred) before the compound MX is added, it may be that all the MX will be converted to MCl by reaction with the hydrogen chloride, and thus no X moieties will be available to replace reactive chlorine atoms.

A third way in which metal derivatives are indirectly able to stabilise polyvinyl chloride is by the addition of the HX, formed by reaction with hydrogen chloride, to any carbon–carbon double bonds which are present in the polymer:

$$HX + \cdots-CH\!\!=\!\!CH-\cdots \longrightarrow \cdots-CH_2-\underset{\underset{X}{|}}{CH}-\cdots$$

Mercaptans, generated from metal mercaptides, are particularly effective in this way. The effect of this reaction is two-fold. In the first place, it removes the unsaturation, and thus the tendency of the polymer to discolour is reduced. Secondly, in removing the unsaturation it converts a neighbouring allylic chlorine atom into an alkyl chlorine atom, and thus reduces its lability and reactivity.

Other stabilisers which function in part at least as hydrogen-chloride acceptors include (a) amines and (b) epoxides. The respective reactions are as follows:

$$RNH_2 + HCl \longrightarrow R\overset{\oplus}{N}H_3Cl^{\ominus}$$

and

$$-\underset{\underset{O}{\diagdown\diagup}}{CH-CH}- + HCl \longrightarrow -\underset{\underset{OH}{|}}{CH}-\underset{\underset{Cl}{|}}{CH}-$$

The fact that epoxidised plasticisers also provide some protection against thermal degradation has already been mentioned in Section 8.3. If epoxides are used in conjunction with stabilisers of the metal-derivative type, it seems probable that other reactions besides hydrogen-chloride acceptance are involved; further reference is made to this matter below.

The majority of commercially-available stabilisers for polyvinyl chloride can be conveniently classified into five groups as follows:

1. basic lead compounds;
2. metal derivatives of carboxylic acids, mercaptans and phenols;
3. organo-tin derivatives;
4. amines; and
5. miscellaneous substances such as epoxidised oils, metal-chelating agents, antioxidants, and absorbers of ultraviolet radiation.

Basic lead compounds have been widely used in this application, partly because of their low cost. The more important examples in this class are white lead (basic lead carbonate), tribasic lead sulphate, dibasic lead phosphite, dibasic lead phthalate, basic lead silicate, and dibasic and monobasic lead stearates. There are three major limitations which restrict the use of lead compounds, namely, (a) the opacity which is inevitably imparted to the polymer, (b) the toxicity hazard, and (c) discoloration in the presence of sulphur and mercaptans. A major use for lead stabilisers is in plasticised polyvinyl chloride to be used for the covering of electric cables. Dibasic lead phthalate is recommended for material which is to be used at elevated temperatures, |because| it functions as an antioxidant as well as a stabiliser against dehydrochlorination. The underlying reason why lead stabilisers are so suitable as stabilisers for polyvinyl chloride to be used in electrical applications is thought to be that neither they nor lead chloride are appreciably soluble in water or ionised. The electrical resistivity of the material is not therefore adversely affected by their presence. Furthermore, lead chloride (unlike many other heavy-metal chlorides) is apparently unable to catalyse further chemical reaction of polyvinyl chloride. Lead compounds are normally used alone; their performance is not significantly affected by the presence of auxiliary organic stabilisers.

A wide variety of metal derivatives of carboxylic acids, phenols and mercaptans is available for the stabilisation of polyvinyl chloride. Combinations of cadmium and barium (and sometimes zinc) carboxylates are widely used. Cadmium carboxylates confer excellent heat stability for a short time, but then the polymer darkens rapidly. Barium carboxylates are less effective but longer-lasting as stabilisers. Combinations of the two types of carboxylate are more effective and longer-lasting than either separately, that is, the mixtures are *synergistic*. The explanation of this synergistic effect is probably as follows: Cadmium carboxylates are able to react rapidly with hydrogen chloride and labile chlorine to produce cadmium chloride. However, cadmium chloride is itself an active catalyst of further polymer degradation. Barium carboxylates do not react so readily with hydrogen chloride and labile chlorine, but they do react with cadmium chloride and chloride-carboxylates to regenerate the cadmium carboxylates and produce barium chloride or chloride-carboxylates, e.g.:

$$CdCl(RCO_2) + Ba(RCO_2)_2 \longrightarrow Cd(RCO_2)_2 + BaCl(RCO_2)$$

Barium chloride is much less active as a catalyst of further polymer degradation than is cadmium chloride. A further observation which illustrates the complex interactions which can occur between stabilisers when they are present together in polyvinyl chloride is that epoxide stabilisers are able to promote the synergism which is exhibited by cadmium and barium carboxylates. It seems likely that epoxides are able to catalyse the interaction between cadmium chloride-carboxylates and barium carboxylates by reactions such as the following:

$$CdCl(RCO_2) + -CH\underset{O}{-}CH- \longrightarrow -\underset{Cl}{CH}-\underset{\underset{Cd(RCO_2)}{O}}{CH}-$$

$$+ Ba(RCO_2)_2$$

$$Cd(RCO_2)_2 + -CH\underset{O}{-}CH- + BaCl(RCO_2)$$

A typical example of a stabiliser in the cadmium–barium carboxylate class is a mixture of cadmium and barium laurates.

The organo-tin stabilisers are relatively expensive, but very effective. They are used principally where the retention of high clarity is important. They are usually of the general structure R_2SnX_2, where R is an alkyl group (usually straight-chain and frequently butyl) and X is a moiety derived from an organic acid, which includes mercaptans as well as carboxylic acids. Typical examples are dibutyltin dilaurate (XCI), dibutyltin maleate (XCII), and dibutyltin bis(lauryl mercaptide) (XCIII). The ability of these stabilisers

$$\begin{matrix} C_4H_9 & & OOC.C_{11}H_{23} \\ & Sn & \\ C_4H_9 & & OOC.C_{11}H_{23} \end{matrix}$$
(XCI)

$$\begin{matrix} C_4H_9 & & OOC.CH \\ & Sn & \| \\ C_4H_9 & & OOC.CH \end{matrix}$$
(XCII)

$$\begin{matrix} C_4H_9 & & SC_{12}H_{25} \\ & Sn & \\ C_4H_9 & & SC_{12}H_{25} \end{matrix}$$
(XCIII)

to promote retention of clarity is attributed to the fact that the organo-tin dichlorides, R_2SnCl_2, which are produced by reaction between R_2SnX_2 and hydrogen chloride, are soluble in plasticised polyvinyl chloride.

Amongst the miscellaneous stabilisers are substances which are added to polyvinyl chloride to confer stability to ultraviolet radiation rather than to confer stability to heat. Several such stabilisers are available, but their effectiveness depends to some extent upon the heat stabiliser with which they are used; thus the hydroxybenzophenones are particularly effective when used with combinations of cadmium and barium carboxylates. Dibasic lead phosphite is an example of a heat stabiliser which also confers some degree of resistance to degradation by ultraviolet radiation.

8.5   FILLERS FOR PLASTICISED POLYVINYL CHLORIDE

Fillers are added to plasticised polyvinyl chloride principally to cheapen the material. In general, physical properties such as tensile strength and elongation at break are reduced by the incorporation of fillers, whereas hardness and modulus are increased. The resistance of the compound to

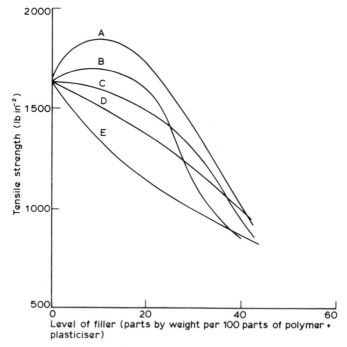

FIG. 8.3.   Effect of various fillers upon tensile strength of plasticised polyvinyl chloride.[1] Compound: polyvinyl chloride 50, tritolyl phosphate 50, white lead 4·5, calcium stearate 1. Fillers: A finely-divided china clay, B lightly-coloured china clay, C coated natural whiting, D precipitated whiting and E asbestos.

ultraviolet light may be increased as a consequence of the opacity of the filler.

The fillers most commonly used with plasticised polyvinyl chloride are the calcium carbonates. Various forms are used, both natural and precipitated, including the grades which are coated with calcium stearate. The addition of calcium carbonate reduces the acid resistance of plasticised polyvinyl chloride compounds. Other fillers which are used include calcined clays, talc, asbestos, slate dust, calcium silicate and barytes. The effects of increasing levels of various fillers upon the tensile strength of polyvinyl chloride plasticised with 100 parts of tritolyl phosphate per 100 parts of polymer are shown in Fig. 8.3. It is interesting to note that, at low levels of addition, some clay fillers increase the tensile strength somewhat.

8.6   OTHER COMPONENTS OF PLASTICISED POLYVINYL CHLORIDE COMPOUNDS

Lubricants are sometimes added to plasticised polyvinyl chloride in order to facilitate processing. Metal soaps are effective in this respect, and it is often the case that a metal carboxylate combination added as a stabiliser also functions as a lubricant. Other substances which are used as lubricants include stearic acid, glyceryl monostearate and paraffin waxes.

Pigments are used for colouring plasticised polyvinyl chloride. Inorganic pigments such as titanium dioxide and carbon black present few problems. Organic pigments are also satisfactory provided that (a) they are stable at processing temperatures, (b) they do not promote thermal degradation of polyvinyl chloride, and (c) they are insoluble in the plasticiser. The latter requirement ensures that subsequent migration of the pigment from the plasticised compound is minimised. Pigments of high opacity are able to contribute significantly to the ability of plasticised polyvinyl chloride to withstand exposure to ultraviolet radiation without undue degradation. This is because they can cause a major proportion of the radiation to be reflected off the surface of the material and then absorb the remainder.

Flame retardants, such as antimony oxide and zinc borate, are used to reduce the flammability of plasticised polyvinyl chloride.

Antistatic agents are sometimes added to reduce the pickup of dirt by the material in service, and also to minimise the possibility of sparking from the discharge of accumulated static electricity. Typical antistatic agents include fatty amines and amides, and esters of polyoxymethylene glycol.

8.7   POLYVINYL CHLORIDE PLASTISOLS

The principles which underlie the technology of polyvinyl chloride plastisols have been explained in the introduction to this chapter. The feasibility of this method of processing plasticised polyvinyl chloride depends upon the fact that, although polyvinyl chloride and ester plasticisers are completely miscible at normal temperatures in the sense that an intimate molecular mixture is thermodynamically stable relative to a state in which the two components are separated, a long time is required for the attainment of equilibrium at room temperature. Thus it is that the finely-powdered polymer can be dispersed in plasticiser to give a plastisol which is reasonably stable at room temperature, although the plastisol does thicken progressively with storage at room temperature. This latter aspect will be discussed in more detail below. If the temperature of the plastisol is raised slowly, the viscosity falls at first because the viscosity of the plasticiser falls with rising temperature. The viscosity of the plastisol reaches a minimum in the region of 45 °C, and then begins to rise with further increase of temperature. The rise in viscosity occurs because the particles of polymer begin to imbibe the plasticiser (thereby becoming 'solvated') and in consequence increase in size. In effect, the volume fraction of the disperse phase in the plastisol increases as a consequence of the imbibition of plasticiser, and so therefore does the viscosity of the plastisol. The exact nature of the variation of viscosity with temperature depends upon both the grade of polymer and the type of plasticiser. In many cases, the viscosity begins to rise sharply by about 60 °C. As the temperature is raised still further, the particles continue to imbibe plasticiser and soften increasingly under the joint influence of the temperature rise and the increasing ratio of plasticiser to polymer. Eventually the particles fuse together, partly because they themselves have become very soft, and partly because the plasticiser which acted as the dispersion medium and separated them from each other has now disappeared. The final product is a homogeneous blend of polymer and plasticiser which cools down to a homogeneous plasticised polyvinyl chloride. The process by which the plasticiser-swollen particles of polyvinyl chloride integrate together to form a coherent, homogeneous mass is called *gelation*, and the plastisol is said to have been *gelled*. Note that, because molecular mixtures of polyvinyl chloride and ester plasticisers are thermodynamically stable at room temperatures, there is no tendency for the gelled plastisol to separate into two phases (polymer and plasticiser) when it is cooled. The gelation of polyvinyl chloride plastisols is an irreversible process.

Polyvinyl chloride plastisols find application in processes such as dipping, wire coating, rotational casting and the production of cellular plasticised polyvinyl chloride. In dipping processes, a heated former (usually metal) is immersed in the plastisol. The rise of temperature in the plastisol in the vicinity of the former surface causes a thin layer of gelled plastisol to form around the outside of the former. After a suitable length of time has elapsed, the former is removed from the plastisol and allowed to cool. The deposit of plasticised polymer is then removed from the former. In this way, thin-walled products having the shape of the former can be produced from a plastisol. The principle underlying coating of wire articles, such as wire baskets, is similar, except that, of course, no attempt is made to remove the plasticised polymer from the wire. A major use for polyvinyl plastisols is in the production of hollow mouldings by rotational casting. A measured quantity of plastisol is charged into a sealed split mould. The mould is then slowly rotated in two planes simultaneously whilst being heated. The objective is to form a uniform layer of gelled plastisol within the mould cavity. When gelation is complete, the mould is cooled and then opened to release the moulding. Cellular plasticised polyvinyl chloride can be produced by incorporating a 'blowing agent' in the plastisol. If the blowing agent and plastisol are such that gas is evolved just before the plastisol finally gels, then the product is a fine dispersion of gas bubbles in plasticised polyvinyl chloride. A variant of this process is one in which a gas (such as carbon dioxide) is dissolved in the plastisol under pressure, and then the pressure is subsequently released as gelation occurs.

The polyvinyl chloride which is to be used for the production of plastisols should have the form of a fine powder. A wide range of special 'paste' grades is available. For a given grade of polymer, the behaviour of the plastisol is determined largely by the plasticiser. The phthalates are the best general-purpose plasticisers for this application, bearing in mind cost as well as performance. The dioctyl phthalates, both straight-chain and branched, are widely used for the production of plastisols. Rather surprisingly, there seems to be an inverse correlation between the ability of a plasticiser to solvate polyvinyl chloride in a plastisol and its ability to impart resistance to stiffening at low temperature in the gelled material. Thus the phosphate plasticisers have relatively high solvating power but are relatively inefficient as low-temperature plasticisers. The adipates and sebacates, on the other hand, have low solvating power but are very effective in imparting resistance to stiffening at low temperatures.

An important matter is the extent to which the viscosity of the plastisol tends to increase with storage at room temperature. For a given grade of polymer, this too is principally determined by the plasticiser. As has been noted above, the reason why the plastisol viscosity tends to increase slowly is that, even at room temperature, the polymer particles imbibe the plasticiser slowly and in consequence the polymer particles gradually increase in size. In general, the greater is the solvating power of the plasticiser for the polymer, the more rapidly does the viscosity increase during storage. Dibutyl phthalate, which has a strong tendency to solvate the polymer, gives a plastisol which rapidly thickens at room temperature. Dialkyl phthalates of higher molecular weight, e.g., the dioctyl phthalates, give plastisols which thicken much less rapidly. A plasticiser which has very little solvating tendency at room temperature, such as triethylene glycol dicaprylate, gives a plastisol which undergoes very little change of viscosity at room temperature.

The viscosity characteristics of plastisols can also be modified by the addition of a so-called *extender* polymer. This is a vinyl chloride polymer or copolymer which is of much larger particle size than the paste polymer, and does not itself form a plastisol with plasticiser. Extender polymers are usually made by suspension polymerisation. The effect of progressive replacement of paste polymer by extender polymer upon the viscosity of

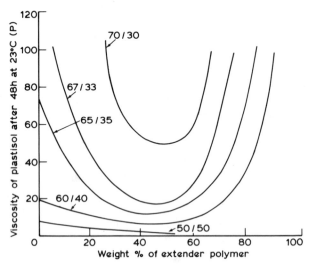

FIG. 8.4.   Effect of replacement of polyvinyl chloride paste polymer by extender polymer upon viscosity of polyvinyl chloride plastisols.[1] The numbers appended to the curves give the weight ratios of polymer to plasticiser in the plastisol.

plastisols of various compositions is shown in Fig. 8.4. The viscosity falls progressively until about half of the paste polymer has been replaced; thereafter it increases with further replacement.

GENERAL BIBLIOGRAPHY

Penn, W. S. (1971). *PVC Technology*, 3rd edition (revised and edited by W. V. Titow and B. J. Lanham), Applied Science Publishers Ltd, London.

Brighton, C. A. (1971). Vinyl chloride polymers (introduction). In: *Encyclopedia of Polymer Science and Technology*, Vol. 14, John Wiley and Sons, New York, pp. 305 f.

Brighton, C. A. (1971). Vinyl chloride polymers (compounding). In: *Encyclopedia of Polymer Science and Technology*, Vol. 14, John Wiley and Sons, New York, pp. 394 f.

Brighton, C. A. (1971). Vinyl chloride polymers (fabrication). In: *Encyclopedia of Polymer Science and Technology*, Vol. 14, John Wiley and Sons, New York, pp. 434 f.

Darby, J. R. and Sears, J. K. (1969). Plasticizers. In: *Encyclopedia of Polymer Science and Technology*, Vol. 10, John Wiley and Sons, New York, p. 228 f.

Nass, L. I. (1970). Stabilization. In: *Encyclopedia of Polymer Science and Technology*, Vol. 12, John Wiley and Sons, New York, p. 725 f.

Ritchie, P. D., Critchley, S. W. and Hill, A. (Eds.) (1972). *Plasticisers, Stabilisers and Fillers*, Iliffe, London, for The Plastics Institute.

Collins, E. A. and Hoffmann, D. J. (1979). Rheology of plastisols of poly(vinyl chloride), *Rubb. Chem. Technol.*, **52**, 676.

REFERENCES

1. Brighton, C. A. (1971). *Encyclopedia of Polymer Science and Technology*, Vol. 14, John Wiley and Sons, New York, p. 394.
2. Darby, J. R. and Sears, J. K. (1969). *Encyclopedia of Polymer Science and Technology*, Vol. 10, John Wiley and Sons, New York, p. 228.

Chapter 9

Other Types of Synthetic Rubber

9.1 CARBOXYLATED RUBBERS (XSBR, XBR, XNBR, XCR, ETC.)

We begin this review of 'miscellaneous' synthetic rubbers by considering a group of rubbers, the members of which are, in effect, special-purpose chemical modifications of other synthetic rubbers. In this group, known collectively as the *carboxylated rubbers*, the main chain of the rubber molecule has been modified in such a way that it carries a minor number of pendant carboxylic-acid groups. There are several types of synthetic rubber which|lend|themselves to modification in this way. The more important are the styrene–butadiene rubbers, the acrylonitrile–butadiene rubbers, the chloroprene rubbers, and the acrylic rubbers (to be considered later in this chapter—see Section 9.2). The concern here is thus not with a single type of synthetic rubber, but with a particular type of synthetic rubber variant. The justification for treating this group of rubbers as a whole is that the advantages which make carboxylation attractive, and the modifications to the processing technology which accrue from carboxylation, are common to the group as a whole. Thus, taking the group as a whole, it is the presence of the additional carboxylic-acid-group functionality which distinguishes them from other rubbers, rather than the nature of the repeat units which make up the greater part of their molecule.

The type of synthetic rubber variant with which we are concerned here is one which contains between, say, 1% and 10% of its weight as units to which carboxylic-acid groups are attached. In most cases, the proportion will be nearer to 1% than to 10%. Furthermore, the carboxylic-acid groups are approximately randomly distributed along the polymer chain. We are not concerned with a special type of carboxylated polymer in which the carboxylic-acid groups are situated exclusively at the ends of the polymer chains, although some reference will be made to this type of polymer in Chapter 10 (Section 10.4).

There are two general methods for producing carboxylated rubbers of the type under consideration here:

1. by copolymerisation of the main monomers with minor amounts of an unsaturated carboxylic acid such as acrylic acid;
2. by modification of an existing polymer chain.

Whilst it is in principle possible to copolymerise unsaturated carboxylic acids with other monomers by a variety of free-radical techniques, emulsion copolymerisation is the method usually employed. Unsaturated carboxylic acids are particularly well suited to incorporation by emulsion copolymerisation if the reaction conditions are appropriately adjusted, not least because, when copolymerised, they tend to increase the stability of the latex which forms. Being active-hydrogen compounds, unsaturated carboxylic acids are not susceptible to polymerisation by cationic or anionic mechanisms, or by Ziegler–Natta catalysis.

As regards carboxylation by the modification of an existing polymer chain, the following are amongst the more obvious possibilities:

1. hydrolysis of cyano groups;
2. hydrolysis of ester groups; and
3. addition of carboxyl-containing compounds to the carbon–carbon double bonds of a diene polymer.

The proportion of the repeat units which require to be modified in any individual case will depend partly upon the degree of carboxylation which is required in the final polymer and partly upon the concentration of modifiable repeat units in the initial polymer. But whatever is the overall degree of modification of repeat units which is necessary in order to attain the requisite degree of carboxylation, it is usually desirable that such modification should occur randomly. However, the distribution of carboxylic-acid groups produced by modification of an existing polymer chain may be rather different from that which is obtained by copolymerisation with an unsaturated carboxylic acid. Differences in distribution may arise in part from the nature of the distribution in the initial polymer of those groups which are to undergo modification.

Of the various production methods which have been outlined above, the only one which has achieved any commercial importance is that of incorporation of carboxylic-acid groups by emulsion copolymerisation. This is interrelated with the fact that carboxylated rubbers currently find little application as dry rubbers, notwithstanding the early hopes that they might form a commercially-important group of special-purpose synthetic rubbers. Although carboxylated rubbers do currently find widespread application, it is almost exclusively in the form of latices that they are used. No attempt will be made in this book to consider the very extensive subject of carboxylated rubber latices. The intention here is to summarise briefly the salient properties which distinguish dry carboxylated rubbers from their non-carboxylated counterparts.

The advantages which are said to accrue from the incorporation of a minor proportion of carboxylic-acid groups into the polymer chain of a rubber are as follows:

1.   It makes the rubber susceptible to vulcanisation by a range of unusual reagents.
2.   It increases the strength of the subsequent vulcanisates.
3.   It increases resistance to swelling in hydrocarbon oils.

In addition to being vulcanisable by whatever reagents are capable of vulcanising the non-carboxylated equivalent (e.g., sulphur and conventional accelerators in the case of a styrene–butadiene rubber), it is claimed that carboxylated rubbers are crosslinkable through reactions involving the carboxylic-acid groups. The possibilities include:

1.   reaction with polyvalent metal oxides to form ionic crosslinks;
2.   reaction with polyamines to form amide crosslinks;
3.   reaction with polyhydric alcohols to form ester crosslinks;
4.   reaction with epoxides to form ester crosslinks;
5.   reaction with methylol-containing resins to give crosslinks of rather uncertain chemical structure; and
6.   reaction between pairs of carboxylic-acid groups to form acid-anhydride crosslinks.

There is also the possibility of crosslinking by reaction with di-functional isocyanates, but, as far as is known, this possibility has never been seriously contemplated. The presence of carboxylic-acid groups has little or no effect upon the ability of a rubber to be vulcanised by heating with organic peroxides.

The following are the mechanisms by which crosslinks are commonly supposed to form by reactions (1)–(6) above. It should be noted, however, that there appears to be little direct evidence that any of these crosslinks actually forms—such evidence as does exist is of a rather circumstantial nature:

1.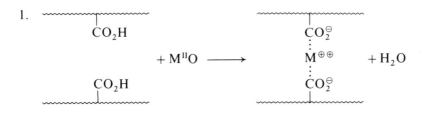

2.

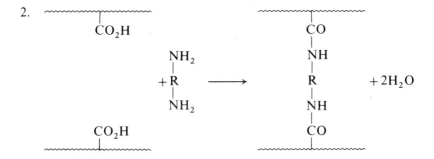

3.

4.

5.

or alternatively a reaction such as:

where $R(CH_2OH)_2$ is of one of the following types:

(A)   (B)   (C)

these structures being derived respectively from reaction between phenol and formaldehyde (A), urea and formaldehyde (B), and melamine and formaldehyde (C).

6.

Reaction (6) is of interest in that it indicates that carboxylated rubbers should be crosslinkable by heating alone. Crosslinking with metal oxides is said to occur so rapidly that problems of scorching during mill mixing can be encountered. For this reason, various substances such as organic acids, organic anhydrides, amines, boric acid and silica have been proposed as retarders or 'controllers' of the reaction between metal oxides and carboxylated rubbers.

The effect of level of carboxylation upon the tensile strength of vulcanisates obtained by heating carboxylated rubbers with excess zinc oxide is shown in Fig. 9.1. The results summarised in this diagram are for copolymers of butadiene and an acid of the acrylic acid type. It appears that, over the range of carboxylation investigated, the tensile strength of

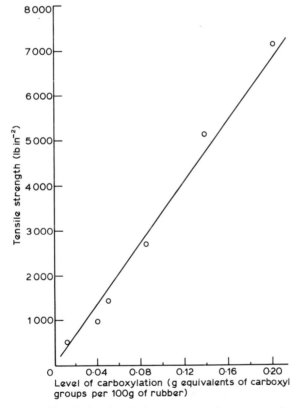

FIG. 9.1. Effect of level of carboxylation upon tensile strength of carboxylated butadiene rubbers vulcanised in presence of excess zinc oxide.[1]

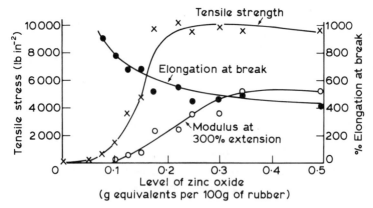

FIG. 9.2.   Effect of level of zinc oxide upon tensile properties of a gum vulcanisate obtained from a carboxylated rubber prepared by polymerising a 35/55/10 acrylonitrile/butadiene/methacrylic acid monomer mixture to 73 % conversion.[1] Carboxyl content of rubber: 0·099 g equivalents per 100 g of rubber. All vulcanisates contained 0·05 g equivalents of phthalic anhydride per 100 g of rubber.

the vulcanisate is roughly directly proportional to the level of carboxylation. Figure 9.2 illustrates that the properties of the vulcanisate depend upon the amount of zinc oxide used in relation to the level of carboxylation. These results refer to gum vulcanisates from a copolymer of acrylonitrile, butadiene and methacrylic acid. Although it might be expected that the optimum vulcanisate properties would be observed when the mole ratio Zn/CO_2H was 1/2, in fact, optimum tensile strength is observed when the ratio is $c.$ 1/1. Further additions of zinc oxide have little further effect upon tensile strength. That the optimum mole ratio Zn/CO_2H for tensile strength is 1/1 may mean that the entity which is effectively responsible for the association between the neighbouring chains is a species such as is shown at (XCIV), rather than the species shown previously. It should be

$$\left\{-CO_2^{\ominus} \cdots \overset{\oplus}{Zn}-O-H \cdots \overset{H}{\overset{|}{O}}-\overset{\oplus}{Zn} \cdots {}^{\ominus}O_2C-\right\}$$

(XCIV)

noted that the vulcanisate modulus increases progressively with Zn/CO_2H ratio up to about 2/1. As expected, the elongation at break falls with increasing level of zinc oxide, but the reduction is slight when the zinc oxide level exceeds the mole ratio 1/1 which is optimum for tensile strength.

The ability of carboxylation to increase the tensile strength and modulus of vulcanisates is also evident in the case of vulcanisates which have been made using a curing system which comprises sulphur, a conventional accelerator, and a metal oxide. In this case, it is presumed that ionic crosslinking by reaction between the acid groups and the metal oxide occurs, in addition to sulphur vulcanisation. Whatever may be the explanation for the observed effects, these effects are illustrated clearly by the data which are summarised in Table 9.1. These data are for gum vulcanisates prepared from carboxylated acrylonitrile–butadiene rubbers. They show clearly that, whatever the state of cure, the vulcanisate modulus increases sharply with increasing level of carboxylation, and the elongation at break falls. In general the tensile strength also increases with level of carboxylation, regardless of the state of cure. Table 9.1 also shows that the rate of stress relaxation of the vulcanisates increases rapidly as the level of carboxylation is increased. It is a general characteristic of vulcanisates produced by reaction between carboxylated elastomers and metal oxides that they tend to flow when subjected to stress. This behaviour is attributed

TABLE 9.1

EFFECT OF LEVEL OF CARBOXYLATION UPON MECHANICAL PROPERTIES OF SULPHUR-VULCANISED CARBOXYLATED ACRYLONITRILE–BUTADIENE RUBBER[1]

Compound	*Parts by weight*			
Acrylonitrile–butadiene rubber	100			
Zinc oxide	5			
Softener	5			
Sulphur	1·25			
Di-2-benzthiazyl disulphide	1·25			
Cure	*Mechanical properties*	*Level of carboxylation*		
(minutes		*(g equivalents per 100 g of rubber)*		
at 300°F)		*0*	*0·02*	*0·08*	*0·18*
10	Tensile strength (lb in^{-2})	1 630	5 300	4 690	6 080
	% Elongation at break	880	700	595	450
	Modulus at 300% (lb in^{-2})	<100	760	900	2 300
20	Tensile strength (lb in^{-2})	2 850	3 650	5 000	6 090
	% Elongation at break	860	595	540	380
	Modulus at 300% (lb in^{-2})	200	700	1 350	3 000
40	Tensile strength (lb in^{-2})	3 490	3 060	2 950	4 200
	% Elongation at break (%)	795	540	460	365
	Modulus at 300% (lb in^{-2})	310	1 000	1 590	3 320
80	Tensile strength (lb in^{-2})	2 400	1 600	2 850	5 090
	% Elongation at break	700	380	395	395
	Modulus at 300% (lb in^{-2})	460	1 110	1 830	4 390
40	Time for loss of 20% of stress at constant extension (h)	1·0	0·56	0·07	0·03

to the ability of ionic crosslinks to rearrange amongst themselves by *bond interchange*. Stress relaxation by this mechanism will be further discussed subsequently in connection with polysulphide rubbers (see Section 9.6). Whatever is the true explanation of the phenomenon, it has the practical consequence that the behaviour of these vulcanisates in respect of properties such as stress relaxation, creep and compression set is poor.

It is true that the carboxylation of rubbers which are not inherently resistant to swelling in hydrocarbon oils does increase their oil resistance. However, the improvements which are brought about by the levels of carboxylation which are commonly used to produce rubbers which react with the reagents described above are often too small to be significant. Although carboxylation has been used as a means of improving the oil resistance of rubbers, there are frequently better ways of achieving such improvement.

The processing behaviour of a rubber can also be affected by carboxylation. Thus, whilst very low levels of carboxylation have little effect upon processing, levels of the magnitude required to introduce sufficient reactivity can increase the 'nerve' of the rubber during milling, the stiffness of the rubber, and the strength of the crumbs. Higher temperatures may also be required in order to achieve sufficient plasticity for mixing and processing.

It is appropriate to conclude this discussion of carboxylated rubbers by pointing out that there are many other 'functional-group' comonomers, besides unsaturated carboxylic acids, which are incorporated into the molecules of rubbery substances in order to confer upon those rubbers properties which they would not otherwise have. Some of these have already been encountered. Thus the dienes which are incorporated in isobutene rubbers and ethylene–propylene rubbers are, in effect, functional-group comonomers. Others will be encountered in the course of this chapter. Two which will not otherwise be mentioned, and which are typical of the diversity of functional-group comonomers which may be incorporated into rubbers which are used in latex form, are acrylamide (XCV) and *N*-methylol acrylamide (XCVI).

$$CH_2 = CH$$
$$|$$
$$CO . NH_2$$
(XCV)

$$CH_2 = CH$$
$$|$$
$$CO . NH . CH_2OH$$
(XCVI)

They are frequently incorporated together with one or more types of carboxylic-acid-bearing unit. The unit derived from *N*-methylol acrylamide is especially interesting in that it is in effect a kind of amide–formaldehyde unit, of the type present in, say, urea-formaldehyde resins, incorporated in the polymer chain. If present in conjunction with units containing a carboxylic-acid group, it is believed to give a polymer which is readily self-vulcanisable by heating.

9.2 ACRYLIC RUBBERS (ACM)

We consider next a family of synthetic rubbers which are obtained by polymerising esters of acrylic and methacrylic acid. The principal repeat units in these rubbers are of the types shown at (XCVII) and (XCVIII) below. In these structures, the moiety R is usually an n-alkyl group.

$$\cdots -CH_2-CH-\cdots \qquad\qquad \cdots -CH_2-\overset{\overset{\displaystyle CH_3}{|}}{\underset{\underset{\displaystyle CO_2R}{|}}{C}}-\cdots$$
$$\underset{\underset{\displaystyle CO_2R}{|}}{}$$

(XCVII)   (XCVIII)

Whether or not the polymer is rubbery at room temperature depends upon (a) the nature of R, and (b) the ratio of acrylic to methacrylic ester units if the polymer contains units of both types. The glass-transition temperatures for polymers from a range of lower n-alkyl acrylates and methacrylates are given in Table 9.2. It can be seen that increasing the alkyl chain length reduces the glass-transition temperature, whereas the replacement of the α-hydrogen atom by a methyl group in going from an acrylate to the corresponding methacrylate raises the glass-transition temperature considerably. Thus it is that poly-n-butyl acrylate is obviously rubbery at room temperature and also at temperatures considerably below room temperature, whereas polymethyl methacrylate is a rigid plastic at room temperature and well above. Polymethyl acrylate and poly-n-butyl methacrylate

TABLE 9.2

GLASS-TRANSITION TEMPERATURES ($^\circ$C) OF VARIOUS POLY-n-ALKYL ACRYLATES AND METHACRYLATES

n-$Alkyl$ $substituent$ (R)	$Polyacrylate$ $-CH_2-CH(CO_2R)-$	$Polymethacrylate$ $-CH_2-C(CH_3)(CO_2R)-$
$-CH_3$	0	72
$-CH_2.CH_3$	-23	47
$-CH_2.CH_2.CH_3$	-51	33
$-CH_2.CH_2.CH_2.CH_3$	-63	17

have glass-transition temperatures which are too close to ambient temperatures to make them useful either as rubbers or plastics in themselves. Copolymers of the various acrylic and methacrylic esters have approximately the glass-transition temperature which would be predicted on the basis of linear variation with the mole fraction of the two types of unit. Increasing the alkyl chain length of a poly-*n*-alkyl acrylate makes the unvulcanised rubber increasingly soft and tacky at room temperature. However, eventually the material becomes waxy at room temperature because of crystallisation of the alkyl side chains. Replacing the *n*-alkyl group of the ester by a branched alkyl group increases the glass-transition temperature, the amount of the increase itself increasing with the extent of branching. The commercially-available acrylic rubbers are based predominantly upon *n*-butyl acrylate and ethyl acrylate, together with, in some cases, acrylic monomers of the alkoxyalkyl type such as methoxyethyl acrylate (XCIX) and ethoxyethyl acrylate (C).

$$CH_2{=}CH$$
$$|$$
$$CO_2CH_2CH_2OCH_3$$

(XCIX)

$$CH_2{=}CH$$
$$|$$
$$CO_2CH_2CH_2OCH_2CH_3$$

(C)

Acrylic ester polymers are readily prepared by emulsion polymerisation of the appropriate acrylic esters. They are not vulcanisable by heating with sulphur and conventional accelerators, because they do not contain any carbon–carbon double bonds. They can, however, be vulcanised by heating with a variety of reagents. Thus in some early compounds based upon polyethyl acrylate rubbers, vulcanisation was effected by heating with alkaline substances such as sodium silicate or potassium hydroxide. According to one interpretation of the changes in physical properties which accompany heating of the rubber under these conditions, crosslinking occurs by way of a Claisen-type condensation reaction involving the ethoxy groups of the ester units and the α-hydrogen atoms in the main chain:

$$\cdots{-}CH_2{-}CH{-}\cdots$$
$$|$$
$$CO$$
$$|$$
$$OC_2H_5$$
$$|$$
$$H$$
$$|$$
$$\cdots{-}CH_2{-}C{-}\cdots$$
$$|$$
$$CO_2C_2H_5$$

$$\xrightarrow{\text{OH}^-}$$

$$\cdots{-}CH_2{-}CH{-}\cdots$$
$$|$$
$$CO$$
$$|$$
$$\cdots{-}CH_2{-}C{-}\cdots$$
$$|$$
$$CO_2C_2H_5$$

$$+ C_2H_5OH$$

However, the more modern types of acrylic rubber contain minor amounts (1–5% by weight) of units derived from a functional-group comonomer whose presence in the polymer chain confers the ability to be vulcanised in more conventional ways. Thus copolymerisation with dienes such as butadiene and isoprene has been used to give rubbers which can be vulcanised in a conventional way with sulphur and accelerators. Introduction of carboxylic-acid groups by copolymerisation with, say, acrylic acid, endows the polymer with the possibility of being crosslinked by all the reactions (and more besides) discussed in the previous section of this chapter. Copolymerisation with allyl glycidyl ether (LXXXI) results in the incorporation in the polymer chain of groups which have the structure (CI). Rubbers containing these units can be crosslinked by reaction with dibasic carboxylic acids, the result of the reaction being an ester crosslink.

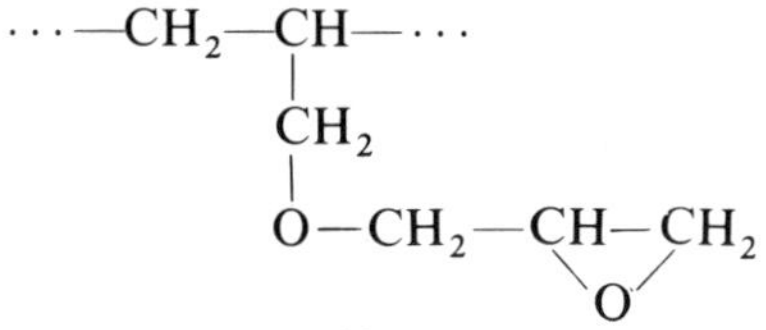

$$\cdots-CH_2-CH-\cdots$$
$$\mid$$
$$CH_2$$
$$\mid$$
$$O-CH_2-CH-CH_2$$
$$\diagdown O \diagup$$

(CI)

However, the most widely-used functional-group comonomers which are used in the production of acrylic rubbers are 2-chloroethyl vinyl ether (CII) and vinyl chloroacetate (CIV). Copolymerisation in the presence of these monomers results in the incorporation of the units shown at (CIII) and (CV) respectively. Crosslinking of the resultant rubbers can be achieved by

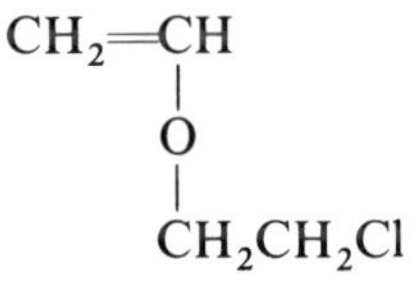

$$CH_2{=}CH$$
$$\mid$$
$$O$$
$$\mid$$
$$CH_2CH_2Cl$$

(CII)

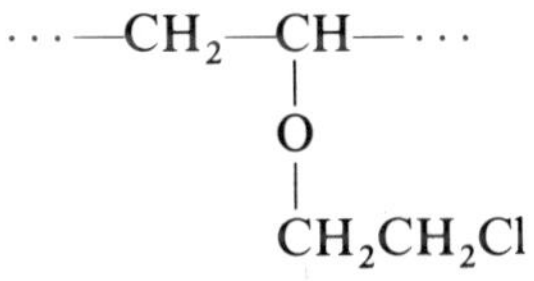

$$\cdots-CH_2-CH-\cdots$$
$$\mid$$
$$O$$
$$\mid$$
$$CH_2CH_2Cl$$

(CIII)

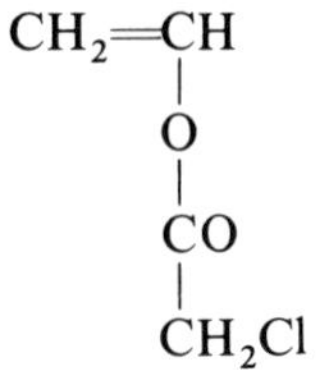

$$CH_2{=}CH$$
$$\mid$$
$$O$$
$$\mid$$
$$CO$$
$$\mid$$
$$CH_2Cl$$

(CIV)

$$\cdots-CH_2-CH-\cdots$$
$$\mid$$
$$O$$
$$\mid$$
$$CO$$
$$\mid$$
$$CH_2Cl$$

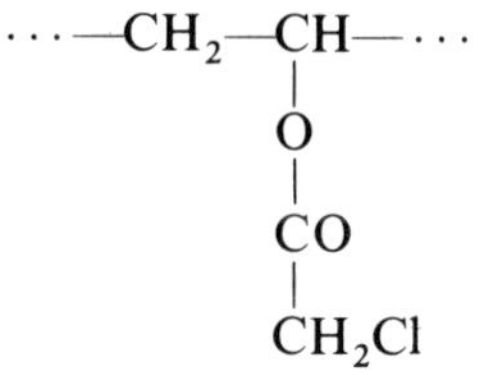

(CV)

heating with difunctional reagents which react with the pendant chloro-ethyl or chloromethyl groups. These reagents include polyamines such as triethylenetetramine, and red lead. Ancillary curatives tend to be used as well. Thus sulphur and sulphur-containing accelerators such as di-2-benzthiazyl disulphide have been used with polyamines, and 2-mercap-toimidazoline (ethylene thiourea) has been used with red lead. The function of the sulphur and sulphur-bearing accelerator in the amine/sulphur vulcanisation system seems to be two-fold: they act as cure-retarders, thereby making the stock less prone to scorching than would otherwise be the case; and they improve the ageing resistance of the vulcanisate. In respect of the latter attribute, the use of sulphur and sulphur-bearing accelerators in conjunction with amine curatives is said to give vulcanisates which are stable as regards hardness and which retain their tensile strength and elongation at break during exposure to oils at high temperature, and also during heat ageing in air. A combination comprising triethylene-tetramine and di-2-benthiazyl disulphide is especially recommended for the vulcanisation of acrylic rubbers containing units derived from 2-chloroethyl vinyl ether. This curing system is said to confer good initial vulcanisate properties, in particular, low compression set, and to give a vulcanisate which retains these properties well during ageing.

As might be expected, the chlorine atom in a copolymerised unit derived from vinyl chloroacetate is considerably more reactive than is that in a unit derived from 2-chloroethyl vinyl ether, because of the activating influence of the adjacent carboxyl group. Additional vulcanising systems are therefore available for acrylic rubbers which contain vinyl chloroacetate units. Ammonium benzoate was initially recommended as a vulcanising agent for these rubbers; it offers the advantages over the older type of amine curatives of greater processing safety and longer storage life of the unvulcanised stock. More recently, other ammonium salts such as ammonium adipate have been recommended as vulcanising agents. An important advance has been the development of a new type of vulcanising system which comprises a combination of an alkali-metal salt of a fatty acid plus sulphur. The primary vulcanising agent is the metal salt, which is typically sodium or potassium stearate. The sulphur functions as an accelerator. The ratio of metal salt to sulphur is typically 10:1. The vulcanising effect seems to be specific to sodium and potassium salts, the latter being more active than the former. This vulcanising system is said to be the one which is most commonly used for acrylic rubbers at the present. The subject of the vulcanisation of acrylic rubbers is rather complex; as the foregoing remarks suggest, a wide range of curing combinations has been proposed over the years.

The useful properties which the acrylic rubbers offer can be summarised as follows:

1. They retain their rubbery properties over long periods of time at a wide range of temperatures (*c.* $-10\,°C$ to $200\,°C$).
2. They are resistant to oxidation at normal and elevated temperatures.
3. They have excellent flex life.
4. Resistance to swelling in oils is good, especially to sulphur-containing oils at high temperatures.
5. Resistance to ozone is excellent.
6. Resistance to discoloration in sunlight is good.
7. Resistance to weathering is good.

One of the principal disadvantages of acrylic rubbers is that they are prone to hydrolysis in the presence of water and steam. In alkaline media, the ester linkages in the polymer side chains hydrolyse, resulting in some degradation of the rubber. Acid solutions cause the rubbers to swell.

9.3 FLUORINATED RUBBERS (FKM)

It has long been known that the fluorine analogues of hydrocarbons (that is, compounds which are derived from hydrocarbons by replacing some or all of the hydrogen atoms by fluorine) are thermally very stable and resistant to oxidative attack. Their stability is due primarily to the strength of the carbon–fluorine bond. High-temperature stability and resistance to oxidation is similarly observed in the fluorinated analogues of hydrocarbon polymers, e.g., polytetrafluoroethylene. In addition, such polymers show good resistance to attack by chemicals, to swelling by oils and solvents, and, of course, they also show low flammability. These further characteristics (other than low flammability) have been attributed to the existence of strong secondary valence forces between the various parts of the polymer molecules, and also to steric hindrance, as well as to the inherent strength of the carbon–fluorine bond.

In view of the foregoing remarks, it is hardly surprising that a great deal of effort has been directed towards the development of fluorinated polymers for application at elevated temperatures and under conditions which are hostile in respect of exposure to oxidative conditions and to oils and solvents. Polytetrafluoroethylene was first discovered in 1938. It is a rigid plastic whose unusual properties have attracted continued interest for specialist applications since then. (Its unusual properties include not only thermal stability and resistance to hostile environments, but also a very low coefficient of friction both to itself and to other surfaces.) The first fluorinated rubbers were developed in the early 1950s, and became commercially available in 1958. Many types of fluorinated rubber have been described in the literature, and many more have no doubt been the subject of hitherto unpublished research and development. It would be quite impossible in the limited space available here to review even briefly all the types whose existence is publicly documented. Attention will be restricted here to two types:

1.   those which can be regarded as being derived principally from saturated hydrocarbon polymers by partial or complete replacement of hydrogen by fluorine—these can be conveniently called 'fluorocarbon rubbers'; and
2.   those which can be regarded as being similarly derived from acrylic rubbers.

Note will subsequently be taken of the existence of fluorinated silicone rubbers when the subject of silicone rubbers is reviewed later in this chapter (Section 9.8).

The names and chemical structures of the monomers which are used for the production of rubbers in the first of the above two categories are given at (CVI)–(CXI). Also given are the chemical structures of the repeat units which result from the enchainment of each of the monomers. Fluorocarbon rubbers of current commercial importance include (a) a 30/70 weight ratio copolymer of hexafluoropropylene and vinylidene fluoride, (b) 50/50 and 30/70 mole ratio copolymers of chlorotrifluoroethylene and vinylidene fluoride, (c) copolymers of 1-hydropentafluoropropylene and vinylidene fluoride, and (d) terpolymers of 1-hydropentafluoropropylene, vinylidene fluoride and tetrafluoroethylene. Fluorocarbon rubbers are produced by free-radical emulsion polymerisation of the appropriate monomers, using an initiator such as ammonium persulphate. If a surfactant is required, it should be fully fluorinated; an example is ammonium perfluoroöctanoate. However, it appears that it not always necessary to use a surfactant.

A variety of reagents can be used for the vulcanisation of fluorocarbon rubbers. These include (a) diamines and their derivatives, (b) dithiols, (c) peroxides, and (d) aromatic polyhydroxy compounds in combination with a strongly basic organic co-agent. It is normally necessary to use a basic metal oxide with all these systems. Fluorocarbon rubbers can also be crosslinked by exposure to high-energy radiation. A combination which

$$CH_2{=}CF_2$$
vinylidene fluoride
(CVI)

$$\cdots{-}CH_2{-}CF_2{-}\cdots$$
(CVIa)

$$CF_2{=}CF_2$$
tetrafluoroethylene
(CVII)

$$\cdots{-}CF_2{-}CF_2{-}\cdots$$
(CVIIa)

$$CFCl{=}CF_2$$
chlorotrifluoroethylene
(CVIII)

$$\cdots{-}CFCl{-}CF_2{-}\cdots$$
(CVIIIa)

$$CF_2{=}CF{-}CF_3$$
hexafluoropropylene

(CIX)

$$\cdots{-}CF_2{-}\underset{\underset{CF_3}{|}}{CF}{-}\cdots$$
(CIXa)

$$CHF{=}CF{-}CF_3$$
1-hydropentafluoropropylene

(CX)

$$\cdots{-}CHF{-}\underset{\underset{CF_3}{|}}{CF}{-}\cdots$$
(CXa)

$$CF_2{=}CF{-}OCF_3$$
perfluoromethylvinyl ether

(CXI)

$$\cdots{-}CF_2{-}\underset{\underset{OCF_3}{|}}{CF}{-}\cdots$$
(CXIa)

appears to have been quite widely used is hexamethylenediamine carbamate plus magnesium oxide. Another example of a diamine derivative which has been recommended for the vulcanisation of fluorocarbon rubbers is N,N'-dicinnamylidene-1,6-hexanediamine. The presence of the metal oxide is essential to the development of a high state of cure in these rubbers. The metal oxide also helps to stabilise the vulcanisate by acting as an acceptor of the small amounts of acidic substances which are released when the rubber is heated to high temperatures. Of the various metal oxides available, magnesium oxide gives the best overall balance of vulcanisate properties, and is probably the most widely-used oxide in conjunction with diamine vulcanising agents. The type of magnesium oxide used affects the rate and state of vulcanisation, the heat ageing characteristics of the vulcanisate, and its compression set. Magnesium oxide is not recommended where optimum water-, steam- or acid-resistance is required. In such cases, lead compounds such as litharge, or dibasic lead phosphite plus zinc oxide should be used. Calcium oxide can be used in place of magnesium oxide to reduce the shrinkage of the rubber during vulcanisation. In order to develop a satisfactory state of cure in fluorocarbon rubbers, it is normally necessary to subject them first to vulcanisation in a press, and then to a period of post-vulcanisation in an oven.

The formulation of compounds based upon fluorocarbon rubbers is relatively simple, because the compounds contain rather fewer ingredients than do normal rubber compounds. In addition to the vulcanising system, the compound usually contains a small amount of a processing aid and up to 30 pphr of a filler. As regards plasticisers and processing aids, it is not normally advisable to use hydrocarbon oils and organic esters, because of problems of limited compatibility with the rubber and volatilisation during post-vulcanisation or when exposed to high temperatures during service. Fluorosilicone fluids of high viscosity (c. 10^4 cPs) are recommended as plasticisers for fluorocarbon rubbers, being thermally stable and of low volatility. A low-molecular-weight copolymer of hexafluoropropylene and vinylidene fluoride is available as a processing acid for these rubbers.

The fluorocarbon rubbers are used in speciality applications where exceptionally high resistance to oils and chemicals at elevated temperatures is required. Their major disadvantage is their very high price; this inevitably limits their application to those areas where their unusual properties provide sufficient mitigation for the high material costs which attend their use. The serviceability of fluorocarbon rubbers at high temperatures can be illustrated by reference to vulcanisates from a hexafluoropropylene–vinylidene fluoride copolymer. The resistance to dry heat is such that vulcanisates are serviceable for continuous periods in excess of three years at 200 °C, for about 1000 h at 260 °C, and for about 48 h at 315 °C. The lower temperature limit of serviceability in dynamic applications is about −20 °C for this type of rubber. The brittle temperature is about −45 °C.

Under the general heading of fluorocarbon rubbers, it is appropriate to draw attention to the existence of a comparatively new type of rubber which is principally a copolymer of tetrafluoroethylene and perfluoromethylvinyl ether, but which also contains a minor proportion of units derived from what is described as a 'perfluorinated cure-site monomer'. It is stated that the very high chemical resistance of the rubber precludes it being vulcanisable by reaction with di-functional reagents of the type used for the more conventional types of fluorocarbon rubber. For this reason, it is necessary to incorporate a small amount of a reactive third monomer in order to provide sites for crosslinking. Terpolymers have also been developed recently which are particularly responsive to vulcanisation by organic peroxides. The vulcanisation system comprises an aliphatic peroxide, a co-agent such as triallyl isocyanurate, and a fixed-base acid acceptor. These terpolymers have been developed specifically for the manufacture of products the vulcanisation of which is carried out under low pressure (such as in liquid curing media or continuously in hot air). The use of peroxide-vulcanised terpolymers is said to obviate the problem of porosity which occurs when more conventional fluorinated rubbers are used; the porosity is attributed to the formation of volatile by-products during the vulcanisation of fluorinated rubbers by the more conventional methods.

The fluoroacrylic rubbers warrant only very brief mention. Those concerning which most information has so far been published are polymers of the 1,1-dihydroperfluoroalkyl acrylates, such as 1,1-dihydroperfluoro-butyl acrylate (CXII) and 3-perfluoromethoxy-1,1-dihydroperfluoropropyl acrylate (CXIII). These rubbers can be produced by emulsion polymeri-sation of the appropriate monomers using potassium persulphate as

$$
\begin{array}{cc}
CH_2{=}CH & CH_2{=}CH \\
| & | \\
CO & CO \\
| & | \\
O & O \\
| & | \\
CH_2 & CH_2 \\
| & | \\
CF_2 & CF_2 \\
| & | \\
CF_2 & CF_2 \\
| & | \\
CF_3 & O \\
 & | \\
 & CF_3 \\
(CXII) & (CXIII)
\end{array}
$$

initiator and sodium lauryl sulphate as surfactant. They appear to be vulcanisable by heating with mixtures of amines and sulphur. Although the resistance of these rubbers to swelling in aromatic hydrocarbon fuels is much superior to that of vulcanised acrylonitrile–butadiene rubbers, it is inferior to that of rubbers based upon vinylidene fluoride. It is this latter consideration which has probably limited markedly the application and commercial development of this type of rubber.

9.4   URETHANE RUBBERS (AU AND EU)

In the group of rubbers to be considered in this section, the high molecular weight which is necessary for the development of satisfactory mechanical properties in a rubber is achieved by the *chain extension* of polymers of relatively low molecular weight. These low-molecular-weight polymers are terminated with active hydrogen atoms, and chain extension is achieved by reaction with polyfunctional organic isocyanates, that is, with compounds which contain the functional group $-N:C:O$. The linkages which are produced by the chain-extension reaction are of the urethane type, and from this fact is derived the generic name of these rubbers.

The reaction between an isocyanate and an active-hydrogen compound can be represented generally as follows:

$$R.N{=}C{=}O + HX \longrightarrow \left[\begin{array}{c} R.N{=}C{-}OH \\ | \\ X \end{array} \right] \longrightarrow \begin{array}{c} R.N{-}C{=}O \\ |\ \ | \\ H\ X \end{array}$$

Thus, for example, reaction between an isocyanate R.NCO and an alcohol R'OH gives the urethane $R.NH.CO_2R'$ in the first instance. Further reaction with isocyanate can then take place by way of the active-hydrogen atom of the urethane compound. The product of this further reaction is an *allophanate* compound, that is, a compound of the type $R.NH.CO.NR.CO_2R'$. Reaction with an amine $R'NH_2$ gives, in the first place, the compound $R.NH.CO.NHR'$, which is a di-substituted urea; reaction with an amide $R'CONH_2$ gives the acyl urea $RNH.CO.NH.COR'$; and reaction with a urea $HR'N.CO.NHR''$ gives the biuret $RNH.CO.NR'.CO.NHR''$. The low-molecular-weight polymers which are employed for chain extension by reaction with isocyanates are usually hydroxyl-terminated polyesters or hydroxyl-terminated polyethers; the elastomers derived from these two classes of hydroxyl-terminated precursor are known as AU and EU types respectively. Hydroxyl-terminated polyester-amides were also used formerly. Thus the chain-extension reaction by which urethane elastomers are manufactured can be represented generally as follows:

$$HO{\sim}OH + OCN.R.NCO + HO{\sim}OH$$
$$\downarrow$$
$$HO{\sim}O.CO.NH.R.NH.CO.O{\sim}OH$$

where OCN.R.NCO denotes a di-functional organic isocyanate.

The term *urethane elastomer* embraces a wide range of synthetic rubbers. These rubbers can be conveniently classified according to the way in which they are processed. The principal groups are as follows:

1. millable gums, which are processed using conventional rubber processing machinery;
2. liquid elastomer prepolymers which are processed by casting techniques;
3. rubbers which are processed by techniques used for thermoplastics (e.g., injection moulding), and which may by processed either as thermoplastic rubbers, or subsequently post-cured by heating;
4. rubbers which are applied as solutions;
5. rubbers which are applied as latices; and
6. liquid elastomer prepolymers which are used for the manufacture of the so-called 'urethane foams'.

The concern here is with the millable gums exclusively, although some reference will be made in Chapter 10 to the urethane rubbers which are used as thermoplastic rubbers (Section 10.2), and to the use of urethane-extended hydroxyl-terminated polymers as fluid rubbers (Section 10.4). Although brief mention is also made in Chapter 10 (Section 10.4) of the use of castable urethane-extended fluid rubbers for the manufacture of elastomeric urethane foam materials, this aspect of the technology of synthetic rubbers has been largely excluded from this survey.

There are many hydroxyl-terminated polymers which have been proposed as candidates for chain extension by reaction with di-functional isocyanates to form millable elastomer gums. The two most important types in current use are:

1. hydroxyl-terminated polyesters, and
2. hydroxyl-terminated polyethers.

The most commonly-used polyesters are polyethylene glycol adipate† (CXIV), polyethylene–propylene glycol adipate (CXV), and polytetramethylene glycol adipate (CXVI). Blends of hydroxyl-terminated polyesters can be used in order to reduce any excessive tendency to crystallisation in the subsequent elastomer. Hydroxyl-terminated polyethers are used in preference to the polyesters where a urethane elastomer of improved water

---

† The word 'glycol' is now sometimes omitted from the chemical names of these polyesters. Thus (CXIV) is now sometimes referred to as 'polyethylene adipate'; likewise the other polyesters.

resistance is required. The polyether urethane elastomers are more resistant to water principally because ether linkages are more hydrolytically stable than are ester linkages. In the case of the polyether urethane elastomers, hydrolysis can occur only at the urethane linkages, these being less prone to hydrolysis than are ester linkages, and also at such hydrolytically-unstable linkages as may be introduced during vulcanisation. The polyether urethane elastomers are also rather more resistant to low-temperature stiffening than are the polyester types. Polytetramethylene glycol (CXVII) is typical of the polyethers which can be used for this application. Whilst the polyethylene glycols give high-strength urethane elastomers, the products suffer from the disadvantage that they are hydrophilic and have very poor resistance to water. By contrast, the polytetramethylene glycols give elastomers which have excellent resistance to water. Unsaturation can be introduced into both polyester and polyether urethane elastomers by incorporating in the elastomer units derived from an unsaturated diol such as glycerol monoallyl ether (CXVIII). Such diols are usually incorporated

$$\cdots-OCH_2CH_2O.CO.CH_2CH_2CH_2CH_2.CO-\cdots$$

(CXIV)

$$\cdots-OCH_2CH_2O.CO.(CH_2)_4.CO-\cdots$$

$$\cdots-OCH_2CH(CH_3)O.CO.(CH_2)_4.CO-\cdots$$

(CXV)

$$\cdots-O(CH_2)_4O.CO.(CH_2)_4.CO-\cdots$$

(CXVI)

$$\cdots-OCH_2CH_2CH_2CH_2-\cdots$$

(CXVII)

$$CH_2OCH_2CH{=}CH_2$$
$$|$$
$$CHOH$$
$$|$$
$$CH_2OH$$

(CXVIII)

by using them as chain extenders. As a matter of historical interest, it may be noted that the two earliest urethane elastomers which were developed commercially were based upon (a) polyethylene glycol adipate ('Vulkollan', developed by I. G. Farbenindustrie) and (b) a polyester-amide derived from ethylene glycol, ethanolamine and adipic acid ('Vulcaprene', developed by I.C.I.).

The isocyanates which are most commonly used for the production of urethane elastomers are 4,4′-diphenylmethane diisocyanate (CXIX) (MDI), and tolylene diisocyanate (CXX) (TDI). The latter is used either as the pure 2,4-isomer (CXXa), or as an 80/20 mixture of the 2,4-isomer and the 2,6-isomer (CXXb). Other aromatic diisocyanates which have been used for the production of urethane elastomers include 1,5-naphthalene diisocyanate (CXXI) (NDI) and 3,3′-dimethyldiphenyl-4,4′-diisocyanate (CXXII) (TODI).

Several methods are available for effecting the vulcanisation of urethane elastomers. These methods can be conveniently classified under four headings:

1. reaction between isocyanate-terminated polymer molecules and polyfunctional active-hydrogen compounds;
2. reaction between active-hydrogen-terminated polymer molecules and polyfunctional isocyanates;
3. reaction with sulphur and conventional accelerators; and
4. reaction with organic peroxides.

Amongst the various polyfunctional active-hydrogen compounds which have been used for the vulcanisation of isocyanate-terminated urethane elastomer gums are 1,4-butanediol (CXXIII), trimethylol propane (CXXIV), 4,4'-methylene bis (o-chloroaniline) (CXXV) (MOCA), and 3,3'-dichlorobenzidine (CXXVI) (DCB). In connection with the latter two

$$CH_2\!\!-\!\!CH_2\!\!-\!\!CH_2$$
$$CH_2\quad CH_2\quad CH_2$$

HO . CH₂CH₂CH₂CH₂ . OH

OH   OH   OH

(CXXIII)

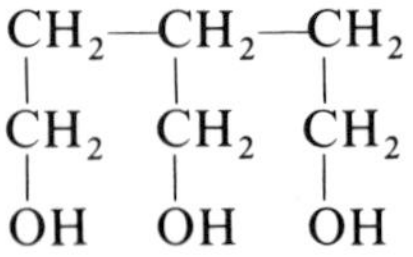

(CXXIV)

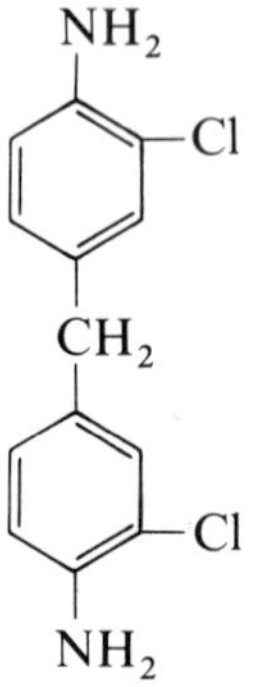

(CXXV)

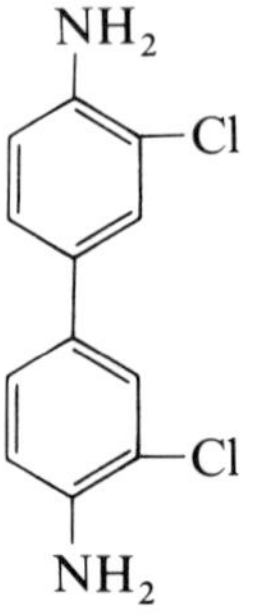

(CXXVI)

compounds, it should be noted that, whereas the reaction between aliphatic diamines and isocyanates is almost instantaneous, reaction between these compounds and isocyanates is moderated to some extent, firstly because they are aromatic diamines, and secondly because they are sterically hindered. For the vulcanisation of active-hydrogen-terminated polymers by reaction with isocyanates, dimerised isocyanates are often used.

Various combinations of sulphur, accelerators and activators have been developed for the vulcanisation of those urethane rubbers which are susceptible to vulcanisation by reaction with sulphur because they contain unsaturated groupings in the rubber molecule. One such vulcanisation combination comprises sulphur, di-2-benzthiazole disulphide, 2-mercapto-benzthiazole, a complex of zinc chloride and di-2-benzthiazole disulphide, and cadmium stearate. The following combinations in pphr of each of these additives respectively is recommended as giving a reasonable balance of vulcanisate physical properties: 0·75, 4, 1, 0·35 and 0·5. The effect of increasing the sulphur level is to increase both the rate of vulcanisation and the extent of vulcanisation which is attainable, but at the expense of the retention of the vulcanisate mechanical properties on ageing. The processing safety is determined, in part at least, by the ratio of di-2-benzthiazole disulphide to 2-mercaptobenzthiazole; a higher proportion of the former favours processing safety. The importance of the zinc activator is evident from Fig. 9.3. Whilst other zinc compounds can be used as activators, it appears that the complex formed between zinc chloride and di-2-benzthiazole disulphide is particularly useful in that the rate of vulcanisation is enhanced without sacrifice of processing safety. The effect of increasing levels of cadmium stearate co-activator is illustrated in Fig. 9.4. Zinc stearate can be used as an alternative to cadmium stearate.

Certain urethane rubbers can be vulcanised by heating with organic peroxides, such as dicumyl peroxide. The necessary prerequisite for a urethane rubber to be peroxide-vulcanisable is that it should contain

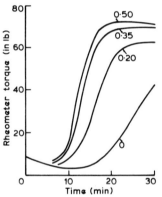

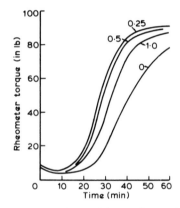

FIG. 9.3. Effect of level of zinc activator upon sulphur vulcanisation of a sulphur-vulcanisable urethane rubber.[2] Compound: urethane rubber 100, HAF carbon black 30, coumarone–indene resin 15, di-2-benzthiazyl disulphide 4, 2-mercaptobenzthiazole 1, sulphur 0·75, cadmium stearate 0·5, zinc activator variable. Vulcanisation carried out at 154°C. Numbers appended to curves indicate level of zinc activator in pphr. The ordinates are the torque developed in an oscillating-disc rheometer.

FIG. 9.4. Effect of level of cadmium stearate upon sulphur vulcanisation of a sulphur-vulcanisable urethane rubber.[2] Compound: urethane rubber 100, HAF carbon black 30, coumarone–indene resin 15, di-2-benzthiazyl disulphide 4, 2-mercaptobenzthiazole 1, sulphur 0·75, zinc activator 0·35, cadmium stearate variable. Vulcanisation carried out at 154°C. Numbers appended to curves indicate level of cadmium stearate in pphr. The ordinates are the torque developed in an oscillating-disc rheometer.

hydrogen atoms which can be fairly readily removed by reaction with the radicals derived from the peroxide. If the urethane rubber has been chain-extended with 4,4'-diphenylmethane diisocyanate (CXIX), then this condition is met by the presence of the methylene group in the diisocyanate. Urethane rubbers can also be made peroxide-vulcanisable by the inclusion of unsaturation.

Urethane rubbers give gum vulcanisates of high tensile strength. Apart from the novelty of certain of the vulcanising systems used with them, the compounding principles for the millable gums are similar to those which apply to other synthetic rubbers of, say, the chloroprene and acrylonitrile–butadiene types. The outstanding properties of urethane rubbers are high resistance to abrasion and high resistance to tearing. They are therefore of particular interest for articles which are to be subjected to adverse wearing conditions in service. Resistances to heat, ozone cracking, and swelling in oils and solvents are good; the oil resistance of urethane rubbers as a family is similar to that of the acrylonitrile–butadiene rubbers. The weakness of the urethane rubbers is their susceptibility to hydrolysis; it has already been noted that those derived from polyethers are less susceptible to hydrolytic degradation than are those derived from polyesters. An unusual feature of urethane rubbers as a family is that gum vulcanisates covering a very wide range of hardness can be obtained from them.

9.5 POLYPENTENAMER AND PLASTICISED POLYNORBORNENE

We now give very brief consideration to two newer types of hydrocarbon rubber in which some interest has been shown in recent years. The first of these is known as *polypentenamer*. It is obtained by the ring-opening polymerisation of cyclopentene (CXXVII), the ring-opening occurring via cleavage of the carbon–carbon double bond. The repeat unit of the polymer sequence is shown at (CXXVIIa). The polymerisation, which is an example of an 'olefin metathesis' reaction, can be represented as follows:

$$CH=CH \atop CH_2 \quad CH_2 \longrightarrow \cdots -CH_2CH=CHCH_2CH_2 \cdots \atop CH_2$$

(CXXVII)   (CXXVIIa)

Polymerisation is effected by means of Ziegler–Natta catalysts.

Polypentenamer is one member of a wider family of polyalkenamers. Of this family, it is the one which has attracted most interest, partly because its monomer is readily available from the petrochemical industry, and partly because it is a readily-vulcanisable rubber which is potentially of interest as a general-purpose synthetic rubber. The configuration of the carbon–carbon double bond in the polymer can be controlled by varying the polymerisation conditions. Polymers of various structures have been prepared. A virtually 100% *cis* form is known; it possesses excellent resistance to low-temperature stiffening. However, the polymer which is of principal interest is one in which approximately 85% of the carbon–carbon double bonds have the *trans* configuration. This polymer has a glass transition temperature in the region of $-95\,°C$ (near to that of *cis*-1,4-polybutadiene), and a crystalline melting point just below room temperature. This combination of properties results in vulcanisates which have excellent resistance to low-temperature stiffening, good resilience, good tensile strength, especially in gum vulcanisates, and good resistance to abrasion.

The second of the synthetic rubbers to be considered here is that which is based upon polynorbornene. Norbornene (bicyclo-2,2,1-heptene-2) (CXXIX) is obtained by the Diels–Adler addition of ethylene to cyclopentadiene (CXXVIII). It can be polymerised by ring-opening polymerisation to give an unsaturated polymer chain in which the repeat units are as shown at (CXXIXa). The carbon–carbon double bonds can have both the *cis* configuration and the *trans* configuration. The formation of norbornene and its polymerisation can be represented as follows:

It is understood that polymerisation is effected by means of an alkyllithium initiator.

Polynorbornene is unusual amongst hydrocarbon synthetic rubbers in that it is a glassy polymer at room temperature, its glass-transition temperature being c. 35 °C. In order to acquire rubbery properties, it is necessary that it be plasticised with a suitable oil. In this respect, it resembles polyvinyl chloride. Oils of the paraffinic, naphthenic and aromatic types are recommended for the plasticisation of polynorbornene. Paraffinic oils give the best resistance to low-temperature stiffening, whereas aromatic oils give the best mechanical properties and resistance to ageing. Blends of the various types of oils can be used in order to give a balance of properties. Polynorbornene can be compounded with large quantities of both oil and filler; typical levels would be 200 pphr of oil (possibly of more than one type) plus 200 pphr of filler. Vulcanisation of polynorbornene rubbers is achieved by reaction with sulphur and conventional accelerators. The areas of interest for the application of polynorbornene rubbers are those which call for a rubber having high mechanical damping characteristics.

9.6   POLYSULPHIDE RUBBERS (TR)

Polysulphide rubbers are produced by reaction between sodium polysulphide and polychloro derivatives of aliphatic compounds. The latter are mainly dichloro derivatives (giving linear sequences of repeat units), but small amounts of compounds containing more than two chlorine atoms are also used in order to impart branching. In the case of a dichloro derivative, the reaction by which the polysulphide rubber is formed can be represented as follows:

$$ClRCl + Na_2S_x \longrightarrow \cdots -R-S_x- \cdots + 2NaCl$$

The principal organic dihalides used for the commercial production of polysulphide rubbers are ethylene dichloride (CXXX) and bis(2-chloroethyl) formal (CXXXI). 1,2,3-Trichloropropane (CXXXII) is used in small amounts (c. 2%) to introduce branching. The sulphur 'rank' of the rubber (defined as the number of sulphur atoms per polysulphide link—x in the above general structure) depends upon that of the sodium polysulphide which was used for its preparation. The usual range of values is c. 2–4. The earliest polysulphide rubber was a polysulphide of rank 4 derived from ethylene dichloride. It had extremely good resistance to swelling in hydrocarbon liquids, this property deriving in part at least from its very high sulphur content (c. 84% by weight). However, this rubber was difficult to process, had a bad odour which was evident in the vulcanised rubber, and gave off irritating fumes during processing. It was subsequently found

$$CH_2Cl . CH_2Cl$$
(CXXX)

$$CH_2 \Big\langle \begin{array}{l} OCH_2CH_2Cl \\ OCH_2CH_2Cl \end{array}$$
(CXXXI)

$$CH_2Cl . CHCl . CH_2Cl$$
(CXXXII)

$$ClCH_2CH_2OCH_2CH_2Cl$$
(CXXXIII)

that bis(2-chloroethyl)ether (CXXXIII) gave rubbers which were less objectionable in respect of odour and processing behaviour; polysulphide rubbers of sulphur rank 2 and 4 were produced from this dihalide. However, the resistance of these rubbers to swelling in hydrocarbon solvents was not as great as that of the rubbers derived from ethylene dichloride. Bis(2-chloroethyl)formal has been found to be a more satisfactory organic dihalide for the manufacture of polysulphide rubbers, and this now appears to be the most generally useful monomer for this purpose. Besides having the advantages noted above, this organic dihalide condenses readily with sodium polysulphides to give polysulphide elastomers which have remarkable resistance to stiffening at low temperatures, bearing in mind their high resistance to swelling in hydrocarbon liquids. A mixture of roughly equal parts of bis(2-chloroethyl)formal and ethylene dichloride is used to produce a polysulphide rubber of greater solvent resistance than that obtained from bis(2-chloroethyl)formal alone. Inclusion of a small amount of a monochloro compound, such as butyl chloride, can be added to the polymerisation system in order to control the molecular weight of the product.

An outline of the manufacturing process for polysulphide rubbers is as follows: A sodium polysulphide of appropriate sulphur rank is prepared by reacting sodium hydroxide with the appropriate amount of sulphur at elevated temperature. An aqueous solution of the sodium polysulphide is then mixed with the appropriate chloro compound(s), and the mixture heated and stirred. In order to obtain a product of high molecular weight, it is necessary to employ a substantial excess (c. 30%) of sodium polysulphide; this is because the polysulphide anions tend to hydrolyse. Wetting and dispersing agents are present in the aqueous solution in order that the rubber may be formed as small particles. The dispersion of polymer particles is then washed to free it of soluble salts. The solid rubber is obtained by coagulating the dispersion with acid. It appears that the polysulphide rubbers produced by this process are hydroxyl-terminated, the hydroxyl groups being formed by slight hydrolysis of the organic chloro compound during polymerisation. Unless the rubber is subjected to a further treatment, it is presumed that the molecules remain hydroxyl-terminated. In the case of polysulphide rubbers in which branching has been deliberately introduced by the use of 1,2,3-trichloropropane, a further treatment (such as heating with excess sodium polysulphide and a small quantity of butyl chloride) is given to the water dispersion just prior to coagulation. The purpose of this treatment is to cleave some of the polysulphide links in the polymer chain, and thereby to plasticise the rubber. Without this treatment, the rubber is too tough to be satisfactorily processable on conventional rubber mills. A secondary consequence of this treatment is that the terminal hydroxyl groups become converted to mercaptan groups.

The hydroxyl-terminated polysulphide rubbers are vulcanised by heating them with zinc oxide. The reaction is obscure. It is necessary to choose the grade of zinc oxide with care, because of the adverse effects of impurities. The Grade XX4 from the New Jersey Zinc Company is specifically recommended. The level of zinc oxide is usually about 10 pphr. The rate of vulcanisation is not significantly affected by increasing or decreasing the amount of zinc oxide in the compound; any zinc oxide in excess of about 10 pphr merely acts as a filler. Organic accelerators are not used to enhance the rate of vulcanisation. If fast-curing stocks are required, it is possible to add 0·5–1·0 pphr of sulphur as an accelerator, although the use of sulphur is not normally recommended because vulcanisation tends to occur too quickly if it is present.

The vulcanisation of mercaptan-terminated polysulphide rubbers is effected rather differently. In this case, vulcanisation is brought about by heating with oxidising agents. Presumably chain extension occurs because the terminal mercaptan groups become oxidised to disulphide linkages:

$$\text{---SH} + \text{HS---} \longrightarrow \text{---S---S---} + H_2O$$

If the rubber molecules are initially branched, then chain extension will clearly result in the formation of infinite molecular networks if the chain-extension reaction occurs to a sufficient extent. The usual oxidising agent used for bringing about chain extension by this reaction is zinc peroxide; the level is usually c. 5 pphr. Alkaline-earth hydroxides such as calcium hydroxide (1 pphr) are used as accelerators of the chain-extension reaction. Other metal peroxides have also been used as vulcanising agents. Amongst the non-peroxidic oxidising agents which have been used for the vulcanisation of mercaptan-terminated polysulphide rubbers is p-benzoquinone dioxime (LI). A typical vulcanising combination is 1·5 pphr of *p*-benzoquinone dioxide plus 0·5 pphr of zinc oxide.

Whilst conventional accelerators of sulphur vulcanisation do not function as accelerators for the vulcanisation of polysulphide rubbers, certain classes, notably the thiazoles, thiurams and dithiocarbamates, do function as peptising agents. The effect of di-2-benzthiazyl disulphide in peptising a hydroxyl-terminated polysulphide rubber during mill mixing is illustrated in Fig. 9.5. The peptising action of di-2-benzthiazyl disulphide by itself is very slow. A small amount of an activator, such as diphenyl guanidine, is required in order to achieve speedy plasticisation; activators such as diphenyl guanidine are ineffective by themselves. Peptisation is presumed to occur by way of cleavage of polysulphide linkages. Use is made of the effect in order to improve the processing behaviour of polysulphide rubber stocks. A combination of 0·3 pphr of di-2-benzthiazyl disulphide plus 0·1 pphr of diphenyl guanidine is suitable. As is evident

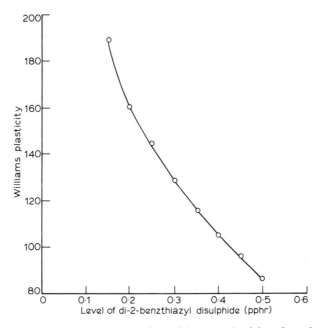

FIG. 9.5. Effect of di-2-benzthiazyl disulphide upon plasticity of a polysulphide rubber.[3] Compound: polysulphide rubber 100, zinc oxide 10, stearic acid 0·5, carbon black 60, diphenyl guanidine 0·1, di-2-benzthiazyl disulphide variable. Conditions of Williams plasticity test: original height of plug 0·5 in; readings taken after 10 min at 158 °F under 5 kg load.

from Fig. 9.5, close control of the amount of di-2-benzthiazyl disulphide is essential if control of the plasticity of the rubber is to be achieved.

The outstanding property of polysulphide rubbers is their resistance to swelling in hydrocarbon solvents. Of the commercially-available organic rubbers, these show the highest degree of resistance to swelling. The swelling resistance increases with the sulphur content of the rubber. It has already been noted that the rubber obtained from ethylene dichloride and having a sulphur rank of 4 contains approximately 84% of sulphur, and is extremely resistant to swelling in hydrocarbon solvents.

There are several disadvantages which attend the use of polysulphide rubbers. Bad odour is a problem, the odour being inherent in the sulphidic nature of the polymer. As might be expected, detailed studies have been undertaken of the effect of polymer structure upon the odour of organic polysulphides. The conclusion has been reached that the odour diminishes as the number of atoms separating the polysulphide units increases. However, other factors have had to be balanced against diminishing odour, such as problems associated with the condensation polymerisation of the dihalide and the tendency of the polysulphide rubber to stiffen by crystallisation when cooled. Polysulphide rubbers process poorly and give vulcanisates having poor mechanical properties. In particular, stress relaxation and creep occur very quickly in polysulphide rubber vulcanisates. This is illustrated in Fig. 9.6, where the stress-relaxation behaviour of a

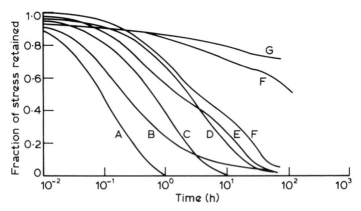

FIG. 9.6.   Illustrating stress-relaxation behaviour of various vulcanised rubbers in air at 130 °C as follows:[4] A polysulphide rubber, B chloroprene rubber, C natural rubber, D acrylonitrile–butadiene rubber, E styrene–butadiene rubber, F polyester rubber, G polyethyl acrylate rubber.

polysulphide rubber is compared with that of other vulcanised rubbers. It will be noted that the stress relaxation of the polysulphide rubber is virtually complete over two decades of time. There are cogent reasons for believing that rapid stress relaxation and creep occur in polysulphide rubbers because of the existence of an unusual molecular stress-relieving mechanism which is not available to most other rubbers. This is the

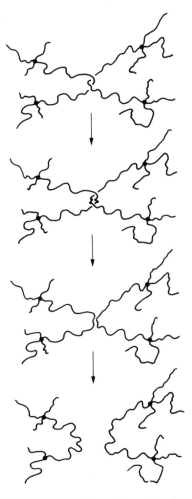

FIG. 9.7. Illustrating stress relaxation in a rubber vulcanisate by rearrangement of molecular entanglements through bond interchange.

mechanism of *bond interchange*. Bond interchange is believed to occur via the polysulphide linkages, and can be represented as follows:

$$\begin{array}{c} \text{\sim\!S—S\!\sim} \\ \text{\sim\!S—S\!\sim} \end{array} \rightleftharpoons \begin{array}{c} \text{\sim\!S} \quad \text{S\!\sim} \\ | \quad | \\ \text{\sim\!S} \quad \text{S\!\sim} \end{array}$$

This provides a means whereby stresses can be rapidly relieved by, say, the rearrangement of molecular entanglements (as illustrated in Fig. 9.7), and without change of either average molecular weight or crosslink density. It appears that, like bad odour, rapid stress relaxation and creep are inherent features of a polysulphidic structure.

It is convenient to note here that some interest has also been shown in a sulphur-vulcanisable polysulphide rubber in which the major repeat unit is propylene sulphide (CXXXIV). The ability to be vulcanised by sulphur is conferred by the inclusion of a minor amount (3–10%) of units derived from an unsaturated comonomer such as allyloxymethyl thiirane (CXXXV). As far as is known, these rubbers have not been developed commercially.

$$\cdots\!-S—CH_2—CH—\cdots \qquad\qquad \cdots\!-S—CH_2—CH—\cdots$$
$$\qquad\qquad | \qquad\qquad\qquad\qquad\qquad\qquad |$$
$$\qquad\qquad CH_3 \qquad\qquad\qquad\qquad\qquad CH_2$$
$$\qquad\qquad\qquad\qquad\qquad\qquad\qquad\qquad\qquad | $$
$$\qquad\qquad\qquad\qquad\qquad\qquad\qquad O—CH_2—CH\!=\!CH_2$$

(CXXXIV)   (CXXXV)

9.7 NITROSO RUBBERS

The nitroso rubbers are a family of special-purpose synthetic rubbers which have an —N—O— unit in the main chain of the polymer molecule, together with a second unit derived from an unsaturated carbon compound. They are therefore partly inorganic in nature and partly organic. The units derived from the two types of monomer alternate. The generic structure of the repeat unit of these rubbers is therefore as shown at (CXXXVI), in which x is a small number. Because these rubbers have been developed for application under rather severe conditions (see further

$$\cdots\!-N—O\!\left\{\!\! \begin{array}{c} | \; | \\ C \\ | \; | \end{array} \!\!\right\}_{\!\!x}\!\cdots$$
$$\qquad\; | $$
$$\qquad\; R$$

(CXXXVI)

below), the substituents attached to the carbon atoms are usually either
fluorine or fluorinated hydrocarbon derivatives. Likewise, the group R is
usually a fluorinated hydrocarbon moiety.

The first member of this family of rubbers—often referred to as just
nitroso rubber—has the repeat unit shown at (CXXXVII). It is an
alternating copolymer of high molecular weight prepared from trifluoro-
nitrosomethane (CXXXVIII) and tetrafluoroethylene by reacting the two

$$\cdots -N-O-CF_2-CF_2-\cdots \qquad\qquad CF_3NO$$
$$\quad\;\; | $$
$$\quad CF_3$$

(CXXXVII)   (CXXXVIII)

monomers at temperatures below $0\,^\circ C$. The outstanding characteristics of
this rubber are its resistance to solvents combined with good low-
temperature flexibility (its glass-transition temperature being $c. -50\,^\circ C$),
and its resistance to chemical deterioration, in particular, to oxidation. It
does not burn in pure oxygen, even at high pressures. However, its excellent
resistance to chemical attack is itself the basis of a major problem, namely,
that of bringing about effective crosslinking. Vulcanisation can be achieved
by heating with polyfunctional amines, but the vulcanisates obtained in this
way have poor mechanical properties and reduced resistance to oxidation.

Because of the difficulty of satisfactorily vulcanising the above rubber,
other nitroso rubbers have been developed which contain a small
proportion of reactive functional groups. One such material is known as
carboxy nitroso rubber. Most of the repeat units are of the type (CXXXVII)
above, but a small proportion ($0 \cdot 5–1 \cdot 5\,\%$) are derived by copolymerisation
with nitrosoperfluorobutyric acid (CXXXIX). The resulting repeat units are
as indicated at (CXL). These rubbers, which are known generically by the

$$NO \qquad\qquad\qquad \cdots -N-O-CF_2-CF_2-$$
$$\;| \qquad\qquad\qquad\qquad\quad |$$
$$(CF_2)_3 \qquad\qquad\qquad\quad (CF_2)_3$$
$$\;| \qquad\qquad\qquad\qquad\quad |$$
$$CO_2H \qquad\qquad\qquad\quad CO_2H$$

(CXXXIX)   (CXL)

letters AFMU, are susceptible to crosslinking by most of the reactions
which are available for the crosslinking of the carboxylated versions of
more conventional synthetic rubbers. In particular, these rubbers can be
cured by heating with metal oxides and salts. Heating with chromium (III)

trifluoroacetate (*c.* 5 pphr) is specifically recommended as giving a vulcanisate having excellent chemical resistance, good mechanical properties, and relatively low compression set. The latter feature is surprising in view of the fact that the crosslinks which form are presumably ionic in nature. The structure of the vulcanisate is supposedly represented as shown at (CXLI).

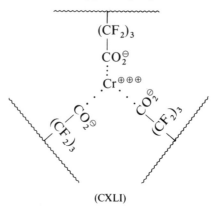

(CXLI)

In order to attain low compression set, it is necessary to subject the vulcanisate to an extended period of postcuring at high temperature. The curing of this rubber with chromium trifluoroacetate can be retarded by the addition of small quantities of an acid (such as salicylic or benzoic acid) or an acid anhydride (such as phthalic anhydride). By this means, the processing safety of the stocks can be improved.

The principal area of application for nitroso rubbers (in particular, the carboxy type) seems to be where extreme resistance to oxidation is required, such as non-flammability in an atmosphere of pure oxygen. One specific application which has been described is that of expulsion bladders to be used within rocket propellant tanks. These bladders serve as containers for the propellant, the propellant being expelled from the bladder into the rocket combustion chamber by means of pressure applied to the exterior wall of the bladder. Dinitrogen tetroxide is typical of the type of oxidiser which it is necessary for the bladder to contain, and it is apparently in specialised applications of this kind that nitroso rubbers have proved successful.

9.8   SILICONE RUBBERS (QR)

We consider now an important class of special-purpose synthetic rubber, the members of which are, like the nitroso rubbers, partly inorganic in nature and partly organic. The 'silicone' rubbers derive their name from the fact that their stoichiometric composition, RR'SiO (where R and R' are organic moieties), is identical with that of the silicon analogue of the ketone RR'CO. Structurally, however, silicones are quite different from ketones in that they are polymeric in nature, the repeat unit being as shown at (CXLII). It is seen that the main chain comprises a sequence of —Si—O— units, and is thus inorganic in nature, and that the groups attached to the

$$\cdots -\underset{\underset{R'}{|}}{\overset{\overset{R}{|}}{Si}}-O-\cdots$$

(CXLII)

silicon atoms are organic in nature. These groups are normally either methyl or phenyl, but in some rubbers a small proportion of vinyl groups is present as well in order to confer the ability to be vulcanised by heating with sulphur and conventional accelerators. As a class, silicone rubbers are conveniently described chemically as polydisubstituted siloxanes. They are denoted generically by the letter Q. The letters M and P are used to denote, respectively, methyl and phenyl substituents on the polysiloxane chain. Thus rubbers which have only methyl substituents on the polysiloxane chain are denoted by the letters MQ. The letters MPQ (sometimes PMQ) denote silicone rubbers which have both methyl and phenyl substituents on the polysiloxane chain. The letters VQ denote silicone rubbers which contain vinyl substituents, the two main types being MVQ and MPVQ (sometimes denoted by VMQ and PVMQ respectively).

Silicone rubbers are also unusual amongst synthetic rubbers in that, being partly inorganic in nature, they are not obtained entirely from petroleum sources. The starting materials are sand or quartz (silicon dioxide) and alkyl or aryl halides. The first step is the reduction of the sand or quartz to elemental silicon by heating with carbon in an electric furnace:

$$SiO_2 + 2C \longrightarrow Si + 2CO$$

The next step is the conversion of the elemental silicon into alkyl- or aryl-substituted silicon chlorides, that is, into compounds of the type R_xSiCl_{4-x} (often referred to as organochlorosilanes). Although this conversion can be brought about in various ways, the procedure which is most widely used is the so-called 'direct process'. In this process, alkyl or aryl chlorides are caused to react directly with elemental silicon. The product is a mixture of organochlorosilanes, in which the value of x in the general formula R_xSiCl_{4-x} ranges from 1 to 3. (Depending upon the reaction conditions, compounds having $x = 0$ or $x = 4$ can also be obtained.) The di-substituted organochlorosilanes ($x = 2$) are the most important of these for silicone rubber manufacture. Their formation can be represented stoichiometrically as follows:

$$2RCl + Si \longrightarrow R_2SiCl_2$$

The reaction is carried out at elevated temperature ($c.\ 275\,°C$ for the production of dimethyldichlorosilane) in the presence of a catalyst. Copper powder is the catalyst which is most frequently used, and the amount is usually approximately $10\,\%$ of the weight of the silicon. However, for the production of the phenylchlorosilanes by the direct process, silver is a more effective catalyst than is copper.

Silicone rubbers are obtained by hydrolysis of mixtures of organochlorosilanes which contain predominantly di-substituted organochlorosilanes. The overall reaction for the hydrolysis of di-substituted organochlorosilanes can be represented as follows:

$$n\underset{\underset{R}{|}}{\overset{\overset{R}{|}}{Cl-Si-Cl}} + nH_2O \longrightarrow \left(\underset{\underset{R}{|}}{\overset{\overset{R}{|}}{Si-O}}\right)_n + 2nHCl$$

It is evident that the structure of the polymer which is obtained depends very much upon the relative amounts of mono-, di- and tri-substituted organochlorosilanes which are present in the mixture which is being hydrolysed. Mono-substituted organochlorosilanes will act as terminators of the growing siloxane chain because they will lead to the formation of the unreactive terminal groups shown at (CXLIII). Tri-substituted organochlorosilanes will lead to the formation of branch points, as shown at (CXLIV). It is therefore necessary to refine the mixture of organochlorosilanes into a blend of controlled composition if a rubber of controlled

$$\cdots -O-\underset{\underset{R'}{|}}{\overset{\overset{R}{|}}{Si}}-R$$

(CXLIII)

$$\cdots -O-\underset{\underset{\underset{\vdots}{|}}{O}}{\overset{\overset{R}{|}}{Si}}-O-$$

(CXLIV)

molecular structure is to be obtained as a result of the subsequent hydrolysis. This refining is achieved by fractional distillation.

The earliest linear silicone polymers of high molecular weight intended for use as rubbers were made from organochlorosilanes by a two-stage hydrolysis process. The organochlorosilane was first hydrolysed to a low-molecular-weight silanol-stopped silicone fluid. The further condensation of these molecules was effected by heating in the presence of iron(III) chloride as catalyst:

$$
\underset{\underset{R}{|}}{\overset{\overset{R}{|}}{\sim\!\!\text{Si}}}\!\!-\!\text{OH} + \text{HO}\!-\!\underset{\underset{R}{|}}{\overset{\overset{R}{|}}{\text{Si}\!\!\sim}} \xrightarrow{\text{FeCl}_3} \underset{\underset{R}{|}}{\overset{\overset{R}{|}}{\sim\!\!\text{Si}}}\!\!-\!\text{O}\!-\!\underset{\underset{R}{|}}{\overset{\overset{R}{|}}{\text{Si}\!\!\sim}} + \text{H}_2\text{O}
$$

Because of difficulties in ensuring the absence of branching and crosslinking in the product, silicone rubber gums are now made by first hydrolysing the organochlorosilanes to cyclic oligomers, and then causing the cyclic oligomers to polymerise to high-molecular-weight linear polysiloxanes:

$$
\left(\underset{\underset{R}{|}}{\overset{\overset{R}{|}}{-\text{Si}}}\!\!-\!\text{O}- \right)_x \longrightarrow \cdots\!\!-\!\!\underset{\underset{R}{|}}{\overset{\overset{R}{|}}{\text{Si}}}\!\!-\!\text{O}\!\!-\!\cdots
$$

Cyclic oligomers with $x = 3, 4, \ldots$ are formed together with linear polymers of low molecular weight when an organochlorosilane is hydrolysed. The cyclic oligomers are separated from the hydrolysate stream and polymerised in bulk using either a strongly acidic or a strongly basic catalyst. Polysiloxanes containing units of more than one kind can be obtained by co-hydrolysing an appropriate mixture of organochlorosilanes.

Silicone rubbers intended for use as millable gums have molecular weights in the range 0.5–1.0×10^6. The earliest rubbers were polydimethylsiloxanes ($R = CH_3$). Although these rubbers have a very low glass-transition temperature ($c. -120\,°C$) (the reason being the high degree of torsional flexibility of the Si–O bond), they in fact undergo serious stiffening long before the glass-transition temperature is reached. This stiffening occurs because the regularity of the polydimethylsiloxane structure permits the polymer to crystallise at $c. -60\,°C$. In order to improve the low-temperature flexibility of the polydimethylsiloxane rubbers, a small proportion ($c. 10\%$) of phenylmethylsiloxane units

$(R = C_6H_5, R' = CH_3)$ can be introduced. Rather paradoxically, the glass-transition temperature of the rubber is increased by the replacement of a few methyl groups by phenyl groups. But because the regularity of the polymer structure is destroyed, crystallisation is inhibited and so low-temperature flexibility is improved. Thus whereas polydimethylsiloxane vulcanisates remain flexible down to about $-55\,°C$, vulcanisates from rubbers containing phenyl groups can remain flexible down to about $-100\,°C$.

Polydimethylsiloxane rubbers, and the phenylmethyl variants, are usually vulcanised by heating with organic peroxides (see further below). The mechanism of the vulcanisation reaction is believed to be abstraction of hydrogen from the methyl groups by radicals derived from the peroxide. Crosslinks between polymer chains are then formed by combination of the $-CH_2\cdot$ radicals produced by the initial hydrogen-abstraction reaction. Silicone rubbers which contain a small proportion ($c.\ 1\,\%$) of vinyl groups ($R = CH:CH_2$ and $R' =$, say, CH_3 or a mixture of CH_3 and C_6H_5) are capable of being vulcanised by sulphur and conventional accelerators. It will be seen subsequently that the crosslinking of silicone rubbers by peroxides is also facilitated by the presence of vinyl substituents.

A further interesting variant of the silicone rubber structure is obtained by introducing a few boron atoms into the main chain by way of the type of structure shown at (CXLV). The presence of boron atoms in the ratio $B:Si = 1:500$ to $1:200$ improves the self-adhesive tack of the rubber, and this is important for building operations using unvulcanised silicone rubber stocks.

$$\begin{array}{ccccc} R & & & R & \\ | & & & | & \\ \sim\!\!\sim\!Si\!-\!O\!-\!B\!-\!O\!-\!Si\!-\!O\!\sim\!\!\sim \\ | & & | & & | \\ R' & & R'' & & R' \end{array}$$

(CXLV)

Several organic peroxides can be used for the vulcanisation of silicone rubbers. The most widely used is probably dibenzoyl peroxide. Vulcanising temperatures in the region of $120\,°C$ are used. If too low a temperature is used, e.g., $80\,°C$, then the peroxide appears to decompose ineffectively, possibly because of the excessive diffusion of atmospheric oxygen into the vulcanising rubber. Other suitable organic peroxides include di-2,4-dichlorobenzoyl peroxide, t-butyl perbenzoate, and dicumyl peroxide. Alkylhydroperoxides and dialkylperoxides tend to give rather poor cures. However, the use of such peroxides is attractive because they do not generate acid by-products. A more satisfactory cure is obtained if they are used with the vinyl-containing silicones, because it appears that the radicals derived from the peroxide are able to add to the vinyl groups and so give radicals which then form crosslinks by addition to a second radical site, or by interaction with a methyl side chain of a second rubber molecule with the formation of a second radical which can then promote further crosslinking:

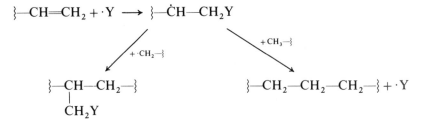

In these reactions, $\cdot$Y denotes a radical derived from the peroxide. Whatever are the exact details of the reaction mechanism, it seems clear that the effect of vinyl substitution is to improve the efficiency of peroxides as curatives for silicone rubbers.

It is generally necessary to post-cure peroxide-vulcanised silicone rubbers in a hot-air oven, in order to develop an optimum balance of vulcanisate properties for a given application. A period of several hours (up to, say, 24 h) heating at 250 °C is typical. Apart from desirable effects upon the initial vulcanisate properties, post-curing stabilises the vulcanisate to subsequent ageing, by eliminating curative decomposition products and also low-molecular-weight silicone polymers.

Silicone rubber gum vulcanisates are extremely weak mechanically. Tensile strengths are of the order of only 50 lb in^{-2}. In order to develop tolerable mechanical strength, it is essential to incorporate fine-particle fillers. Carbon blacks are not widely used as fillers for silicone rubbers, because of their relatively high content of adsorbed volatile matter, leading to the formation of voids in the rubber during vulcanisation. As regards other fillers, the strength obtained depends upon the type of filler and the level at which it is present. Some generalisations are indicated in Table 9.3. For maximum strength (which is still rather low by conventional standards), fine-particle silicas are used as fillers.

Mixtures of unvulcanised silicone rubber and fine-particle silicas present a processing problem which probably has its origin in the interaction between silica particles and rubber which gives rise to the enhancement of the strength of the vulcanisate. It is particularly acute in the case of the

TABLE 9.3.
EFFECTS OF VARIOUS FILLERS IN SILICONE RUBBER VULCANISATES[5]

Type of filler	Average particle diameter (nm)	Specific surface area ($m^2 g^{-1}$)	Mechanical properties of filled vulcanisate	
			Tensile strength ($lb\ in^{-2}$)	% Elongation at break
Silica aerogel	30	110–150	600–1 000	200–350
Fumed silica	15–20	175–200	600–1 800	200–600
Acetylene black	45	78–85	600–900	200–350
Flux-calcined diatomaceous silica	1 000–5 000	< 5	400–800	75–200
Calcined diatomaceous silica	1 000–5 000	< 5	400–800	75–200
Calcined kaolin	1 000–5 000	< 5	400–800	75–200
Precipitated calcium carbonate	30–50	32	400–600	100–300
Ground silica	1 000–10 000		100–400	200–300
Zinc oxide	300	3	200–500	100–300
Iron oxide	< 1 000		200–500	100–300
Zirconium silicate			400–600	100–300
Titanium dioxide	300		200–500	300–400

fumed silicas which give maximum enhancement of strength. This is the problem of so-called *crepe hardening* during storage. As the name implies, the unvulcanised mix undergoes severe hardening during storage, thereby becoming crepe-like in character. Crepe-hardened stocks crumble when put on a two-roll mill, and become processable again only after prolonged milling. The 'bound rubber' content of the mix increases throughout the crepe-hardening process, that is, there is a progressive reduction in the proportion of the rubber which can be extracted by solvents. Both phenomena are interpreted in terms of interaction taking place slowly between the rubber-molecules and the surface of the silica particles, thereby giving rise to what might be described as a 'quasi-vulcanised' material. In the absence of additives which retard crepe hardening, significant changes occur over a period of 1–3 weeks at room temperature. Not only is this phenomenon important because of its implications for processing; the vulcanisate properties are also influenced by crepe hardening. Various procedures have been proposed for overcoming the problem of crepe hardening. Thus *structure-control additives* (sometimes called *antistructure agents*) can be used; pinacoxydimethylsilane (CXLVI) is one such additive.

(CXLVI)

Alternatively, the silica filler particles can be pre-treated with organo-chlorosilanes and then washed to free them from hydrochloric acid which results from interaction between the silane and the hydroxyl groups on the surface of the silica particles. The effect of such treatment is to block some of the reactive sites at the surface of the silica particles, and so inhibit the type of interaction which leads to crepe hardening.

The outstanding property of silicone rubbers as a family is their ability to resist extremes of temperature. Although the mechanical properties of silicone rubber vulcanisates are generally poor at room temperature, these properties are retained at elevated temperatures very much better than are the properties of conventional general-purpose rubbers. As a guide, the useful life of silicone rubber vulcanisates at 150 °C is in the range 5–10 years. At 200 °C, this drops to 2–5 years. The heat resistance of silicone rubber is the attribute which has contributed most to its development. Retention of flexibility at low temperatures is also very good, especially if the regularity of the polydimethylsiloxane structure is disturbed by the presence of a few phenyl groups. Whilst the mechanical properties of silicone rubbers are generally poor, one property—compression set—is remarkably good. Low compression set is presumably the consequence in part of the thermal stability of the Si–O links of the main chain, and in part of the stability of the C–C crosslinks (in the case of a peroxide-cured rubber, at least). Apart from the problem of poor mechanical properties at room temperature, the general application of silicone rubbers is inhibited by their high cost. One area of application in which silicone rubbers are used, and in which cost is of little consequence, is that of body implant materials. Silicone rubbers have the two-fold advantage over many other rubbers that (i) they do not cause untoward reactions with body tissues within which they are in contact, partly, no doubt, because they are very hydrophobic in nature, and (ii) they are not metabolised by the human body. A third factor which is important in some implant applications is that the permeability of silicone rubber to oxygen is unusually high. This is believed to be associated with the high torsional flexibility of the Si–O bond.

A problem which is sometimes encountered with silicone rubbers is that they are susceptible to main-chain cleavage by hydrolysis, the reaction being as follows:

$$\begin{array}{ccc} R & R & R & R \\ | & | & | & | \\ \sim\!\!Si-O-Si\!\!\sim + H_2O \longrightarrow \sim\!\!Si-OH + HO-Si\!\!\sim \\ | & | & | & | \\ R & R & R & R \end{array}$$

This type of degradation is commonly known as *reversion*. It is mainly encountered when the rubber is subjected to heat in confined spaces. Few problems are encountered in this respect if water is able to escape from the vulcanisate and its environment, because the above reaction is reversible.

Some interest has been shown in recent years in the development of fluorosilicone rubbers (which are denoted by the letters FQ). The objective has been to obtain silicone rubbers having improved heat resistance and (in particular) improved resistance to solvents. The rubber in which most interest has been shown is poly(trifluoropropyl)methylsiloxane ($R = CF_3CH_2CH_2$, $R' = CH_3$) (MFQ). An interesting variant has also been developed which is resistant to the hydrolytic degradation which has been referred to in the preceding paragraph. This rubber is a 'hybrid' fluoro-silicone which contains units derived from a fluorocarbon monomer as well as trifluoropropylmethylsiloxane units.

9.9 INORGANIC RUBBERS

The silicone rubbers and nitroso rubbers are partly inorganic in nature and partly organic. We come now to consider briefly those synthetic rubbers which are entirely inorganic in nature. There is only one important family of these at present, namely, the polyphosphazenes. Interest in these materials has arisen from their lack of flammability, and their general resistance to oxidation and charring at high temperatures. A further significant attribute is their resistance to low-temperature stiffening.

It has been known for many years that phosphorus pentachloride reacts with ammonia or ammonium chloride to give a volatile white solid, commonly known as 'phosphonitrilic chloride'. This is now known to have the cyclic structure shown at (CXLVII). When this compound is strongly heated, it melts and changes into a rubbery substance. The cause of this change is a ring-opening polymerisation which leads to the formation of a linear polydichlorophosphazene (CXLVIII) of high molecular weight. These reactions can be represented as follows:

$$3PCl_5 + 3NH_4Cl \longrightarrow \quad + 12HCl$$

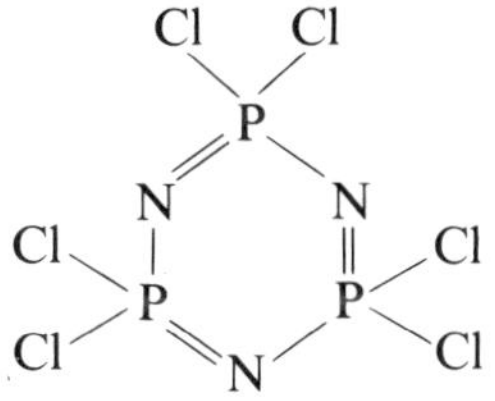

(CXLVII)

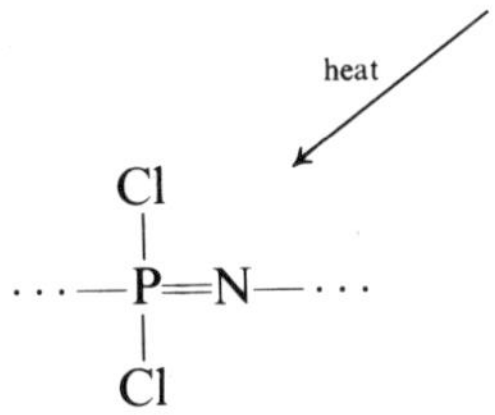

heat

$$\cdots -\!\!\!\!\begin{array}{c} Cl \\ | \\ P\!=\!N \\ | \\ Cl \end{array}\!\!\!\!-\cdots$$

(CXLVIII)

Polydichlorophosphazene slowly hydrolyses in contact with atmospheric moisture, yielding a crusty conglomerate of ammonium phosphate and phosphoric acid. It is principally for this reason that it has remained little more than a laboratory curiosity until very recently.

The problem of making useful polyphosphazene rubbers (general structure (CXLIX)) is that of making flexible linear macromolecules of high

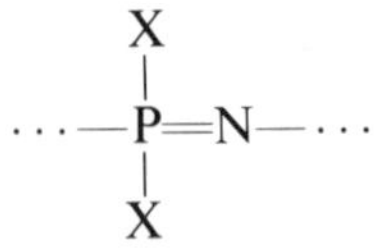

$$\cdots -\underset{\underset{X}{|}}{\overset{\overset{X}{|}}{P}}=N-\cdots$$

(CXLIX)

molecular weight which are resistant to hydrolysis. On general chemical grounds, it might be inferred that the hydrolytic instability of poly-dichlorophosphazene is associated with the presence of the phosphorus–chlorine bonds rather than with the $-P\!=\!N-$ bonds of the main chain. For this reason, attempts have been made to replace the chlorine by other atoms or moieties which give non-hydrolysable bonds with phosphorus. Initially, attempts were made to polymerise thermally organocyclo-phosphazenes of the type (CL), which were expected to give hydrolytically stable polymers. However, the presence of the bulky organic substituents in compounds like (CL) prevents ring-opening polymerisation. A second

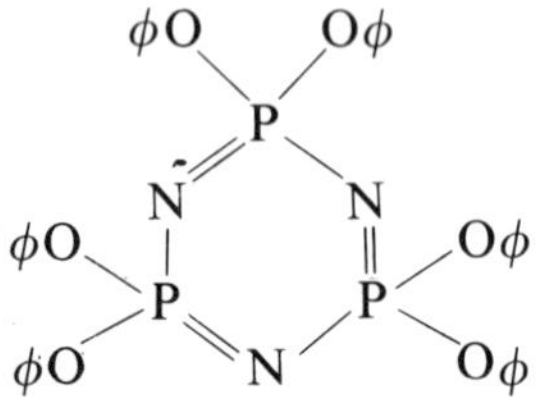

(CL)

approach has proved to be more successful. In this approach, nucleophilic substitutions are carried out on polydichlorophosphazene in order to replace the chlorine atoms by suitable organic moieties. It appears that it is necessary to control the initial polymerisation of the dichlorophosphazene rather carefully if complete nucleophilic substitution is to be achieved and a hydrolytically-stable polymer produced. The problem is that the ring-opening polymerisation of dichlorophosphazene takes place in two steps. The initial polymer which forms is not crosslinked, and so is soluble in organic solvents such as benzene or tetrahydrofuran. Complete nucleo-philic substitution of the chlorine atoms is possible with this polymer.

However, in the second step of the polymerisation, the polymer becomes crosslinked and insoluble in organic solvents, although it does swell. This polymer undergoes incomplete nucleophilic substitution only. It is therefore necessary to carry out the substitution reaction before the polymer reaches the crosslinked state.

The nature of the polymer which is obtained by nucleophilic substitution of the polydichlorophosphazene precursor depends upon the group which has replaced the chlorine atoms. Alkoxy substituents, notably methoxy and ethoxy, give rubbery products having good resistance to low-temperature stiffening. Replacement of chlorine by an alkoxy group is achieved by treatment of the polydichlorophosphazene precursor with an alkali-metal alkoxide, according to the reaction

$$\cdots\!-\!\overset{\displaystyle Cl}{\underset{\displaystyle Cl}{\overset{|}{\underset{|}{P}}}}\!\!=\!\!N\!-\!\cdots + 2NaOR \longrightarrow \cdots\!-\!\overset{\displaystyle OR}{\underset{\displaystyle OR}{\overset{|}{\underset{|}{P}}}}\!\!=\!\!N\!-\!\cdots + 2NaCl$$

In order to discourage any tendency to crystallise when the rubber is cooled, the regularity of the polymer sequence can be destroyed by introducing two or more substituents into the same polymer chain. One way in which this can be done is to allow the polydichlorophosphazene precursor to react simultaneously with two different nucleophilic reagents. Rubbers produced by this procedure are currently being developed commercially. Because they are required to have good heat resistance and also resistance to hydrocarbon fuels and oils, the substituent groups are fluorinated. A typical combination is one in which the OR groups comprise a mixture of trifluoroethoxy groups (OCH_2CF_5) and pentafluoropropoxy groups ($OCH_2CF_2CF_3$). These rubbers are known as 'phosphonitrilic fluoroelastomers'. They also contain a small amount of a reactive substituent attached to the $-P=N-$ backbone to facilitate vulcanisation, which can be effected by means of reaction with organic peroxides or with sulphur and conventional accelerators, or by exposure to high-energy radiation.

Polyphosphazene rubbers can have excellent low-temperature flexibility, and this is of considerable significance in relation to some of their potential applications. Thus the phosphonitrilic fluoroelastomers are described as being solvent resistant with a temperature range of serviceability from about -60 to about $+205\,°C$. It is perhaps rather surprising that the molecules of a substituted polyphosphazene chain remain flexible at low temperatures, partly in view of the presence of the alternating double bonds in the main chain, and partly because of possible polarisation effects. However, whatever may be the reason, polyphosphazene rubbers can have glass-transition temperatures as low as about $-80\,°C$ and, as long as crystallisation is inhibited by, say, multiple substitution, the low glass-transition temperature is reflected in the retention of flexibility when the rubber is cooled to low temperatures. Other ways in which the mixed substitution of a polydichlorophosphazene precursor can be achieved include (i) partial replacement of one organic moiety by another, and (ii) partial substitution of the chlorine atoms by one type of moiety, followed by completion of the substitution by another type of moiety.

Some interest is currently being shown in possible biomedical applications of polyphosphazene rubbers. These rubbers are very hydrophobic in nature, especially those which contain fluorinated alkoxy side groups. It appears that these rubbers show good compatibility with body tissues, in particular, with blood. They also degrade slowly in the body environment, the products of degradation being said to be innocuous. They are therefore of interest for the manufacture of sutures and implants which are required slowly to disappear from the body. Special types of polyphosphazene rubber are said to be being developed for this purpose.

9.10   HYDROGEL RUBBERS

We conclude this survey of miscellaneous types of synthetic rubber by making brief reference to what may fairly be regarded as the most unconventional group of synthetic rubbers which has yet been developed. These are materials which are rigid in the dry state, but which in contact with water become hydrated and, in consequence, rubbery in nature. In some respects, there is a formal analogy with plasticised polyvinyl chloride, in that an otherwise rigid polymer is caused to become flexible by plasticising it with a low-molecular-weight liquid with which it is compatible. However, all similarity ends at this formal resemblance.

Several types of hydrogel rubber have found application principally as materials for the manufacture of soft contact lenses. A typical material is a copolymer of 2-hydroxyethyl methacrylate (CLI) and N-vinyl-2-pyrrolidone (CLII) as the main monomers, and minor amounts of methacrylic acid and a crosslinking monomer such as tetraethylene glycol dimethacrylate. Polymerisation is often carried out in bulk in the absence of water, the polymer being allowed to hydrate after polymerisation is

$$CH_2{=}CH.CH_3$$
$$|$$
$$CO$$
$$|$$
$$OCH_2CH_2OH$$

(CLI)

(CLII)

complete. The function of the crosslinking monomer is, of course, to cause the polymer chains to become lightly crosslinked into an indefinitely large molecular network which swells, but does not dissolve, in water or other solvents. The function of the methacrylic acid is to increase the extent to which the polymer swells when placed in contact with water, especially water which is sufficiently alkaline to bring about ionisation of the copolymerised carboxylic-acid groups. Poly-2-hydroxyethyl methacrylate is a weakly-hydrophilic polymer, and poly-N-vinyl-2-pyrrolidone is a strongly hydrophilic polymer. The tendency of the copolymer to hydrate is principally determined by the ratio of the two types of unit present in it. The effect of the minor amount of copolymerised methacrylic acid is supplementary to that of the major amount of copolymerised N-vinyl-2-pyrrolidone. The equilibrium water content of hydrogel rubbers (typically in the region 45–65 % by weight) is an important variable in biomedical applications because it is the principal factor which determines the oxygen permeability of the hydrogel. The oxygen permeability of this type of hydrogel increases uniformly with increasing water content, approaching the oxygen permeability of pure water at very high water contents.

One of the major disadvantages of hydrogel rubbers is that, as might be anticipated, they are extremely weak mechanically. The strength tends to fall as the water content of the hydrogel increases. The tensile strength of a hydrogel which contains 50 % water is of the order of $5\,\mathrm{lb\,in^{-2}}$.

GENERAL BIBLIOGRAPHY

Brown, H. P. and Gibbs, C. F. (1955). Carboxylic elastomers, *Ind. Eng. Chem.*, **47**, 1006.

Brown, H. P. (1957). Carboxylic elastomers, *Rubb. Chem. Technol.*, **30**, 1347.

Brown, H. P. (1963). Crosslinking reactions of carboxylic elastomers, *Rubb. Chem. Technol.*, **36**, 931.

Whitby, G. S., Davis, C. C. and Dunbrook, R. F. (Eds.) (1954). *Synthetic Rubber*, John Wiley and Sons, New York, Chapter 25.

Fram, P. (1964). Acrylic elastomers. In: *Encyclopedia of Polymer Science and Technology*, Vol. 1, John Wiley and Sons, New York, pp. 226f.

Vial, T. M. (1971). Recent developments in acrylic elastomers, *Rubb. Chem. Technol.*, **44**, 344.

Montermoso, J. C. (1961). Fluorine-containing elastomers, *Rubb. Chem. Technol.*, **34**, 1521.

Arnold, R. G., Barney, A. L. and Thompson, D. C. (1973). Fluoroelastomers, *Rubb. Chem. Technol.*, **46**, 619.

Wall, L. A. (Ed.) (1972). *Fluoropolymers*, Wiley-Interscience, New York. (Especially Chapters 5 and 10.)

Barney, A. L., Kalb, G. H. and Khan, A. A. (1971). Vulcanizate properties from a new perfluoroelastomer, *Rubb. Chem. Technol.*, **44**, 660.

Brown, J. H., Finlay, J. B., Hallenbeck, A., MacLachlan, J. D. and Pelosi, L. F. (1977). *Peroxide-curable fluoroelastomers*, Paper No. 8 presented at International Rubber Conference, Brighton, England.

Frisch, K. C. and Dieter, J. A. (1975). An overview of urethane elastomers, *Polym. Plast. Technol. Eng.*, **4**(1), 1.

Saunders, J. H. (1960). The relations between polymer structure and properties in urethans, *Rubb. Chem. Technol.*, **33**, 1259.

Saunders, J. H. (1960). The formation of urethan foams, *Rubb. Chem. Technol.*, **33**, 1293.

Ofstead, E. A. (1977). Polypentenamer. In: *Encyclopedia of Polymer Science and Technology*, Supplement Vol. 2, John Wiley and Sons, New York, pp. 610f.

Le Delliou, P. (1977). *Properties and applications of polynorbornene: a new elastomer in powder form*, Paper No. 12 presented at International Rubber Conference, Brighton, England.

Bertozzi, E. R. (1968). Chemistry and technology of elastomeric polysulfide polymers, *Rubb. Chem. Technol.*, **41**, 114.

Berenbaum, M. B. (1969). Polysulfide polymers. In: *Encyclopedia of Polymer Science and Technology*, Vol. 11, John Wiley and Sons, New York, pp. 425f.

Gaylord, N. G. (1962). *Polyethers. Part III. Polyalkylene Sulfides and Other Polythioethers*, Interscience, New York, Chapters XIII and XIV.

Levine, N. B. (1971). Carboxy nitroso rubber, chemistry, properties, and applications, *Rubb. Chem. Technol.*, **44**, 40.

Lewis, F. M. (1962). The science and technology of silicone rubber, *Rubb. Chem. Technol.*, **35**, 1222.

Warrick, E. L. (1976). Silicone rubber—a perspective, *Rubb. Chem. Technol.*, **49**, 909.

Warrick, E. L., Pierce, O. R., Polmanteer, K. E. and Saam, J. C. (1979). Silicone elastomer developments 1967–1977, *Rubb. Chem. Technol.*, **52**, 437.

Lichtenwalner, H. K. and Sprung, M. N. (1970). Silicones. In: *Encyclopedia of Polymer Science and Technology*, Vol. 12, John Wiley and Sons, New York, p. 464 f.

Pierce, O. R. and Kim, Y. K. (1971). Fluorosilicones as high temperature elastomers, *Rubb. Chem. Technol.*, **44**, 1350.

Allcock, H. R. (1976). Polyphosphazenes: new polymers with inorganic backbone atoms, *Science*, **193**, 1214.

Kyker, G. S. and Antkowiak, T. A. (1974). Phosphonitrilic fluoroelastomers—a new class of solvent resistant polymers with exceptional flexibility at low temperature, *Rubb. Chem. Technol.*, **47**, 32.

REFERENCES

1. Brown, H. P. (1957). *Rubb. Chem. Technol.*, **30**, 1347.
2. Golder, R. I. (1973). *Adiprene CM: A Sulphur Curable Urethane Rubber*, Elastomer Chemicals Department, E.I. Du Pont de Nemours & Co. (Inc.), Wilmington, Delaware.
3. *FA Polysulphide Rubber*, Thiokol Chemical Corporation publication TD No. 921R(6-73).
4. Tobolsky, A. V. (1956). *J. Appl. Phys.*, **27**, 673.
5. Lewis, F. M. (1962). *Rubb. Chem. Technol.*, **35**, 1222.

Chapter 10

Novel Developments in the Technology of Synthetic Rubbers

10.1 INTRODUCTION

This chapter is concerned not so much with different chemical types of synthetic rubber as with novel forms of synthetic rubber which in turn have led to the development of unusual processing technologies as judged by the standards of conventional rubber technology. It is in this latter sense that the adjective 'novel' is used. Of course, it has sometimes been the case that novel chemical variations have been necessary in order to produce a rubber which is capable of being processed in a particular way. But these chemical variations are largely secondary to the processing technology by which the rubber is intended to be fabricated into the final product. Indeed, it has often been the case that the chemical variations have been dictated by the processing requirements.

The first, and most important, of these novel developments which we consider is that of the so-called *thermoplastic rubbers*, also known variously as *thermoplastic elastomers*, *rubberlike thermoplastics*, *plastomers*, and *elastoplastics*. These are materials which can be processed at elevated temperatures as thermoplastics, but which cool down to give rubbery materials at normal temperatures. Unlike conventional rubbers, the change from thermoplastic melt to rubbery solid is not brought about by the formation of chemical crosslinks between the polymer molecules, but instead arises from certain reversible physical changes which occur within the material as it is cooled down. The change from thermoplastic melt to rubbery solid is therefore reversible; on heating, the rubber reverts to a thermoplastic melt, and so can be reprocessed without difficulty.

The second novel development which will be considered is that of *powdered rubbers*. Here the idea is to produce the rubber in the form of a powder which can then be mixed with other powdered compounding

320

ingredients and the mixture processed directly as a powder blend. The intention is to eliminate the necessity for the conventional mixing step, with its requirement for heavy equipment and energy input.

The third and final novel development to be considered is that of *fluid rubbers*. As with powdered rubbers, the intention underlying this development is the elimination of the conventional mixing step. The idea here is to produce a low-molecular-weight version of the rubber in the first instance, and then to link together these molecules subsequently by suitable chemical reaction in order to produce a molecular network which exhibits reversible elasticity. The initial polymer molecular weight is sufficiently low for the material to be a viscous liquid rather than an elastic gum. It can therefore, in principle at least, be readily mixed with other compounding ingredients, and then fabricated into the final product by casting in suitable moulds. The chemistries of the various systems are designed to be such that the linking of the polymer molecules together to form a network takes place within the mould cavity.

10.2   THERMOPLASTIC RUBBERS

Several types of polymer have been developed for use as thermoplastic rubbers. A convenient classification of the more important thermoplastic rubbers is as follows:

1.   styrene–diene block copolymer types;
2.   polyolefin blend types;
3.   urethane types; and
4.   polyester types.

Thermoplastic rubbers of these types became commercially available in (1) 1965, (2) 1972, (3) 1955–60, and (4) 1972 respectively. Other types which will not be considered here include blends of polyvinyl chloride and an acrylonitrile–butadiene rubber, synthetic *trans*-1,4-polyisoprene, and a graft copolymer of isoprene rubber and ethylene. It might be argued with some force that plasticised polyvinyl chloride is also in effect a thermoplastic rubber. Indeed, it could be claimed that this material is the earliest elastomer which can be, and commonly is, processed as a thermoplastic. However, largely as a consequence of history and tradition, it is not usual to regard plasticised polyvinyl chloride as a thermoplastic rubber.

10.2.1 Thermoplastic Rubbers of the Styrene–Diene Block Copolymer Type

The most important class of general-purpose hydrocarbon thermoplastic rubbers comprises materials which are block copolymers of styrene and either butadiene or isoprene. The simplest types have molecules which consist of a centre block of diene units and two terminal blocks of styrene units, as illustrated in Fig. 10.1(a). These are known as styrene–diene–styrene triblock copolymers. Rather more complex structurally are the radial type of block copolymers illustrated in Fig. 10.1(b). These consist of three or more arms of styrene–butadiene diblock copolymers radiating from a centre hub; in each diblock, the butadiene portion is innermost, and the styrene portion forms the terminal block.

The ability of styrene–diene block copolymers to behave like a vulcanised rubber at normal temperatures depends upon a phenomenon to which reference has been made in Chapter 5 (Section 5.2.2), namely, that polymers containing repeat units of different chemical types tend to be incompatible at the molecular level. Thus copolymer molecules normally form a material which is homogeneous at the molecular level only if the various repeat units in the copolymer are reasonably uniformly dispersed

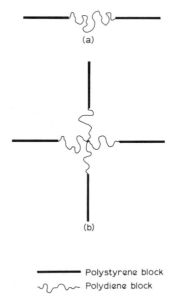

(a)

(b)

━━━━━━ Polystyrene block
∿∿∿∿ Polydiene block

FIG. 10.1.   Schematic illustration of structure of (a) triblock copolymers and (b) radial block copolymers of the styrene–diene type used as thermoplastic rubbers.

amongst each other. Any tendency to blockiness in the copolymer tends to lead to phase separation into domains which are relatively rich in one or the other type of repeat unit. It has been noted in Chapter 5 that the styrene and butadiene units in a styrene–butadiene copolymer are not inherently miscible with each other, although, if the two types of unit are reasonably uniformly distributed in a copolymer molecule, then phase separation of the various parts of the copolymer molecule becomes physically impossible. It is otherwise with block copolymers of styrene and a diene. The material tends to phase separate into domains which are rich in styrene units and domains which are rich in butadiene units. This phase separation into domains is illustrated schematically in Fig. 10.2(a) for a styrene–diene–styrene triblock copolymer, and in Fig. 10.2(b) for a radial block copolymer. Since the styrene domains are well below the glass-transition temperature of polystyrene at normal temperatures, the styrene domains are rigid. By contrast, the diene blocks are well above their glass-transition temperature at normal temperatures, and so the diene domains are rubbery in nature. Furthermore, because the styrene blocks are chemically bonded to the diene block, the rigid styrene domains are not only dispersed within a matrix of rubbery material, but are also very firmly bonded to that matrix. Thus, on the one hand, the styrene domains can be regarded as providing particulate reinforcement to the rubber, and on the other hand, because of the chemical bonding which exists between the styrene and diene phases, they can also be regarded as providing crosslinks which bind the polydiene segments into an indefinitely large molecular network and thus prevent the occurrence of irreversible flow when the material is subjected to stress. The thermoplasticity of these materials arises from the fact that the polystyrene domains soften when the material is warmed. The above description might lead one to suppose that the material would resemble a reinforced vulcanised rubber until the glass-transition temperature of polystyrene (c.

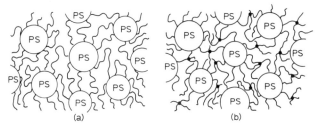

Fig. 10.2. Illustrating phase separation in styrene–diene block-copolymer thermoplastic rubbers of (a) the triblock type and (b) the radial block type.

90 °C) was approached, when the material would suddenly soften and become a thermoplastic melt. In fact, as results which will be given subsequently show, the rubbery properties begin to disappear at temperatures well below the glass-transition temperature of high-molecular-weight polystyrene.

It is clear from the above description of these materials why it is essential that a styrene–diene block copolymer should have the form of either a styrene–diene–styrene triblock, or the form of a radial block copolymer with the styrene segments outermost. It would not be expected, for instance, that a styrene–diene diblock copolymer or a diene–styrene–diene triblock copolymer would be suitable as a thermoplastic rubber. It is essential that each polydiene segment should be terminated either by a polystyrene segment or by the centre point of a radial structure, otherwise the polydiene segment cannot be immobilised by the rigid polystyrene domains as the material cools from the melt. It is clearly a further requirement of the molecular structure that (a) the polydiene segments should be sufficiently long for the matrix of the material to have rubbery character, and (b) the copolymer should be sufficiently blocky for effective phase separation of the two types of units to occur. An indication that requirement (b) has been satisfied is that the material shows two glass-transition temperatures, one of which is approximately that of polystyrene and the other of which is approximately that of the polydiene, rather than one glass-transition temperature which has a value intermediate between those of the two homopolymers.

Block copolymers of styrene and dienes which are suitable for application as thermoplastic elastomers can be synthesised by at least four important processes. These are:

1.　the difunctional-initiator process;
2.　the three-stage sequential addition process;
3.　the coupling process; and
4.　the tapered block process.

The *difunctional-initiator process* uses a difunctional ionic initiator such as the adduct which forms between sodium and naphthalene. In the presence of a diene monomer, an active species is formed which propagates at both ends. The chemistry of the reaction is similar to that which has been described in Chapter 5 (Section 5.5.2) for the polymerisation of dienes by alkyllithiums (see also Section 4.2.5 of Chapter 4). As has been pointed out in Chapter 4 (Section 4.2.5), in the absence of impurities, the propagating polymer chains cannot terminate because, being anionic in nature, no mechanism exists whereby the propagating chains can terminate each other. The propagating chains therefore retain their activity, even although all the diene monomer has been consumed. For this reason, they have become known somewhat inappropriately as 'living' polymers. They have the ability to add further monomer units of themselves if a fresh supply of monomer is provided. If this fresh monomer happens to differ from the initial monomer, then a block copolymer is formed. Furthermore, this will be of the triblock type unless one end of the initial polymer chain became deactivated, in which case a diblock copolymer is formed. In the production of a styrene–diene triblock copolymer, the diene is first polymerised, and then styrene is admitted to the reaction system. The formation of the triblock copolymer can be represented schematically as follows:

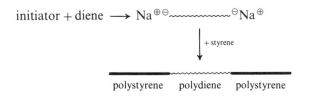

The principal disadvantage of this process is that the active ends of the initial polydiene species are susceptible to deactivation by impurities. If such deactivation occurs, then the final product will contain diblock copolymer and polydiene homopolymer, as well as the desired triblock copolymer. A further disadvantage is that some at least of the difunctional initiators function only in solvents such as aliphatic ethers, thereby reducing the proportion of 1,4-polydiene units which are formed. A high proportion of 1,4-polydiene microstructure is desirable in order to impart as high a degree of rubberiness to the final product as possible.

In the *three-stage sequential addition process*, a hydrocarbon-soluble organo-lithium monofunctional initiator is commonly used. The initial step is the polymerisation of styrene in a hydrocarbon solvent, such as benzene, using an organo-lithium initiator. When the polymerisation is complete, diene monomer is added to the polymer, which is still active. Polymerisation of the diene occurs to form the central polydiene block. The preparation of the copolymer is completed by the addition of a further quantity of styrene, equal in amount to that used initially, when the polymerisation of the diene is complete, followed by deactivation of the ends of the copolymer when the polymerisation of the second aliquot of styrene is complete. The process can be represented as follows:

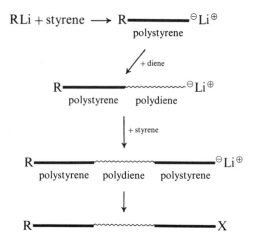

Again, the chief problem is premature termination of the active chain ends by adventitious impurities. Such premature termination can result in the formation of polystyrene homopolymers, diblock copolymers, and triblock copolymers having a shortened terminal polystyrene block.

The two processes described above clearly produce triblock copolymers. The third process, the so-called *coupling process*, is capable of producing radial block copolymers as well as the linear types. The principle underlying this method is to couple together suitable reactive diblock copolymers using a coupling agent whose functionality is appropriate to the structure of the block copolymer which it is desired to produce. The coupling agent can be a polyfunctional halogen compound. The diblock copolymer can be one which is produced by anionic polymerisation and which is still active at the butadiene end. A process for the production of a styrene–diene–styrene triblock copolymer is as follows:

If a coupling agent of functionality greater than two is used, then the product is a radial block copolymer. An example of such a coupling agent is silicon tetrachloride:

$$4 \quad \text{━━━━━━} \sim\sim\sim\sim\sim^{\ominus}\text{Li}^{\oplus} + \text{SiCl}_4 \longrightarrow$$

$$\text{━━━━━} \sim \text{Si} \sim \text{━━━━} + 4\text{LiCl}$$

Apart from the inherent versatility of the process, in that both linear and radial block copolymers can be produced, its main advantage is that the opportunities for unwanted structural variations brought about by adventitious impurities are minimised because only two monomer additions are required. The main disadvantage is the need for high precision and efficiency in the coupling reaction.

The *tapered block process* depends upon the fact that, when an attempt is made to copolymerise styrene and butadiene in hydrocarbon solvents using an alkyllithium initiator, the butadiene polymerises preferentially to the styrene. No appreciable polymerisation of the styrene occurs until the butadiene concentration is reduced to a low level. This phenomenon is illustrated in Fig. 10.3, which shows conversion–time curves for the polymerisation of a 25/75 styrene/butadiene mixture in various hydrocarbon solvents, the reaction being initiated by an alkyllithium. Initially the polymerisation proceeds until about 80 % of the monomer is polymerised. The rate of polymerisation then falls to a low level. Throughout this stage of the reaction, the solution is colourless. But during the period of apparent quiescence, it becomes amber. Polymerisation then recommences and is rapidly completed. The colour change suggests that styryl anions are formed during the period of quiescence. In fact, the colour change, the rapidity of the subsequent polymerisation, and the nature of the product are all consistent with the interpretation that the butadiene polymerises first followed by the styrene. The essential steps in the production of

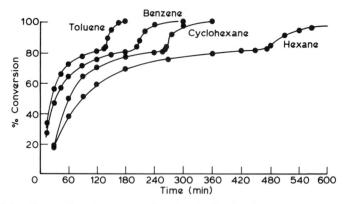

FIG. 10.3.   Conversion–time curves for the alkyllithium-initiated copolymerisation of a 25/75 styrene/butadiene monomer mixture at 122 °F in various solvents.[1]

styrene–butadiene–styrene triblock copolymers by the tapered block process are as follows: The polymerisation of styrene by an alkyllithium initiator is first allowed to proceed to about 50 % conversion. Butadiene is then introduced into the reaction system. The polymerisation of styrene then virtually ceases, and that of the butadiene occurs until almost all the butadiene has reacted. The remaining styrene then begins to polymerise, and the final product is a styrene–butadiene–styrene triblock copolymer having styrene blocks of approximately equal lengths. The copolymer is described as being 'tapered' because the butadiene component does not polymerise completely before the polymerisation of the residual styrene commences. What happens is that, towards the end of the butadiene polymerisation, styrene units become copolymerised with the butadiene units randomly but in increasing amounts as the concentration of styrene monomer rises relative to that of butadiene monomer. Thus the composition 'tapers' gradually from that of almost pure polybutadiene to that of almost pure polystyrene. One advantage of the process is that only two additions of monomer are required; opportunities for unwanted effects arising from the presence of adventitious impurities are therefore minimised.

Considerable interest has been shown in the morphological structures of the domains which form within thermoplastic rubbers of the styrene–diene block copolymer type. In accordance with expectation, the shape and nature of the domains which form the disperse phase depend upon the relative amounts of the two phases. Some generalisations are summarised in Table 10.1. When the two components are present in roughly equal

TABLE 10.1

GENERALISATIONS CONCERNING EFFECT OF VOLUME RATIO OF TWO PHASES UPON MORPHOLOGY OF STYRENE–DIENE BLOCK-COPOLYMER MATERIALS

Volume ratio	Morphology
Increasing A content ↕ Increasing B content ↑	Spheres of A in matrix of B
	Rods of A in matrix of B
	Alternate lamellae
	Rods of B in matrix of A
	Spheres of B in matrix of A

amounts, a system comprising alternating lamellae tends to develop. If one component is present in excess, it becomes the disperse phase or matrix. The other component forms discrete domains which are spherical if the volume fraction of domains is small, but tend to become rodlike as the volume fraction of domain increases. The weight fraction of styrene in commercially-available styrene–diene–styrene block-copolymer thermoplastic rubbers varies over a wide range (c. 10–50 %). The styrene content of some of the more widely used materials is in the region of 25–30 %.

The styrene–diene–styrene block-copolymer thermoplastic rubbers are usually compounded with fillers, resins and oils in order to modify their properties and to lower costs. Reinforcing fillers do not improve the tensile strength of these rubbers, this being already very high (see below). They do, however, increase the modulus, tear resistance, hardness, flex life and abrasion resistance. The presence of fillers increases the viscosity of the melt, whereas the addition of oils reduces melt viscosity.

Unmodified styrene–diene–styrene block-copolymer thermoplastic rubbers have high tensile strengths (up to $5000 \, lb \, in^{-2}$) at normal temperatures, and elongations at break of up to 1000 %. The high tensile strength of these materials is perhaps surprising at first sight when it is recalled that gum vulcanisates from conventional styrene–butadiene rubbers have very low tensile strengths. The high tensile strength of the styrene–diene–styrene block copolymers is attributed primarily to the presence of the glassy polystyrene domains. As has already been pointed out these domains not only function as crosslinking sites; they also function as particulate reinforcing fillers, the reinforcing effect being supplemented by the excellent adhesion which exists between the rubber phase and the glassy phase. Another factor which will promote high tensile strength is the ability of the rubber phase to crystallise on stretching. The separation of the polystyrene as a distinct phase will encourage this, as will stereoregularity of the polydiene sequences. The hardness varies over the approximate range 35–95 IRHD. The hardness of the unmodified rubbers tends to increase with increasing styrene content, but the hardness is determined by other factors besides styrene content. The increase of hardness with styrene content is also attributed to the glassy polystyrene domains functioning as a particulate filler.

The principal shortcoming of these thermoplastic rubbers is that mechanical properties such as tensile strength fall off rapidly as the temperature increases above normal, whilst properties such as compression set and rates of stress relaxation and creep increase sharply with increasing temperature. The effect of temperature upon the tensile strength of several unmodified commercially available thermoplastic rubbers of this type is shown in Fig. 10.4; results for a conventional styrene–butadiene rubber vulcanisate containing HAF black as a reinforcing filler are included for purposes of comparison. A family of stress-relaxation curves for a single unmodified thermoplastic rubber of this type at various temperatures is

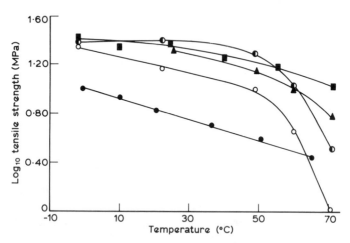

FIG. 10.4.   Effect of temperature upon tensile strength of various unmodified styrene–butadiene block copolymer thermoplastic rubbers as follows:[1] ▲ radial 40 % styrene, ◑ linear 30 % styrene, ○ radial 30 % styrene, ● sulphur-vulcanised radial 30 % styrene; ■ conventional sulphur-vulcanised styrene–butadiene rubber containing 40 pphr of HAF carbon black.

given in Fig. 10.5; it is evident that the rate of stress relaxation increases rapidly with the temperature, especially above about 50 °C.

Much information is available concerning the effects of structural variations upon the mechanical properties of thermoplastic elastomers of the styrene–diene–styrene block-copolymer type. The following effects are particularly interesting:

1. For materials of given overall styrene content, the tensile stress–strain behaviour is little affected by the overall molecular weight of the polymer, provided that it is high (say > 70 000).
2. For materials of given overall styrene content, the tensile strength falls with reduction in the molecular weight of the styrene blocks. The tensile strength is very low if the blocks have molecular weights below about 7000, and this is attributed to inability of the polystyrene blocks to separate as well-defined phases if the block lengths are short.
3. The tensile strength increases progressively with increasing styrene content.
4. Polydispersity of the polystyrene blocks has little effect upon tensile strength, but increasing the polydispersity of the poly-butadiene blocks reduces the tensile strength.

In order to achieve an optimum balance of mechanical properties and processability, it has been suggested that (a) the polystyrene blocks should

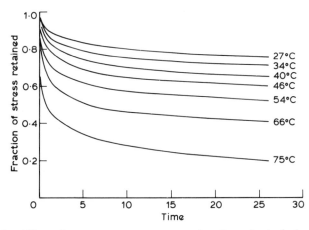

FIG. 10.5.   Effect of temperature upon stress relaxation of a typical unmodified styrene–diene block-copolymer thermoplastic rubber.[2]

have a molecular weight within the range 1×10^5–2×10^5, (b) the polydiene blocks should have a molecular weight within the range 4×10^5–1×10^6, and (c) the overall styrene content should be within the range 20–40 % by weight. The lower end of the molecular-weight range for the polystyrene blocks is a consequence of the need for adequate phase separation of the polystyrene domains. The upper ends of the molecular-weight ranges for both blocks are consequences of the need to avoid unduly high melt viscosity, which makes processing difficult and hinders the separation of the polystyrene domains.

One of the principal areas of application for this type of thermoplastic rubber is in the footwear industry, where these rubbers are used for the production of direct-moulded soles and heels. It is significant that this is a large-scale application where the loss of rubbery properties when the temperature rises is not of great importance. In general, these types of rubber find application in areas where temperatures are not likely to rise much above normal, where properties such as compression set are not of great importance, and where ozone resistance is not required.

An interesting variant of the styrene–diene block-copolymer type of thermoplastic rubber is the product which is obtained by hydrogenating the polydiene centre block to give what is essentially a styrene–olefin block copolymer. If the centre block of the initial block copolymer consisted exclusively of 1,4-butadiene units, then the centre block of the hydro-genated block copolymer would be essentially a linear polyethylene sequence. The centre block would therefore tend to crystallise, and the product would not be rubbery at ambient temperatures. If, on the other hand, the conditions under which the centre blocks have been formed are such that the butadiene units are enchained in the 1,2 mode as well as in the 1,4 mode (e.g. by adding a suitable ether to the reaction system), then the hydrogenated centre block will be in effect a random copolymer of ethylene and butene units. The centre block will not now tend to crystallise, and the elastomeric character of the initial block copolymers will be preserved in the hydrogenated material. An obvious advantage which the hydrogenated styrene–diene block copolymers have over their precursors is the absence of carbon–carbon double bonds in the centre block. They are therefore more resistant to degradation by oxygen and ozone than are the styrene–diene copolymers.

10.2.2 Thermoplastic Rubbers of the Polyolefin Blend Type

These are materials which are produced by blending a polyolefin rubber with a polyolefin plastic. The usual combination is an ethylene–propylene rubber and a polypropylene thermoplastic. When the blend is cooled from processing temperatures down to normal temperatures, the polypropylene molecules tend to separate into rigid domains which are embedded in a rubbery matrix of ethylene–propylene rubber. The formation of rigid polypropylene domains is a consequence of the crystallisation of polypropylene segments, rather than of vitrification because the polymer has been cooled below its glass-transition temperature. Presumably good adhesion between the rubbery matrix and the rigid polypropylene domains is achieved through mixing of propylene-rich segments of the rubber phase with the surface layers of the polypropylene domains. In order to obtain a satisfactory material, it is sometimes necessary to introduce a few crosslinks into the rubber matrix. This is achieved by adding a small amount of an organic peroxide at the blending stage; crosslinking then occurs during processing. The function of these crosslinks is obscure. Their presence does not appear to destroy the thermoplastic character of the rubber. It is possible the primary function of these crosslinks is to ensure that some chemical bonding occurs between the polypropylene and ethylene–propylene rubber phases. The choice of peroxide and of crosslinking conditions is important if concurrent degradation of the polypropylene component is to be avoided.

At the time of writing, increasing interest is being shown in these materials. The major area of application seems to be the manufacture of flexible components for road vehicles, e.g., parts which are required to survive impact during low velocity collisions. The mechanical properties of this type of thermoplastic rubber are not so good as are those of the styrene–diene–styrene block-copolymer type. However, one very important advantage which this type has over the styrene–diene block-copolymer type is that they are inherently resistant to attack by ozone, being largely chemically saturated. Resistance to oxidation and weathering is excellent, as is chemical resistance generally. Electrical properties in both wet and dry environments are excellent.

Turning now to consider the production of these blends in a little more detail, it is convenient to classify the more important production methods under three headings:

1. physical blending of a partially-cured ethylene–propylene rubber with a polyolefin thermoplastic;
2. physical blending of a high-viscosity ethylene–propylene rubber with a polyolefin thermoplastic; and
3. physical blending of an ethylene–propylene copolymer having a high degree of crystallinity in the unstretched state with a polyethylene.

The blending of a partially-cured ethylene–propylene rubber with a polyolefin thermoplastic is probably the method which has been most widely used for the production of this type of thermoplastic rubber. Blending can be carried out either as a *one-step* process or as a *two-step* process. In the one-step process, as the name implies, the blending and crosslinking are carried out as a single operation. In the two-step process, the ethylene–propylene rubber is first lightly crosslinked by mixing with a peroxide and then heating. The lightly-crosslinked rubber is then blended with the thermoplastic olefin. The principal advantage of the two-step process over the one-step process is that the possibility of the polyolefin thermoplastic being degraded under the influence of the decomposing organic peroxide is minimised. As regards the rubber component, ethylene–propylene rubbers which contain a small proportion of unsaturated units are preferred to fully saturated ethylene–propylene rubbers, presumably because of the greater ease with which they are crosslinked with organic peroxides. As regards the polyolefin thermoplastic, highly-crystalline isotactic or syndiotactic polypropylenes are preferred to, say, crystalline polyethylenes. The extent to which the rubber component has been crosslinked is usually assessed by means of the gel content of the rubber as determined by extraction with cyclohexane. The normal range for gel content appears to be 55–90 %, so that the rubber phase is in fact very lightly-crosslinked ethylene–propylene rubber plasticised by uncrosslinked rubber of the same type. Various organic peroxides have been used for effecting crosslinking of the rubber phase. A typical example is dicumyl peroxide; using this peroxide, a satisfactory degree of crosslinking is achieved by working the material for 5–10 min at a temperature in the range 150–200 °C, the level of peroxide being *c.* 0·5 pphr. It may be desirable to terminate the action of the residual peroxide by adding an antioxidant as a radical scavenger at the end of the crosslinking step.

The other two types of process listed above for the production of this type of thermoplastic rubber require the use of special types of rubber. The second group of methods uses an ethylene–propylene rubber of high zero-shear viscosity, thus obviating the need for a crosslinking step. In the third group of methods, the requirement for crosslinking is eliminated by using an ethylene–propylene rubber which forms a high proportion of crystalline domains in the unstretched state.

10.2.3 Thermoplastic Rubbers of the Urethane Type
Thermoplastic urethane rubbers can be regarded as being linear block copolymers of the type $(AB)_x$, in which one block of the polymer chain is a

relatively long, flexible polyester or polyether sequence, and the other block is formed by reaction between an aromatic diisocyanate, the terminal hydroxyl groups of the polyester sequence, and a low-molecular-weight diol or triol chain-extender. The average molecular weight of the polyester or polyether sequences is typically in the range 1000–3000. It is these sequences which impart rubbery character to the material at normal temperatures. It is essential that not only should these sequences be flexible at normal temperatures, but also that they should be virtually amorphous. Because these sequences impart to the material its rubbery character, it has become customary to refer to them as *soft segments*. The second block of the copolymer is commonly known as the *hard segment*. It is these latter blocks which provide the physically effective 'pseudo' crosslinks and also, to a certain extent, give the material something of the character of a rubber containing a strongly reinforcing particulate filler. (It may be noted in passing that the terms *soft segment* and *hard segment* can also sensibly be applied to the diene and styrene blocks respectively of thermoplastic rubbers of the styrene–diene–styrene block-copolymer type. However, it appears that only latterly is this nomenclature being used in this connection, and then only infrequently. This nomenclature was developed in the first place to describe the molecular structure of thermoplastic rubbers of the urethane type, and has subsequently been used to describe the molecular structure of thermoplastic rubbers of the polyester type (see Section 10.2.4).)

It is necessary to consider in a little more detail the way in which the hard segments function as pseudo crosslinks and as a particulate filler. The hard segments are rich in urethane linkages, consisting as they do of the products of reaction between the hydroxyl end-groups of the polyester or polyether soft segments, the diisocyanate and the chain-extending diol or triol. They are therefore polar in nature, and hence there exist strong mutual attractions between the various hard segments. As a consequence, the hard segments tend to aggregate into ordered crystalline domains within the rubbery matrix of flexible chains. An important factor which encourages this aggregation is extensive hydrogen bonding between the hydrogen atoms of the urethane group and electronegative atoms such as carboxyl and ether oxygens and the nitrogen atoms of other urethane groups. It appears that hydrogen bonding of this kind is able to restrict the mobility of the urethane chain segments in the domains, and hence to restrict the ability of these segments to organise extensively into fully-crystalline lattices. The result is semi-ordered regions to which the adjective *para-crystalline* is sometimes applied. Other forces which also promote association of the hard segments include (a) interactions between the π-electron systems of the aromatic rings of the diisocyanate, and (b) van der Waals' forces of attraction. The general nature of the association of hard segments into domains is illustrated in Fig. 10.6, leaving open the question of the exact nature of the forces which bind the hard segments into rigid domains. The effect of the association of the hard segments into rigid domains is to cause the polymer chains comprising the soft segments to be bound into molecular networks such that irreversible flow under stress cannot occur. The thermoplasticity of this type of of thermoplastic rubber arises from the circumstance that, whatever may be the forces which bind the hard segments together, these forces are sufficiently weak to be disrupted when the material is heated to processing temperatures. It is then possible for the soft segments to flow irreversibly under stress, because they are not any longer bound into a molecular network. On cooling, the hard segments reassociate into rigid domains, and the material once again displays behaviour characteristic of a crosslinked rubber. Solvents which are able to solvate the whole of the polymer chains (including the hard segments) are also able to destroy the pseudo crosslinks by destroying the association between the hard segments. The hard segments re-associate into rigid domains when the solvent is removed. It is therefore possible to apply thermoplastic rubbers of this type as solutions, and use is made of this possibility in various coating applications.

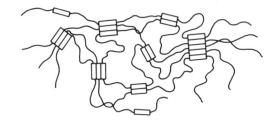

Virtually crosslinked / extended network
of polymer primary chains

↑ ↓ Heat or solvent

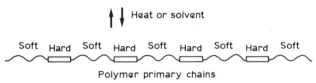

Soft Hard Soft Hard Soft Hard Soft Hard Soft

Polymer primary chains

FIG. 10.6.   Illustrating formation of hard segments in a urethane thermoplastic rubber.[3]

Two obvious compositional variables which could be expected to affect the physical properties of a urethane thermoplastic rubber of given chemical type are:

1. the average molecular weight of the soft segments; and
2. the length of the hard segments.

Tables 10.2(a) and 10.2(b) summarise results which have been reported for the effects of each of these variables respectively. The thermoplastic rubbers to which these results refer were prepared from hydroxyl-terminated polyesters of a given chemical type (polycaprolactone (PCL) (CLIII)) but of various molecular weights. The rubbers were prepared by reaction of the polyesters with 4,4'-diphenylmethane diisocyanate (MDI) (CXIX) and 1,4-butanediol (BDO) (CXXIII) as chain-extender.

$$H\!-\!\{O(CH_2)_5CO\}_n\!-\!OR\,OH$$

(CLIII)

The results given in Table 10.2(a) refer to a series of materials in which the number-average molecular weight of the PCL soft segments was varied over the range 340–3130. For each polymer, the expected sequence length of the hard segments was kept constant by fixing the molar ratio of the intermediates at $PCL/MDI/BDO = 1/2/1$. Thus although the composition of the hard segments was expected to remain constant throughout the series, the weight fraction of hard segments (and also the number concentration) decreased inversely as the molecular weight of the soft segments increased. The physical properties of the various thermoplastic rubbers at normal temperatures vary essentially in accordance with expectation. As the sequence length of the soft segments increases, so the hardness, tensile modulus and tear strength decrease, but the tensile strength and elongation at break increase. It is also evident that the glass-transition temperature of these rubbers (and hence their expected behaviour at low temperatures) depends strongly upon the length of the soft segments. (Strictly speaking, of course, it is the glass-transition temperature of the rubbery phase of the material to which reference is being made here.) Unlike homopolymers and random copolymers, the glass-transition temperature increases sharply as the sequence length is decreased. In fact, as Fig. 10.7 shows, the glass-transition temperature increases linearly with reciprocal average sequence length. The increase in glass-transition temperature with decrease in sequence length is attributed to immobilisation of the ends of the soft segments through chemical attachment to the

TABLE 10.2(a)

EFFECT OF AVERAGE MOLECULAR WEIGHT OF SOFT SEGMENTS UPON PHYSICAL PROPERTIES OF URETHANE THERMOPLASTIC RUBBERS PREPARED FROM HYDROXYL-TERMINATED POLYCAPROLACTONE (PCP), 4,4'-DIPHENYLMETHANE DIISOCYANATE (MDI) AND 1,4-BUTANEDIOL (BDO) (MOLE RATIO PCP/MDI/BDO = 1/2/1)[4]

$\bar{M}_n$ of polycaprolactone	340	530	830	1250	2100	3130
Weight fraction MDI/BDO	0·63	0·53	0·43	0·32	0·22	0·16
Hardness (Shore A)		95	90	80	65	50
Modulus at 100% (lb in^{-2})		2 000	500	400	300	1 000
Modulus at 300% (lb in^{-2})			2 500	1 000	600	1 500
Tensile strength (lb in^{-2})		4 000	6 000	6 000	5 000	4 500
% Elongation at break		250	400	500	600	700
Tear strength (Graves, lb in^{-1})		700	500	350	250	500
Reduced viscosity at 30°C (0·2 g dl^{-1} in dimethylformamide)	1·16	1·44	1·36	1·57	1·57	0·98
Glass-transition temperature (°C)						
(i) from in-phase modulus	53	25	−10	−27	−40	−35
(ii) from loss tangent	65	35	−5	−20	−35	−20

TABLE 10.2(b)

EFFECT OF LENGTH OF HARD SEGMENTS UPON PHYSICAL PROPERTIES OF URETHANE THERMOPLASTIC RUBBERS PREPARED FROM HYDROXYL-TERMINATED POLYCAPROLACTONE (PCP) ($\bar{M}_n = 2100$), 4,4'-DIPHENYLMETHANE DIISOCYANATE (MDI) AND 1,4-BUTANEDIOL (BDO)

| Mole ratio PCP/MDI/BDO | 1/2/1 | 1/3/2 | 1/4/3 | 1/5/4 | 1/6/5 | 1/8/7 | 1/10/9 | 1/15/14 | 1/20/19 |
Weight fraction MDI/BDO	0·22	0·31	0·38	0·43	0·49	0·56	0·61	0·70	0·76
Hardness (Shore A)	65	85	90	95					
Hardness (Shore D)		30	40	45	55	60	60		
Modulus at 100% ($\mathrm{lb\,in}^{-2}$)	300	600	1000	2000	2500	5500			
Modulus at 300% ($\mathrm{lb\,in}^{-2}$)	600	1000	2000	3500	5000				
Tensile strength ($\mathrm{lb\,in}^{-2}$)	5000	5000	7500	7000	6500	6000	3000		
% Elongation at break	600	600	500	450	350	150	25		
Tear strength (die C, $\mathrm{lb\,in}^{-1}$)	250	400	500	600	750	1000	1200		
Glass-transition temperature (°C)									
(i) from in-phase modulus	−40	−40	−32	−30	−30	−30	−30	30	30
(ii) from loss tangent	−35	−30	−23	−17	−14	−5	30	55	65

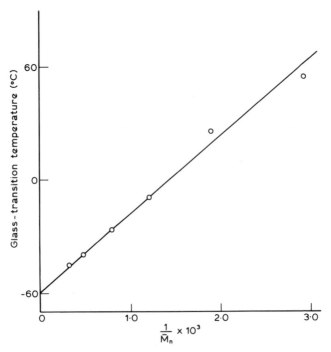

FIG. 10.7.   Effect of sequence length, as quantified by reciprocal number-average molecular weight, upon glass-transition temperature for a urethane thermoplastic rubber.[4] See text for details of polymer.

hard segments. As the sequence length of the soft segments increases, the restrictive influence of the hard-segment domains becomes less important, and the glass transition therefore occurs at progressively lower temperatures. A practical consequence of the results for glass-transition temperature given in Table 10.2(a) is that, in the case of thermoplastic urethane polymers of the type to which these results refer, a useful rubber is obtained only if the average molecular weight of the soft segments exceeds about 800, i.e., the average sequence length exceeds about seven repeat units. However, if the soft segments are too long, then, although the inherent glass-transition temperature of the material is suitably low, the rubbery character is reduced by crystallisation of the soft segments. In the case of the polymers to which the data of Table 10.2(a) refer, soft-segment crystallisation begins to occur when the average molecular weight of the soft segments exceeds about 3000 (i.e., about 26 repeat units). The effect of soft-segment crystallisation is to increase the actual glass-transition temperature of the material somewhat above the value which would be expected on the basis of the molecular weight of the soft segments.

The results summarised in Table 10.2(b) refer to a series of materials in which the sequence length of the soft segments was kept constant at a value corresponding to an average molecular weight of 830, and the average length of the hard segments was varied by varying the proportion of MDI and BDO to PCL. The weight fraction of hard segments (MDI + BDO) was varied over the range 0·42–0·76. As the proportion of hard-segment material increases, so the hardness and tear strength increase, but the tensile strength and elongation at break decrease. The effect upon elongation at break is particularly marked. The length of the hard segments also has a marked effect upon the glass-transition temperature, in that this temperature increases sharply with increase in hard-segment length. At first sight, this result is surprising, because the glass-transition temperature of these materials is determined by the mobility of the soft segments, and this mobility would not be expected to depend upon the length of the hard segments. However, it seems that the overall mobility of the soft segments is affected by the length of the hard segments, as is evidenced by the glass-transition temperature being dependent upon the length of the hard segments.

In addition to the truly thermoplastic urethane rubbers which have been described above, various so-called 'thermoplastic–thermosetting' urethane rubbers have also been developed. One group of these contains an excess of the diisocyanate component; as a consequence, crosslinking occurs when the material is subjected to a post-curing step. It is to be doubted whether materials of this type should be described as thermoplastic rubbers at all, because, although they can be initially processed as thermoplastics, they lose their ability to revert to thermoplastic melts on heating once chemical crosslinking has taken place.

10.2.4 Thermoplastic Rubbers of the Polyester Type

Polyester thermoplastic rubbers resemble urethane thermoplastic rubbers in that they are block copolymers of the type $(AB)_x$ in which one type of block is a flexible soft segment and the other type is a rigid hard segment. Again, the hard segments can associate into rigid domains, the mechanism of association being believed to be crystallisation. The hard segment is a polyester of a short-chain alkylene glycol and terephthalic acid. The structure is as shown at (CLIV). Typically the glycol can be ethylene glycol (in which case $x = 2$) or 1,4-butanediol (in which case $x = 4$). The soft

$$\cdots -OC \text{—} \underset{}{\bigcirc} \text{—} CO.O(CH_2)_xO \text{—} \cdots$$

(CLIV)

segments are described as being polyether polyesters. By this is meant that they are polyesters of polyether glycols, that is, of hydroxyl-terminated polyethers. The acid is usually terephthalic acid, and the glycol component is a polyoxyalkylene glycol. The general structure of the soft segments is therefore as shown at (CLV). Typically (i) $y = 2$, $C_yH_{2y} = $ —CH_2CH_2—, and $z \cong 20$; (ii) $y = 3$, $C_yH_{2y} = $ —$CH(CH_3)CH_2$—, and $z \cong 15$; and (iii) $y = 4$, $C_yH_{2y} = $ —$(CH_2)_4$—, and $x \cong 14$.

$$\cdots -OC \text{—} \underset{}{\bigcirc} \text{—} CO \text{—}(OC_yH_{2y})_z O \text{—} \cdots$$

(CLV)

These copolyesters are made by the melt-transesterification of a mixture which contains a phthalate ester, a low-molecular-weight glycol, and a polyoxyalkylene glycol. A large number of different products can be made in this way. The molecular weight of the polyoxyalkylene glycol can be varied, as can the nature of the low-molecular-weight glycol. It is also possible to use the various phthalate ester isomers and mixtures thereof. In a typical process for the preparation of a polyester thermoplastic rubber, dimethyl terephthalate and polyoxytetramethylene glycol of molecular weight c. 1000 are placed together in a reaction vessel with a stoichiometric excess of 1,4-butanediol. A small amount of a catalyst (e.g., a titanate) is added and the mixture slowly heated to 160 °C, at which temperature methanol begins to distil from the mixture. The temperature is raised to 250 °C over a period of about one hour in order to remove most of the methanol. The product at this stage is a prepolymer. This is then chain-extended to a product of high molecular weight by reducing the pressure and allowing 1,4-butanediol to distil off. The extent of polymerisation is conveniently monitored by means of the viscosity of the reacting mass as evidenced by the torque required to drive the stirrer. When the viscosity levels off or begins to decrease, the reaction is terminated by cooling the reactor.

The overall stoichiometry of the reaction system is chosen in such a way that the final product will contain a preponderance of repeat units of the hard-segment type. On purely statistical grounds, it will then be the case that the polymer produced contains relatively long sequences of crystallisable hard-segment repeat units. The feasibility of the polymer melt giving a thermoplastic rubber when cooled depends upon the presence of these relatively long sequences of hard-segment units. The average length of the hard segments increases with the total content of hard-segment units. The expected relationship for polyesters synthesised from dimethyl terephthalate, 1,4-butanediol, and polyoxytetramethylene glycol of molecular weight 1000 is shown in Fig. 10.8.

For a given type of polymer, both the glass-transition temperature of the rubbery phase and the melting point of the crystalline phase depend upon the proportion of hard-segment units in the material. The effect of this

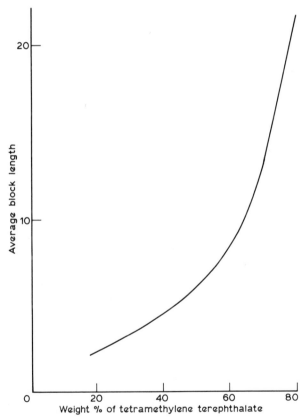

FIG. 10.8.   Average length of hard segments as a function of total content of hard-segment units for product formed by transesterification of dimethyl terephthalate, 1,4-butanediol, and polyoxytetramethylene glycol of molecular weight 1000.[5]

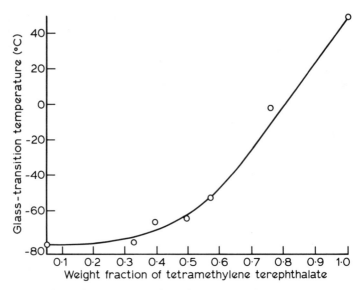

FIG. 10.9.   Effect of proportion of hard-segment units upon glass-transition temperature of rubbery phase for a polyester thermoplastic rubber containing tetramethylene terephthalate hard segments.[5]

variable upon the glass-transition temperature of the rubbery phase is shown in Fig. 10.9. As the proportion of hard-segment units increases, so the glass-transition temperature of the rubber phase increases, as is observed for the urethane-type thermoplastic rubbers. The explanation may be partly that the average length of the soft segments falls as the proportion of hard-segment units increases, and partly that some of the hard-segment units are in fact dispersed throughout the rubbery phase, perhaps because they are in segments which are too short to crystallise or because they are prevented from crystallising by having become entrapped through molecular entanglements. As expected, the melting point of the crystalline phase increases sharply with increasing proportion of hard-segment units. This is illustrated in Fig. 10.10, where at (a) the crystalline melting point is shown as a function of weight fraction of hard-segment units, and at (b) the same data are plotted as a function of average sequence length of the hard segments.

   The polyester thermoplastic rubbers offer an interesting combination of properties, bearing in mind that these are not covalently-crosslinked materials. They are usable over a wide range of temperature, the exact range of usefulness of any particular rubber depending upon its composition. Thus, whilst retaining their flexibility at low temperatures, they

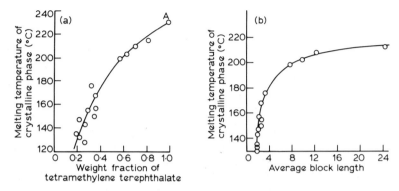

FIG. 10.10.   (a) Effect of weight fraction of hard segments upon melting temperature of crystalline phase for a polyester thermoplastic rubber containing tetramethylene terephthalate hard segments.[5] Point A gives the melting point of tetramethylene terephthalate homopolymer. (b) Effect of average block length of hard segments upon melting temperature of crystalline phase for a polyester thermoplastic rubber containing tetramethylene terephthalate hard segments.[5]

also retain mechanical properties such as tensile strength and resistance to creep when the temperature is raised. Other useful characteristics include good resilience, good abrasion resistance, excellent flex-fatigue resistance, and good resistance to chemicals, oils and solvents.

10.3   POWDERED RUBBERS

The term *powder* is generally used to describe a solid material which has been comminuted into particles less than about 1 mm in size. However, the term *powdered rubber* has tended to become used to describe all rubbers which are available as fine granules, even although these granules may be considerably larger in size than 1 mm.

The possibility of using rubber in powder form rather than as bales has interested rubber technologists ever since the first patent was granted for making powdered rubber in 1930. Powdered rubbers are of interest principally because they offer the prospect of simplifying the rubber-mixing process, and, in particular, because they offer the possibility of continuous mixing and processing with pneumatic transport of rubber and additives. The feasibility of a new technology such as powdered-rubber technology depends almost entirely upon the cost of the process, given that the technology is able to produce a product which is at least as good as that which is produced by conventional technology. Whatever process is used for producing rubber in powdered form, powdered rubber is more expensive than is the corresponding rubber in bale form because of the additional production steps required. Only if more than this premium can be saved in processing costs will the use of powdered rubbers become commercially attractive. It appears that any cost savings which are enjoyed when powdered rubbers are used in conventional rubber processes are rather marginal. The prediction has been made that the use of powdered rubbers only becomes commercially attractive if used in continuous processes in which open and internal mixers are eliminated and extruders or injection-moulding machines are continuously fed with material in powdered form. The commercial feasibility of using powdered rubbers at the present time can be judged from the claim that in 1975 less than 0·05% of rubber throughout the world was utilised in powder form. It remains to be seen whether powdered rubbers will ever become widely used industrially.

There are at least four types of method which have been developed for the production of powdered rubbers:

1.   cryogenic grinding;
2.   latex spray drying and flash drying;
3.   latex coagulation; and
4.   latex freeze drying.

In the *cryogenic grinding process*, the rubber in solid form is frozen and then ground to a coarse powder. It is not possible to produce a fine powder by this technique, and clearly the powdered rubber produced by this method must always be significantly more expensive than the corresponding bale form. Powdered rubbers produced by this method have the advantage that their content of non-rubber constituents is no greater than that of the corresponding bale form. This is not the case with powdered rubbers produced by *latex spray drying*, where the powder contains all the non-rubber constituents (expect water and other volatiles) which were present in the latex. In this type of process, a latex is introduced into a drying chamber by means of a high-speed rotating disc. The rotating disc converts the latex into small droplets which meet a stream of hot air in the chamber. The water in the latex rapidly volatilises and the dried rubber is produced in the form of a fine powder suspended in the gas stream. The rubber powder is separated from the stream by passing through a cyclone collector. The process of latex flash drying is similar, but drying occurs more rapidly. In the *latex coagulation process*, the latex is coagulated under conditions such that fine particles of coagulum are produced and do not agglomerate together during the subsequent filtration and drying steps. It is necessary to choose the coagulating agent and the conditions of coagulation carefully. It is also necessary in many cases to include a *partitioning agent* in the latex, whose function is to co-precipitate with the particles, coat their surfaces, and thereby discourage them from fusing together during filtration, drying and storage. Powders such as calcium carbonate, silica and zinc stearate in proportions ranging from 1 to 15 pphr have been used for this purpose. Starch xanthate has also been used as a partitioning agent. It is added to the latex as an aqueous solution of pH 12. It co-precipitates with the rubber particles as starch xanthide (a crosslinked version of starch xanthate) when the latex is acidified under oxidising conditions, and encourages the formation of particles of uniform size which do not agglomerate with one another. The level of starch xanthate used is typically 3 pphr. The method is said to be applicable to all rubbers which are available in latex form. It has also been used for elastomers like ethylene–propylene rubbers which are not produced as latices in the first place. For such rubbers, an artificial latex is first prepared by emulsifying a solution of the rubber in an organic solvent. The artificial latex is then co-precipitated with the starch xanthate by acidification under oxidising conditions as previously. It is claimed that the rubber particles become encased in a network of starch xanthide, and that this prevents the particles from coalescing. The starch xanthide network is said to break up on shearing, and to coalesce into small reinforcing particles. Both the processing of the rubber and the physical properties of the subsequent vulcanisate are slightly affected by the presence of the starch xanthide. In the *latex freeze drying process*, the latex is sprayed into a cold chamber (below $-30\,°C$) under vacuum. The water sublimes off, and dry rubber in the form of very fine particles is produced. It has been claimed that particles of sizes down to $0·1\ \mu m$ can be produced by this method.

Of all the rubbers which have been prepared in powder form at various times, the greatest interest seems to have been shown in powdered acrylonitrile–butadiene rubbers. Dry powder blends of powdered rubbers and compounding ingredients can be prepared very simply using conventional low-shear blenders. A problem may be encountered if attempts are made to add oils and plasticisers as such, because the rubber particles become sticky and coalesce. One way of overcoming this problem is first to absorb the oil or plasticiser in silica and thus, in effect, transform them into powders.

Apart from the fact that the user has to pay a premium for the privilege of having his rubber prepared in powder form as compared to bale form, there are at least four other problems which attend the use of rubbers in powder form. In the first place, there is the problem of the agglomeration and fusion of the particles during transportation and storage. This phenomenon is often known as *compaction*. It results in a material which no longer has the form of a free-flowing powder. The presence of an efficient partitioning agent will help to prevent compaction, but compaction will be encouraged by storage for long periods at elevated temperatures and under pressure. Furthermore, although the powdered rubber may not be under any obvious excess pressure, it must be remembered that the powder at the bottom of a container has to support the weight of the powder which is above it.

The second problem is that, because of the absence of a conventional mixing step, the reinforcing potential of particulate fillers may not be fully developed. For the development of reinforcement, it seems essential that the rubber–filler blend should be subjected to very high shearing of the type which is encountered in conventional mixing processes.

The third problem is concerned with transportation. Powdered rubbers are only about half as dense as bales. Shipping and storage costs are therefore proportionately higher. The user has to pay these additional costs as well as the incremental cost which is incurred in producing the rubber as a powder in the first place.

The fourth problem arises from the fact that only a small range of types of rubber is yet available commercially in powder form. It has been predicted that widespread adoption of powdered rubbers for general-purpose use will only become feasible when a wide range of types is available. Of course, it can also be argued with some truth that the range of available types will probably increase significantly only if the current usage is such as to provide the producer with a reasonable expectation of general-purpose application.

10.4   FLUID RUBBERS

The concept of producing a rubber precursor as a viscous liquid which can be mixed with compounding ingredients in this form, and then converted into a rubber vulcanisate in a mould, embodies what is probably potentially the single most revolutionary development in modern rubber technology. If this type of technology were ever widely used, then almost all sectors of the rubber manufacturing industry would undergo drastic change.

Like rubber in powder form, the idea of using rubber precursors which are viscous liquids is quite old. It has long been recognised that high molecular weight in a rubber is essential only in the final product. As regards processing, high molecular weight can be disadvantageous because it means high viscosity and consequent difficulty in inducing flow. In fact, natural rubber—the rubber for which conventional rubber technology was evolved—as encountered in bale form is of such high molecular weight and high viscosity as to be effectively a solid until its molecular weight has been reduced by some means, such as a combination of mechanical and chemical action as in mastication. A logical consequence of these considerations is to produce and process the rubber in the form of a precursor which is of sufficiently low molecular weight to be fluid, and then subsequently to link the precursor molecules by suitable chemical reactions into a macro-molecular network.

The earliest approach to developing a rubber technology of this type was that of using a low-molecular-weight version of a conventional rubber, and then relying upon the normal reactivity of the rubber molecule, as manifest in vulcanisation, for the linking together of the small chains into a macromolecular network. The principle which underlies this approach is illustrated in Fig. 10.11(a). An example of a material whose technology depends upon this principle is depolymerised natural rubber. The material is produced by drastically reducing the molecular weight of natural rubber. It is processed by mixing with conventional curatives and fillers, casting into moulds, and then causing the molecules of depolymerised rubber to be linked together by reaction with sulphur. A second example of such a material is a synthetic rubber which has been produced by free-radical emulsion polymerisation in the presence of sufficient modifier to give a product of much lower molecular weight than usual. Fluid rubbers of this type have found very limited application only, principally because the vulcanisates have poor mechanical properties. The poor mechanical properties are attributed to the high concentration of free chain ends which these vulcanisates inevitably contain, as is evident from Fig. 10.11(a).

The more recent approach has been to develop low-molecular-weight castable rubber precursors whose molecules have functional end-groups which have been introduced to confer reactivity towards specific reagents. By means of these end-groups, the low-molecular-weight molecules can be linked together (*chain-extended*) by means of difunctional reagents to form linear sequences. Crosslinking can be brought about by reaction with reagents of functionality higher than two. Crosslinking can also be brought about by utilising the normal chemical reactivity of the polymer sequence as a whole. The concentration of free chain ends in vulcanisates produced

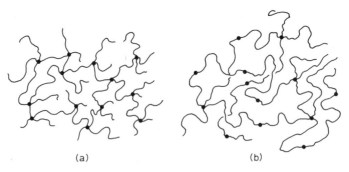

(a)   (b)

FIG. 10.11.   (a) Illustrating molecular structure of vulcanisate from a fluid rubber prepared utilising normal reactivity of rubber molecule. (b) Illustrating molecular structure of vulcanisate from a fluid rubber prepared by chain extension.

in this way is very much lower than in the case of vulcanisates produced from low-molecular-weight precursors which are not capable of chain extension. This should be clear from Fig. 10.11(b), which illustrates the principle which underlies this more recent approach to the production of vulcanisates from fluid rubbers. The name *telechelic polymer* has been adopted for the low-molecular-weight precursors for this type of network. This name is derived from two Greek words, namely *tele* meaning 'far off', and *chele* meaning 'claw'. The name has its origin in the fact that the molecules to which it is applied have two reactive sites which are widely separated.

The principal types of rubber which have been available in recent years in functionally-terminated fluid form are urethane rubbers, butadiene rubbers, acrylonitrile–butadiene rubbers, and polysulphide rubbers. Liquid silicone rubbers have also become available. The following are the principal types of reactive end-group which have been used:

1. hydroxyl
2. carboxyl
3. halogen
4. isocyanate
5. thiol (mercaptan, sulphydryl).

Other miscellaneous reactive end-groups which have been used include epoxide, olefin and aziridinyl.

Hydroxyl-terminated polymers can be chain-extended by reaction with diisocyanates, in the same way that solid urethane rubbers are produced from hydroxyl-terminated polyester and polyether prepolymers. *Carboxyl-*

terminated polymers can be chain-extended by reaction with di-epoxides as follows:

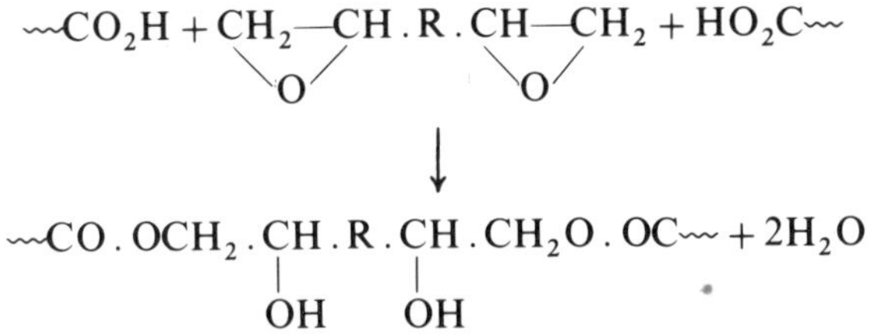

$$\sim\!CO_2H + CH_2\!\!-\!\!CH.R.CH\!\!-\!\!CH_2 + HO_2C\!\sim$$
$$\underset{O}{\diagdown\!\diagup}\underset{O}{\diagdown\!\diagup}$$

$$\downarrow$$

$$\sim\!CO.OCH_2.\underset{\underset{OH}{|}}{CH}.R.\underset{\underset{OH}{|}}{CH}.CH_2O.OC\!\sim + 2H_2O$$

The hydroxyl groups which are formed in this reaction will also react with other carboxyl and epoxide groups to form crosslinks. A typical di-epoxide which has been used for this application is the diglycidyl ether of 4,4′-dihydroxydiphenyl-2,2-propane (CLVI) (the latter compound being com-

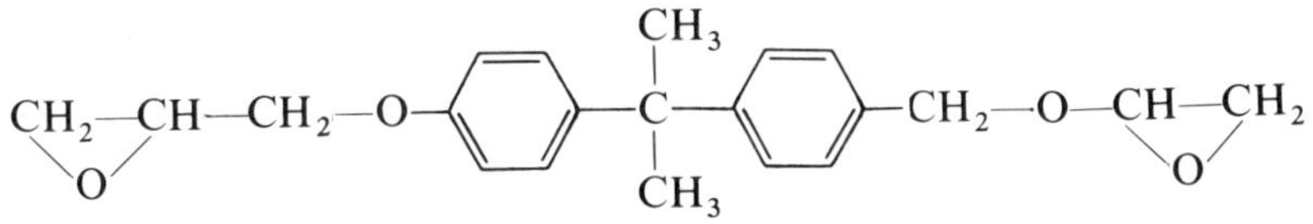

(CLVI)

monly known as bisphenol A). Carboxyl-terminated polymers can also be chain-extended by reaction with aziridinyl compounds

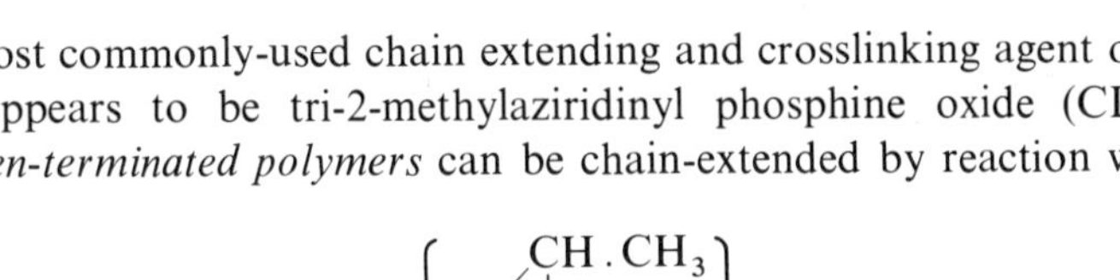

$$\sim\!CO_2H + R.\underset{\underset{CH_2}{|}}{CH}\!\!-\!\!N.R'.N\!\!-\!\!\underset{\underset{CH_2}{|}}{CH}.R + HO_2C\!\sim$$

$$\downarrow$$

$$\sim\!CO.O.CHR.CH_2.NH.R'.NH.CH_2.CHR.O.OC\!\sim$$

The most commonly-used chain extending and crosslinking agent of this type appears to be tri-2-methylaziridinyl phosphine oxide (CLVII). *Halogen-terminated polymers* can be chain-extended by reaction with a

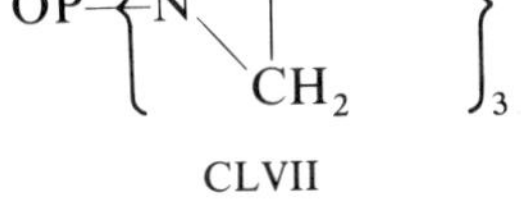

CLVII

diamine. The terminal halogen atom is usually bromine, and so the chain-extension reaction can be represented as follows:

$$\sim\!Br + H_2N.R.NH_2 + Br\!\sim$$

$$\downarrow$$

$$\sim\!HN.R.NH\!\sim + 2HBr$$

Isocyanate-terminated polymers can be chain-extended by reaction with a variety of difunctional active-hydrogen compounds, of which low-molecular-weight diols are perhaps the most obvious. Several types of reagent are available for the chain-extension of *thiol-terminated polymers*. Reaction with suitable oxidising agents gives chain-extension by way of disulphide linkages:

$$\sim\!SH + HS\!\sim \xrightarrow{\;O\;} \sim\!S\!-\!S\!\sim$$

Reaction with an aldehyde gives chain-extension by way of the formation of thioacetal linkages:

$$\sim\!SH + R.CHO + HS\!\sim$$

$$\downarrow$$

$$\sim\!S\!-\!CHR\!-\!S\!\sim + H_2O$$

Furfural (CLVIII) has been used for this purpose. Thiol-terminated

$$\text{(CLVIII)}$$

polymers can also be chain-extended by reaction with diisocyanate, the reaction being as follows:

$$\sim\!SH + OCN.R.NCO + HS\!\sim$$

$$\downarrow$$

$$\sim\!S\!-\!\underset{\underset{O}{\|}}{C}\!-\!NH.R.NH\!-\!\underset{\underset{O}{\|}}{C}\!-\!S\!\sim$$

In each case, crosslinking can be accomplished by simultaneous reaction with a reagent of the same type whose functionality is greater than two.

One of the major problems which has undoubtedly hindered the development of chain-extendable fluid rubbers is that of adequate control of the structure of the prepolymer. The molecular weight of the prepolymer is clearly an important variable. It affects not only the ease with which the fluid rubber can be processed, but also the properties of the vulcanisate which is obtained subsequently. Not only is the average molecular weight important, but also the molecular weight distribution. Of even greater importance, however, is the matter of average functionality and distribution of functionality. Ideally, all the prepolymer molecules should have a functionality of exactly two. In practice, average functionalities are within the range 1·5–2·8, with most prepolymers having an average functionality rather greater than two. Almost as important as average functionality is distribution of functionality, and in many cases very little seems to be known about this. Prepolymer molecules of zero functionality will not be incorporated into the elastomer network at all, but will remain in the vulcanisate as a plasticiser or sol fraction. Prepolymer molecules having a single reactive terminal group will become combined in the elastomer network but will be incapable of participating in further chain extension. They will therefore contribute to the network defects in the form of loose ends. Prepolymer molecules having functionalities of three or more will cause chain-branching and crosslinking. It is perhaps significant that by far the most successful group of fluid rubbers so far developed has been the castable liquid urethane rubbers. In this case, the functionality of the prepolymer molecule can be quite accurately controlled at two. Furthermore, some control over molecular weight distribution can be exercised.

As an example of the preparation of low-molecular-weight functionally-terminated fluid rubbers, we may take the polybutadienes. These can be made by anionic polymerisation in a hydrocarbon solvent, using a difunctional lithium initiator. The polymerisation is terminated when the degree of polymerisation of the butadiene has achieved the desired value. The termination reaction is chosen so that the propagating anionic sites are destroyed and the required terminal functional groups are incorporated simultaneously. Thus, for instance, carboxyl groups can be introduced by reaction with carbon dioxide followed by treatment with acid:

$$Li^{\oplus \ominus} \text{\scriptsize{\textasciitilde\textasciitilde\textasciitilde}} ^{\ominus} Li^{\oplus} + 2CO_2 \longrightarrow Li^{\oplus \ominus} O_2C \text{\scriptsize{\textasciitilde\textasciitilde\textasciitilde}} CO_2^{\ominus} Li^{\oplus}$$

$$+ 2HX \diagup$$

$$HO_2C \text{\scriptsize{\textasciitilde\textasciitilde\textasciitilde}} CO_2H + 2LiX$$

The conversion of the growing polymer chain to the dilithium dicarboxylate is accompanied by a marked rise in the viscosity of the solution and the formation of a stiff gel, presumably because of strong association between the ionic ends of the polymer chains. On subsequent acidification, the viscosity falls to almost that of the corresponding hydrogen-terminated polybutadiene. The feasibility of producing difunctional molecules by this method depends upon the absence of adventitious impurities which react with the propagating ionic species to produce non-carboxyl end-groups. Water is an example of such an impurity. If such reactions occur, then the product may be a mono- or zero-functional polymer molecule, according as one or both of the propagating ions has reacted. Alternatively, carboxyl-terminated low-molecular-weight polybutadienes can be produced by free-radical polymerisation using carboxyl-containing initiators. Two examples of such initiators are azobis(4-cyanopentanoic acid) (CLIX) and glutaric acid peroxide (CLX), of which the second is preferred on both economical

$$\left\{ =N-\underset{\underset{CN}{|}}{\overset{\overset{CH_3}{|}}{C}}-CH_2CH_2CO_2H \right\}_2 \qquad \left\{ HO_2C(CH_2)_3-\underset{\underset{O}{\|}}{C}-O \right\}_2$$

(CLIX)   (CLX)

and technological grounds. In both cases, the possibility of producing difunctional carboxylated polymers depends upon (a) the initiator fragments effecting initiation by combination with monomer rather than by, say, abstraction of hydrogen atoms, (b) termination of the propagating radicals occurring by combination and not by disproportionation, and (c) chain transfer occurring to a negligible extent only. It has been claimed that it is possible by this technique to produce a difunctional carboxyl-terminated polybutadiene of narrow molecular-weight distribution. The molecular weight of the polymer produced depends upon the ratio of monomer to initiator. Apparently, in the case of butadiene, the initiator is consumed more rapidly than the monomer, and the addition of the initiator to the reaction system is programmed in such a way as to maintain the monomer/initiator ratio at a constant level.

Other types of functionally-terminated polybutadienes can be produced by the reaction of lithium-initiated polymers with suitable reagents. Thus, in addition to the production of carboxyl-terminated polymers by reaction with carbon dioxide, hydroxyl-, thiol-, amine-, epoxide- and aziridinyl-terminated polymers can be formed by reactions which can be summarised as follows:

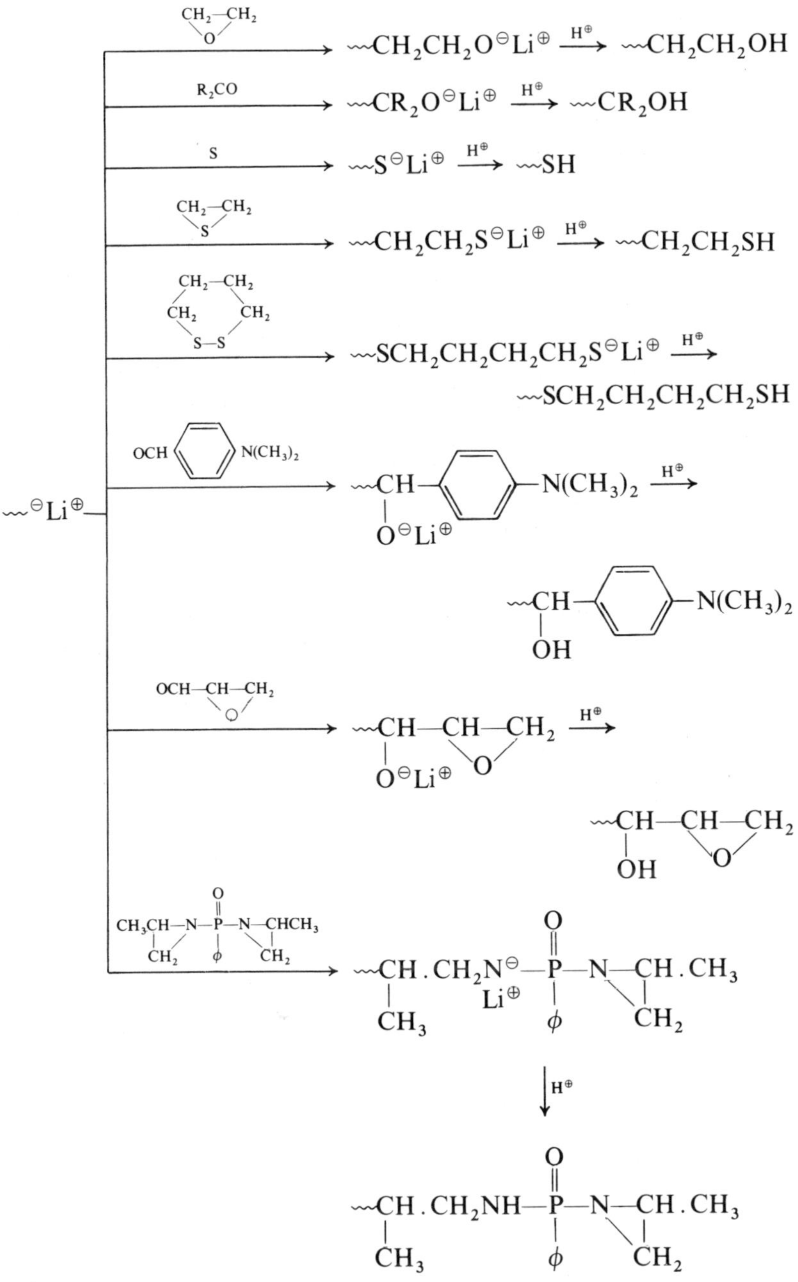

In order to illustrate something of the technology of fluid rubbers, the urethane-extended hydroxyl-terminated polybutadienes will be considered as an example. Results are shown in Fig. 10.12 for the effect of prepolymer molecular weight upon (a) the hydroxyl content of the polymer, and (b) the viscosity of the prepolymer. As expected, the hydroxyl content is inversely proportional to the number-average molecular weight. The logarithm of the bulk viscosity increases approximately linearly with molecular weight over the range for which data are available; thus the bulk viscosity itself increases exponentially with increasing molecular weight. Results are shown in Fig. 10.13 for the effect of prepolymer molecular weight upon (a) tensile strength, (b) elongation at break, (c) modulus, (d) hardness, and (e) tear strength of the final vulcanisate. The rubbers to which these results refer were made from hydroxyl-terminated polybutadienes (HTPB) having number-average molecular weights covering the range 1000–8000 (degree of polymerisation c. 18–150) which were chain-extended and crosslinked by reaction with 4,4'-diphenylmethane diisocyanate (MDI) (CXIX) in the presence of di-n-butyltin dilaurate (XCIII) as catalyst. Each of the graphs in Fig. 10.13 gives results for two types of rubber:

1. rubbers produced by the reaction of hydroxyl-terminated polybutadiene with 4,4'-diphenylmethane diisocyanate and catalyst alone; and
2. rubbers produced by the reaction of hydroxyl-terminated polybutadiene with 4,4'-diphenylmethane diisocyanate, catalyst and a short-chain diol, namely N,N-bis(2-hydroxypropyl)aniline (HPA) (CLXI).

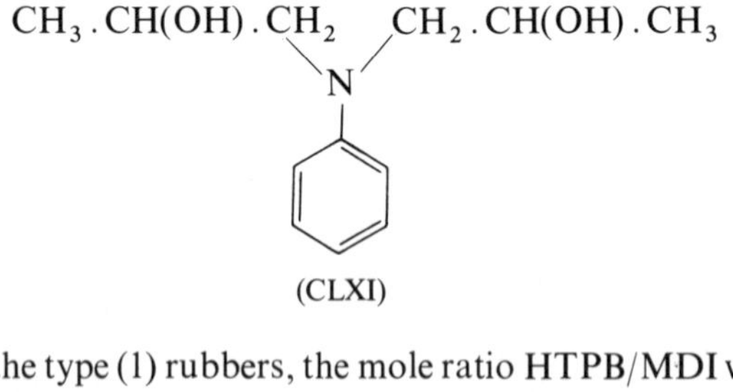

$$CH_3 . CH(OH) . CH_2 \quad CH_2 . CH(OH) . CH_3$$

N

(CLXI)

In the case of the type (1) rubbers, the mole ratio HTPB/MDI was 1/1, that is, the ratio OH/NCO was also 1/1. The rubbers therefore had the structure shown at (CLXII).

$$\sim CO_2 . CO_2 . NH . R . NH . CO_2 . CH_2 \sim$$

(CLXII)

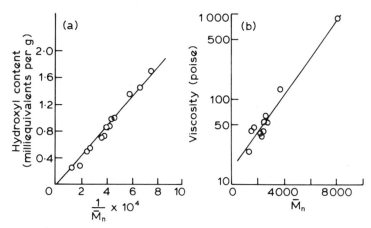

FIG. 10.12.   Effect of prepolymer molecular weight upon (a) hydroxyl content of polymer and (b) viscosity of polymer for hydroxyl-terminated polybutadiene fluid rubbers.[6]

In the case of the type (2) rubbers, the composition of the reaction mixture in all cases was such that the mole ratio HTPB/MDI/HPA was 1/2/1, that is, the ratio HTPB/HPA was 1/1 and the ratio OH/NCO was also 1/1. The rubbers produced by these reaction mixtures therefore had the structure shown at (CLXIII).

$$\sim\!\!CH_2.CO_2.NH.R.NH.CO_2.R'.CO_2.NH.R.NH.CH_2.CH_2\!\!\sim$$

(CLXIII)

In these structures, R is the moiety derived from 4,4'-diphenylmethane, and R' is the moiety derived from N,N-bis(2-hydroxypropyl)aniline. In both cases, the rubbers were made by the so-called *one-shot* process, in which the hydroxyl-terminated rubber precursor is blended with the diisocyanate, catalyst and short-chain diol, and then cast.

The results summarised in Fig. 10.13 show two interesting features. Firstly, they show that properties such as tensile strength, modulus, hardness and tear strength *decrease* as the molecular weight of the rubber precursor *increases*. Furthermore, the effect is particularly marked in those rubbers which were prepared using the short-chain diol (type (2) above). Secondly, the presence of the short-chain diol brings about very marked increases in tensile strength and tear strength, especially when the molecular weight of the rubber precursor is low, and it also brings about increases in modulus, hardness and elongation at break. Thus the short-chain diol is clearly functioning as a reinforcing agent in these rubbers. The

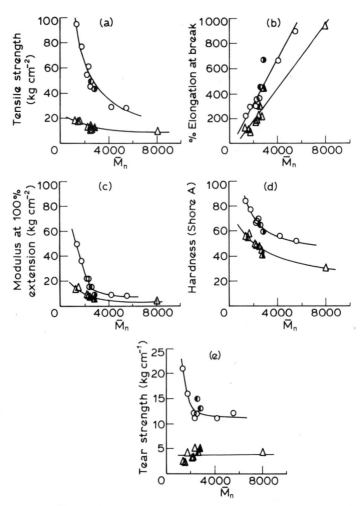

FIG. 10.13.   Effect of prepolymer molecular weight upon mechanical properties of vulcanisates prepared by 'one-shot' urethane-extension of hydroxyl-terminated polybutadiene fluid rubbers: (a) tensile strength, (b) elongation at break, (c) modulus at 100% extension, (d) hardness, (e) tear strength.[6] The triangular points refer to rubbers produced by reaction of hydroxyl-terminated polybutadiene with 4,4'-diphenylmethane diisocyanate and catalyst alone. The circular points refer to rubbers produced by reaction of hydroxyl-terminated polybutadiene with 4,4'-diphenylmethane diisocyanate, catalyst and a short-chain diol (N,N-bis(2-hydroxy-propyl) aniline). The different styles of triangular and circular points distinguish data for vulcanisates based upon different types of hydroxyl-terminated polybutadienes.

See text for further details of the rubbers to which these results refer.

first of these observations can be explained by attributing the strength and stiffness of these rubbers primarily to the presence of polar amide units embedded in sequences of non-polar hydrocarbon units, and to the existence of hydrogen bonding and dipole interaction between those polar units. The chain-stiffening effect of the aromatic rings of the diisocyanate may also be important. The plausibility of this view is demonstrated by the fact that, for both types of rubber, both the modulus and the tensile strength are approximately inversely proportional to reciprocal number-average molecular weight (see Fig. 10.14(a) and (b) respectively). Thus both properties of the vulcanisate are directly proportional to the concentration of chain ends in the rubber precursor. The second of these observations can be explained by arguing that the effect of the short-chain diol is to introduce a 'harder' short segment into the polymer sequence than is introduced by the diisocyanate alone. Not only is the number of amide units per inter-precursor linkage doubled; the stiffness of the segments linking the precursor molecules will also be markedly increased because the number of aromatic rings derived from the diisocyanate is doubled, and the stiffening effects of these is supplemented by the presence of the aromatic ring of the N,N-bis(2-hydroxypropyl) aniline. Both these effects of increased polarity and increased segmental stiffness will be expected to enhance strength and stiffness, and this is what is observed.

It has already been noted that the term *one-shot* is commonly applied to processes for the production of cast urethane rubbers such as the one

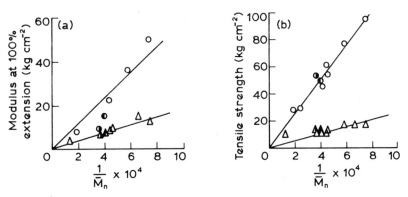

FIG. 10.14.   Modulus at 100 % extension (a), and tensile strength (b), as functions of reciprocal prepolymer molecular weight for vulcanisates prepared by 'one-shot' urethane extension of hydroxyl-terminated polybutadiene fluid rubbers.[6] See caption to Fig. 10.13 for significance of different styles of point.

described above. In these processes, a hydroxyl-terminated rubber precursor is reacted with a polyfunctional isocyanate and, possibly, with a
short-chain diol. An alternative process is available in which the hydroxyl-
terminated rubber precursor is first reacted with a diisocyanate to form a so-
called *prepolymer*, which is terminated with reactive isocyanate groups.
These prepolymer molecules are then chain-extended and crosslinked by
reaction with, say, polyhydroxy compounds. Techniques of this kind are
described as *prepolymer processes*. (The term *prepolymer* applied specifically to this type of process is rather inappropriate, because the rubber
precursor of the one-shot process is no less a prepolymer than is that of the
prepolymer process.) Prepolymer processes seem to be more widely used in
the field of cast urethane rubber technology than do one-shot processes.
The initial hydroxyl-terminated rubber precursor is usually a polyester or a
polyether. The vigorous exclusion of moisture from all the reagents,
especially the rubber precursors, is essential, because water rapidly reacts
with isocyanates to produce, in the first instance, a substituted carbamic
acid which then decomposes to give a primary amine and carbon dioxide:

$$R.NCO + H_2O \longrightarrow R.NH.CO_2H \longrightarrow R.NH_2 + CO_2$$

The primary amine can then react with further isocyanate by reactions of
the kind:

$$R.NCO + R.NH_2 \longrightarrow R.NH.CO.NH.R$$

Whatever may be the precise chemistry of the interaction between the
isocyanate and adventitious moisture, the effect upon the mechanical
properties of the resultant vulcanisate is adverse, e.g., because of wastage of
isocyanate groups in non-chain-extending reactions, and because of
porosity arising from the evolution of carbon dioxide.

In the case of hydrocarbon-type fluid rubbers, it is frequently necessary
to incorporate reinforcing fillers, such as carbon black, in order to obtain
satisfactory mechanical properties in the eventual vulcanisate. Consequently, considerable effort has been expended in investigating the degree
of dispersion required to achieve satisfactory reinforcement, and in
developing satisfactory techniques for attaining the necessary degree of
dispersion. It seems that the fundamental problem is how to break down
the aggregates of a filler such as carbon black, rather than that of the
dispersion of the carbon black aggregates themselves. Not only is
breakdown of the aggregates more difficult in a medium of low viscosity
(such as a fluid rubber) than in one of high viscosity (because the shearing

forces at a given shear rate are lower); it also seems to be the case that the energy which has to be expended in achieving adequate dispersion of a reinforcing filler is essentially independent of the viscosity of the medium in which dispersion is being attempted, because the amount of work which is required to effect dispersion is virtually constant. For this latter reason, some of the alleged advantages of fluid rubber technology over conventional rubber technology—notably a reduced energy requirement for mixing—have proved to be largely illusory.

For reasons such as poor vulcanisate mechanical properties and failure to realise the expected energy savings during mixing, fluid rubber technology as an alternative to conventional rubber technology has failed as yet to achieve much significance. Application of most of the fluid rubbers so far developed has been restricted to rather specialised areas such as mastics, sealants and rollers produced by casting. A notable exception to this generalisation is that of the cast urethane rubbers. So potentially successful have these materials been that serious interest has been shown in recent years in the production of urethane rubber passenger tyres by casting. There is also another area in which what are in effect castable fluid urethane rubbers have been outstandingly successful. This is the area of elastomeric urethane foam materials. Although this review of synthetic rubbers has been confined almost exclusively to rubbers which are intended for application as solid, non-expanded materials, it is appropriate to note here that urethane foam rubbers are produced by the chain-extension and crosslinking of low-molecular-weight polyester and polyether precursors in much the same way that solid urethane rubbers are produced from similar low-molecular-weight precursors by casting. The major point of difference is that, in the case of foam production, use is made of the generation of carbon dioxide by interaction between the isocyanate and water to ensure that foaming takes place concurrently with chain extension and crosslinking.

GENERAL BIBLIOGRAPHY

Anon. (1975). Thermoplastic rubbers, *Eur. Rubb. J.*, **157**(8), 38.

Houston, A. M. (1975). Thermoplastic elastomers, *Mater. Eng.*, **82**(7), 47.

Sorensen, B. H. (1976). Thermoplastic rubbers, their properties and uses, *Plast. News*, February, pp. 11–16.

Walker, B. M. (Ed.). (1979). *Handbook of Thermoplastic Elastomers*, Van Nostrand Reinhold, New York.

Thorn, A. D. (1980). *Thermoplasic Elastomers*, Rubber and Plastics Research Association, Shawbury.

Morton, M. (1971). Styrene–diene block copolymers. In: *Encyclopedia of Polymer Science and Technology*, Vol. 15, John Wiley and Sons, New York, pp. 508f.

Hsieh, H. L. (1976). Synthesis of radial thermoplastic elastomers, *Rubb. Chem. Technol.*, **49**, 1305.

Morris, H. L. (1974). TDR thermoplastic rubber, *J. Elast. Plast.*, **6**(3), 121.

Schollenberger, C. S. and Dinbergs, K. (1975). Thermoplastic urethane chemical crosslinking effects, *J. Elast. Plast.*, **1**(1), 65.

Whittaker, R. E. (1974). Mechanical properties of segmented polyurethane elastomers, *Rheol. Acta*, **13**, 675 (1183).

Cella, R. J. (1977). Polyesters, elastomeric. In: *Encyclopedia of Polymer Science and Technology*, Supplement Vol. 2, John Wiley and Sons, New York, pp. 485f.

Brown, M. (1975). Thermoplastic copolymer elastomers: new polymers for specific end-use applications, *Rubb. Ind.*, **9**(3), 102.

Anon. (1976). Powdered rubbers—what kind of future?, *Plast. Rubb.*, **1**(1), 6.

Monnot, H., Morrell, S. H. and Pyne, J. R. (1974). Powdered rubber, *R.A.P.R.A. Members J.*, **2**(10), 253.

Byrne, P. S. and Schwarz, H. F. (1973). Powdered rubber, *Rubb. Age*, **105**(7), 43.

Widmer, H. and Milner, P. W. (1974). Powdered rubber, *Rubb. Age*, **106**(11), 41.

Woods, M. E. and Whittington, W. H. (1975). Powdered rubber technology, *Rubb. Age*, **107**(10), 39.

Evans, C. W. (1978). *Powdered and Particulate Rubber Technology*, Applied Science Publishers Ltd, London.

Sheard, E. A. (1973). Commercial status of liquid elastomers, *Chem. Technol.*, May, p. 298.

Pyne, J. R. (1971). Liquid rubbers—general rubber products technology, *Rubb. Chem. Technol.*, **44**, 750.

Berry, J. P. and Morrell, S. H. (1974). Liquid rubbers and the problems involved in their application, *Polymer*, **15**, 521.

French, D. M. (1969). Functionally terminated butadiene polymers, *Rubb. Chem. Technol.*, **42**, 71.

Hoffman, R. F. and Gobran, R. H. (1973). Liquid carboxyl-terminated poly(butadiene), *Rubb. Chem. Technol.*, **46**, 139.

Humpidge, R. T., Matthews, D., Morrell, S. H. and Pyne, J. R. (1973). Processing and properties of liquid rubbers, *Rubb. Chem. Technol.*, **46**, 148.

Rivin, D. and True, R. G. (1973). Filler reinforcement of liquid elastomers, *Rubb. Chem. Technol.*, **46**, 161.

REFERENCES

1. Hsieh, H. L. (1965). *Rubb. Plast. Age*, **46**, 394.
2. Kam, Y. F. (1979). Ph.D. Thesis, C.N.A.A.
3. Schollenberger, C. S. and Dinbergs, K. (1975). *J. Elast. Plast.*, **1**(1), 65.
4. Seefried, C. G., Koleske, J. V. and Critchfield, F. E. (1975). *J. Appl. Polym. Sci.*, **19**, 2493, 2503.
5. Cella, R. J. (1977). *Encyclopedia of Polymer Science and Technology*, S2, p. 485.
6. Ono, K. *et al.* (1977). *J. Appl. Polym. Sci.*, **21**, 3223.

Index